QC **REF** Physics and
801 astrophysics from a
.P45 lunar base.
1990

$60.00

		DATE		

**BUSINESS/SCIENCE/TECHNOLOGY
DIVISION**

© THE BAKER & TAYLOR CO.

PHYSICS AND ASTROPHYSICS FROM A LUNAR BASE

AIP CONFERENCE PROCEEDINGS 202

PHYSICS AND ASTROPHYSICS FROM A LUNAR BASE
FIRST NASA WORKSHOP
STANFORD, CA 1989

EDITORS:

A. E. POTTER & T. L. WILSON
NATIONAL AERONAUTICS AND SPACE
ADMINISTRATION, JOHNSON SPACE CENTER,
HOUSTON, TEXAS

American Institute of Physics　　　　　　　　　**New York**

Authorization to photocopy items for internal or personal use, beyond the free copying permitted under the 1978 US Copyright Law (see statement below), is granted by the American Institute of Physics for users registered with the Copyright Clearance Center (CCC) Transactional Reporting Service, provided that the base fee of $2.00 per copy is paid directly to CCC, 27 Congress St., Salem, MA 01970. For those organizations that have been granted a photocopy license by CCC, a separate system of payment has been arranged. The fee code for users of the Transactional Reporting Service is: 0094-243X/87 $2.00.

Copyright 1990 American Institute of Physics.

Individual readers of this volume and non-profit libraries, acting for them, are permitted to make fair use of the material in it, such as copying an article for use in teaching or research. Permission is granted to quote from this volume in scientific work with the customary acknowledgment of the source. To reprint a figure, table or other excerpt requires the consent of one of the original authors and notification to AIP. Republication or systematic or multiple reproduction of any material in this volume is permitted only under license from AIP. Address inquiries to Series Editor, AIP Conference Proceedings, AIP, 335 E. 45th St., New York, NY 10017.

L.C. Catalog Card No. 90-55073
ISBN 0-88318-646-2
DOE CONF 8905272

Printed in the United States of America.

Contents

Preface ix
Foreword xi

SESSION ON SPACE PHYSICS

Space Physics from a Lunar Base Project 3
    F. Curtis Michel
The Moon and the Magnetosphere and Prospects for Neutral Particle Imaging 9
    John W. Freeman, Jr.
Space Plasma Physics Research at a Lunar Base 17
    Richard Vondrak

SESSION ON COSMIC RAY PHYSICS

The Atmosphere as Particle Detector 29
    Todor Stanev
Low Energy Cosmic Ray Studies from a Lunar Base 33
    Mark E. Wiedenbeck
Measurement of Ultra-Heavy Cosmic Rays at a Lunar Base* 42
    M. H. Salamon, P. B. Price, and G. Tarle
Study of Chemical Composition of Ultra-High Energy Cosmic Radiation
from a Lunar Base* 49
    Gautam D. Badhwar

SESSION ON NEUTRINO PHYSICS

Medium and High-Energy Neutrino Physics from a Lunar Base 53
    Thomas L. Wilson
Search for Proton Decay and Supernova Neutrino Bursts with a Lunar Base
Neutron Detector 88
    David B. Cline
Muons on the Moon 113
    Victor J. Stenger
Lunar Neutrino Physics 119
    John G. Learned
A Lunar-Based Detector to Search for Relic Supernovae Antineutrinos 128
    A. K. Mann and W. Zhang

EXPERIMENTS IN GRAVITATION & GENERAL RELATIVITY

On the Use of Clocks in Satellites Orbiting the Moon for Tests of
General Relativity 143
    Arnold Rosenblum
The Apollo Retroreflector Arrays Revisited: A Lunar Beaconed Array 153
    James E. Faller
Thoughts on a Few Gravitation Experiments .... #
    C. W. F. Everitt

SESSION ON GRAVITATIONAL RADIATION PHYSICS

Gravitational Radiation Antennas: History, Observations, and Lunar
Surface Opportunities 159
    Joseph Weber
The Moon as a Gravitational Wave Detector, Using Seismometers 183
    Warren W. Johnson
A Lunar Gravitational Wave Antenna Using a Laser Interferometer 188
    R. T. Stebbins and P. L. Bender

SESSION ON COSMIC BACKGROUND RADIATION PHYSICS

Cosmic Background Radiation Physics 205
    George F. Smoot

SESSION ON PARTICLE ASTROPHYSICS

The Measurement of Elemental Abundances above 10^{15} eV at a Lunar Base 211
    Simon P. Swordy
Possibilities for Fundamental Particle/Astrophysics Experiments at a
Lunar Base 217
    Serge Rudaz
Probing the Halo Dark Matter γ Ray Line from a Lunar Base* 227
    Pierre Salati, Alain Bouquet, and Joseph Silk

SESSION ON GAMMA-RAY AND X-RAY PHYSICS

The Large Area High Resolution Gamma Ray Astrophysics Facility 243
    E. J. Fenyves *et al.*
Large X-Ray Detector Arrays for Physics Experiments at a Lunar Base 263
    Kent S. Wood and Peter F. Michelson

Arcsec Source Location Measurements in Gamma-Ray Astronomy from a Lunar Observatory* 278
    David G. Koch and E. Barrie Hughes

SESSION ON SURFACE PHYSICS

Surface Physics—Materials Science Research Possibilities on a Lunar Base 285
    Alex Ignatiev

GROUP LEADER SUMMARIES

Possible Cosmic Ray Experiments on the Lunar Base Discussion in Stanford on May 20, 1989 291
    Todor Stanev
Comments on Particle Astrophysics at a Lunar Base 293
    Simon P. Swordy

AFTERWORD

Reprint of "The Moon as the Searching Ground for Proton Decay" 297
    J. C. Pati, Abdus Salam, and B. V. Sreekantan

List of Participants 305
Lunar Photography Credits and Maps 311
Lunar Physics Constants 318
Author Index 323

*Manuscript submitted after or in addition to plenary sessions. These were not speakers.
#Manuscript not available.

Preface

The Workshop occurred at a time of interesting transition in what we consider to be traditional physics. As physicists, we have always considered ourselves to be pushing back the frontiers, so to speak, and it is really a tribute to our spirit and imagination the clever ways in which we seem to overcome the obstacles and ironies Nature sets before us. The last several decades have literally transformed our view of the physical Universe. And yet as Earth-bound observers, tradition has trudged on and only a few of us have played with the thought that some of the basic experiments in physics could be more accurately performed on the Moon than on the Earth—*if* we should ever establish a fundamental physics research facility there. Anyone who has studied particle physics *beneath* the Earth's atmosphere knows that its noise is a serious limitation, and the Moon would provide a welcome sanctuary as a remote, quiet place for certain experiments.

To cope with the Earth's atmosphere, some particle physicists have moved underground in their pursuit of new frontiers such as grand unification and supersymmetry. Likewise, the era of astronomy in space has gotten outside of it but has reconciled itself to being Earthbound, perhaps always. However, a small group of us have kept our eye on the Moon, and we began looking for a community of interest in fundamental physics at a lunar base. This most recent effort started at the 14th Texas Symposium on Relativistic Astrophysics in Dallas, Texas in December 1988. It was there and later that Joseph Weber remarked to this editor, "The Moon is a very interesting object." Indeed, it is, and the organization of this Workshop began.

Within this small band of people were Professor Mason Yearian of Stanford University who offered to host the Workshop and Professor John Freeman of Rice University who helped Dr. Andrew Potter of NASA here in Houston convene it through our Lunar and Planetary Institute. Dr. Michael B. Duke as chairman of the Agency's Exploration Science Working Group (EXSWG) for the Office of Exploration (Code Z), was also instrumental in establishing the funds necessary to bring it about.

As a venture in space initiatives, the story behind the Workshop also takes a special twist. It was the foresight of Dr. Carl B. Pilcher, Administrator for Science in NASA's newly formed Office of Exploration (Code Z) that permitted it to happen. Except for the formation of this independently funded branch of the Agency, such an exploratory workshop in frontier physics would not have taken place. It is a tribute to Dr. Pilcher's leadership that we were permitted to address all of the issues, even those that were new and undeveloped. As with any new program, this process has only just begun. Tidal forces continued to struggle over the view that some of these Stanford proposals appear too costly and may be out of scope. That is the distant cloud on the horizon.

The strategy was not to set a standard for future workshops. The intent was simply to do it, and to focus on the main character of the drama, the Moon. It was to focus not only upon traditional space physics but also upon those other fundamental physics experiments which uniquely require the Moon in order to be performed—not Mars and not Earth-orbiting or interplanetary platforms. We then searched for clever people. It was organized rather quickly, and the plenary sessions lasted one day, May 20 at Stanford's Hansen Laboratory for High-Energy Physics. Introductory talks by Dr. Pilcher and Dr. Potter were given the evening before at the Sunnyvale Hilton in Palo Alto, California, where program organization and future funding were discussed.

The organizing committee owes its gratitude to a number of individuals who could not participate, but who contributed nevertheless. These include Dr. T. K. Gaisser (Bartol Research Institute), Professor Gaurang Yodh (University of California, Irvine), Professor Bernard Sadoulet (University of California, Berkeley), Professor Richard E. Lingenfelter (University of California, San Diego), Professor Ronald Drever (California Institute of Technology), Dr. Robert Spiro (California Institute of Technology), Professor Daniel T. Wilkinson (Princeton University), Dr.

Nobuki Kawashima (Institute of Space & Aeronautical Science, Japan), Professor R. D. Davies (University of Manchester, United Kingdom), Dr. Peter L. Bender (University of Colorado, Boulder), Dr. Don Kniffen (NASA, Goddard), Dr. Floyd Stecker (NASA, Goddard), and Dr. Robert H. Manka (National Academy of Sciences).

Also, special thanks are due to Professor David B. Cline (University of California, Los Angeles) who not only participated but who pointed out the earlier work of Jogesh Pati, Abdus Salam, and B. V. Sreekantan on the potential importance of the Moon to the ongoing search for proton decay. To Professor Pati *et al.*, we owe our gratitude for the privilege to re-publish this work, thus making it more available as well as bringing it together with the proceedings of this Workshop. Dr. K. K. Phua of World Scientific is to be congratulated for publishing that fundamental investigation of the Moon's environment as a testbed for physics research, which we reproduce in the Afterword.

In conclusion, the workshop was not intended to bring about a major change in lunar physics research. Rather, it was a modest effort at dealing with all scientific issues which time would permit. As you read the manuscripts, however, you will see a new point of view emerging which is far from being modest. We do not know what experiments will be most crucial 20 or 25 years from now, nor which ones will be established on the Moon. But it is obvious that the ideas put forth crossed many of the traditional boundaries which lie between NASA as a space agency and the Department of Energy which historically has pressed forward the urgency and the need for Earth-based high-energy physics facilities. As we approach the next century, these traditional boundaries and values need to be re-assessed.

Revolutions have occurred when people no longer look at the world the way they used to see it. Some of these proposals are bold, and some may be unrealistic. However, if they do so much as change our Earth-based view of physics, that in itself is a new initiative for all of us.

Subsequent to this Workshop, as you may know, the President announced a return to the Moon as NASA's next major space initiative beyond Earth and the Space Station. To the extent that this Workshop contributed to that decision, it was a success. For one brief day in May, we "got it off the ground."

It is a pleasure to thank the Lunar and Planetary Institute, the Universities Space Research Association (USRA), and Stanford University for their generous support and hospitality during this Workshop.

Thomas Wilson
Houston, Texas
December 1, 1989

Foreword

When one considers the title of this Workshop, its subject matter seems straightforward enough. Only after some consideration do you begin to realize the enormity of the unexplored realm of physics which we considered at Stanford. We recognized a need to establish new initiatives and studies for an emerging class of experiments in fundamental physics from the frontier of space, specifically from the Moon because it could support such a program as our closest neighbor and nearest opportunity. The Moon likewise provides a large fixed-base platform which can serve as an observatory of the Earth, its magnetosphere, and its fragile but complex dependence on the Sun.

Several notable points became clear. Aside from finishing the mandate of a global mapping of the Moon [H. C. Urey *et al.*, Philos. Trans. R. Soc. London A **285**, 600 (1977)] with a precursor mission and re-establishing studies of the Moon itself (e.g., atmospheric venting, sesimic networks, and magnetic fields), the possibility of a permanent human or man-tended presence at a mature lunar base implied the following possibilities:

- Probable migration of particles and fields research from low Earth orbit to the new frontier at the lunar base.
- Major large-scale arrays.
- A fundamental and high-energy physics facility.
- Particle astronomy and particle astrophysics testbed.
- Cosmic ray and cosmic abundance observatory.
- New vistas in neutrino/antineutrino astronomy.
- An Earth–Moon laboratory.
- A Moon-tethered space physics observatory.

None of these is currently funded by NASA. Further, none refers to physics of former programs. You will discover upon reading the manuscripts that when we speak of detector designs such as seismometers, for example, we are referring to a new breed of seismometer/gravimeter which might possibly search for gravitational radiation as well as probe planetary interiors. There is no such instrument in the Agency's present planetary programs, and the interferometry mentioned is totally new.

As scientific justifications for a return to the Moon, the above possibilities in turn raise several significant physics issues which need to be dealt with:

- There is no standard model of the Moon.
- The *in situ* construction of much of the instrumentation for particle detectors would establish a turning point in the problem (averting Earth-to-Moon transport expense).
- The *in situ* production of liquid scintillators (including water shared with the life-support systems), radiochemical solutions, and heavy gas (e.g., xenon) would likewise establish a turning point in the problem.
- Light-weight calorimeter design (of plastic?), and possibly inflatable detectors.
- Use of lunar material (e.g., dust and regolith) as much as possible (e.g., to pack plastic precision tubing machined and brought from Earth).
- The proposed L2 relay satellite system should accommodate scientific payloads.

These are just some of the multitude of issues which have now been raised, and which need international enabling resource and development as we progress forward.

The Editors
Andrew E. Potter
Thomas L. Wilson

NASA Workshop on Physics from a Lunar Base

ORGANIZING COMMITTEE:

Gautam Badhwar	NASA, Johnson Space Center
John W. Freeman, Jr.	Rice University
Andrew E. Potter	NASA, Johnson Space Center
Thomas L. Wilson	NASA, Johnson Space Center

SPONSORS:

National Aeronautics and Space Administration
Lunar and Planetary Institute
Stanford University

Earthrise, as viewed from Lunar orbit looking westward. This would become a common view for the Moon-tethered satellite system discussed at the Workshop. A return to the Moon would offer a unique observatory for studying the Earth and Moon, and their complex interaction with the solar wind. (AS8-14-2383).

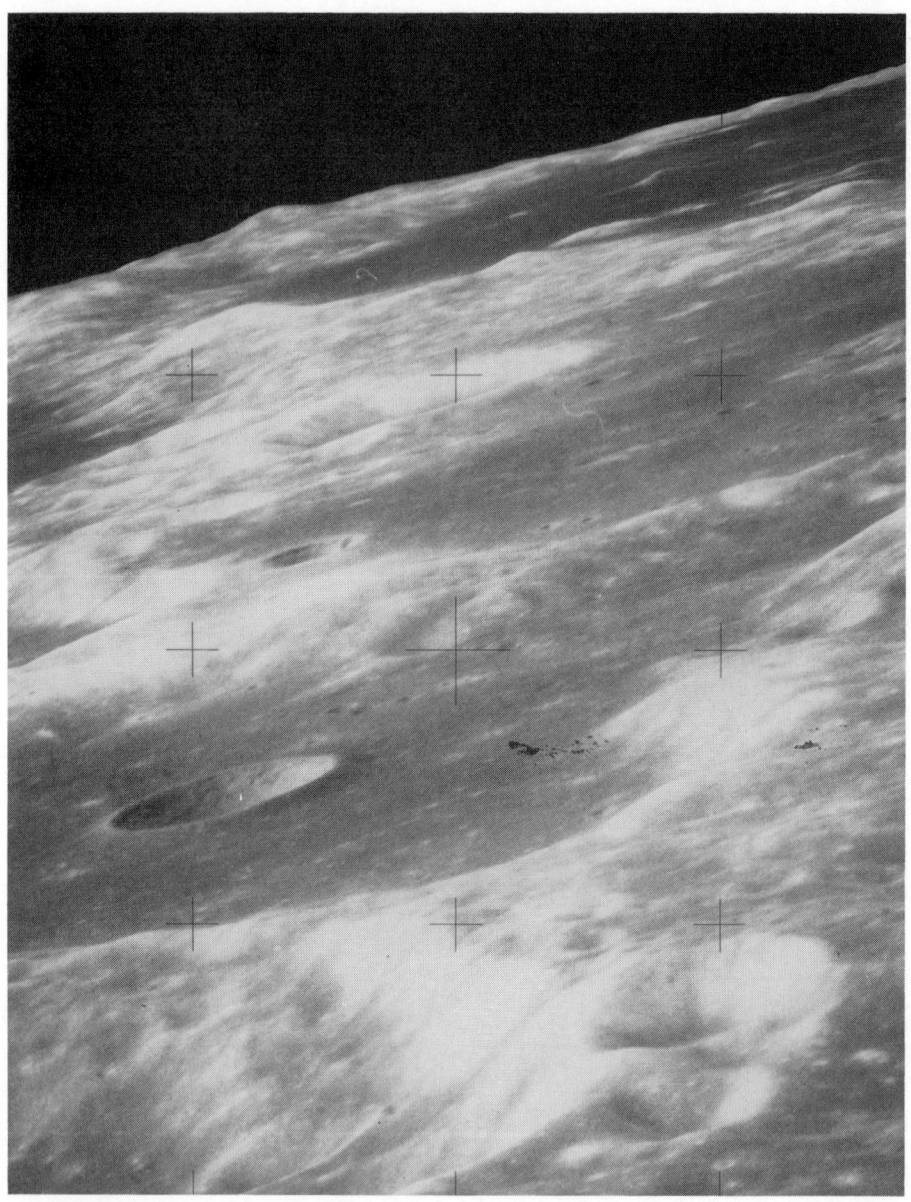

View from Lunar orbit at an altitude of 121 km of craters Maclaurin [1.9^0 south, 68.0^0 east: 50 km diameter] and Dubyago [4.4^0 north, 70.0^0 east: 51 km diameter]. Principal point (the intersection of photographic frame diagonals) is 0.5^0 south, 68.0^0 east. (AS15-81-10924).

A trans-earth coast view of Mare Australe (Southern Sea). The center of this view (the "headphones") is near latitude 34^0 south, longitude 96^0 east. In the upper left is Humboldt [27.2^0 south, 80.9^0 east: 207 km diameter], and in the upper right is Hilbert [18.0^0 south, 17.8^0 east: 170 km diameter]. Scaliger [27.1^0 south, 108.9^0 east: 84 km] is just below right, center, with much larger Milne [30.5^0 south, 112.6^0: 262 km] below it. Fermi and Tsiolkovsky are just east (to the right of the photgraph). Top center of the photo is north-northeast. (AS17-152-23288).

The crater flooded with dark mare material is Tsiolkovsky [20.4^0 south, 129.1^0 east: 180 km diameter], a candidate lunar base site on the far side of the Moon. (Lunar Orbiter III photograph 121M).

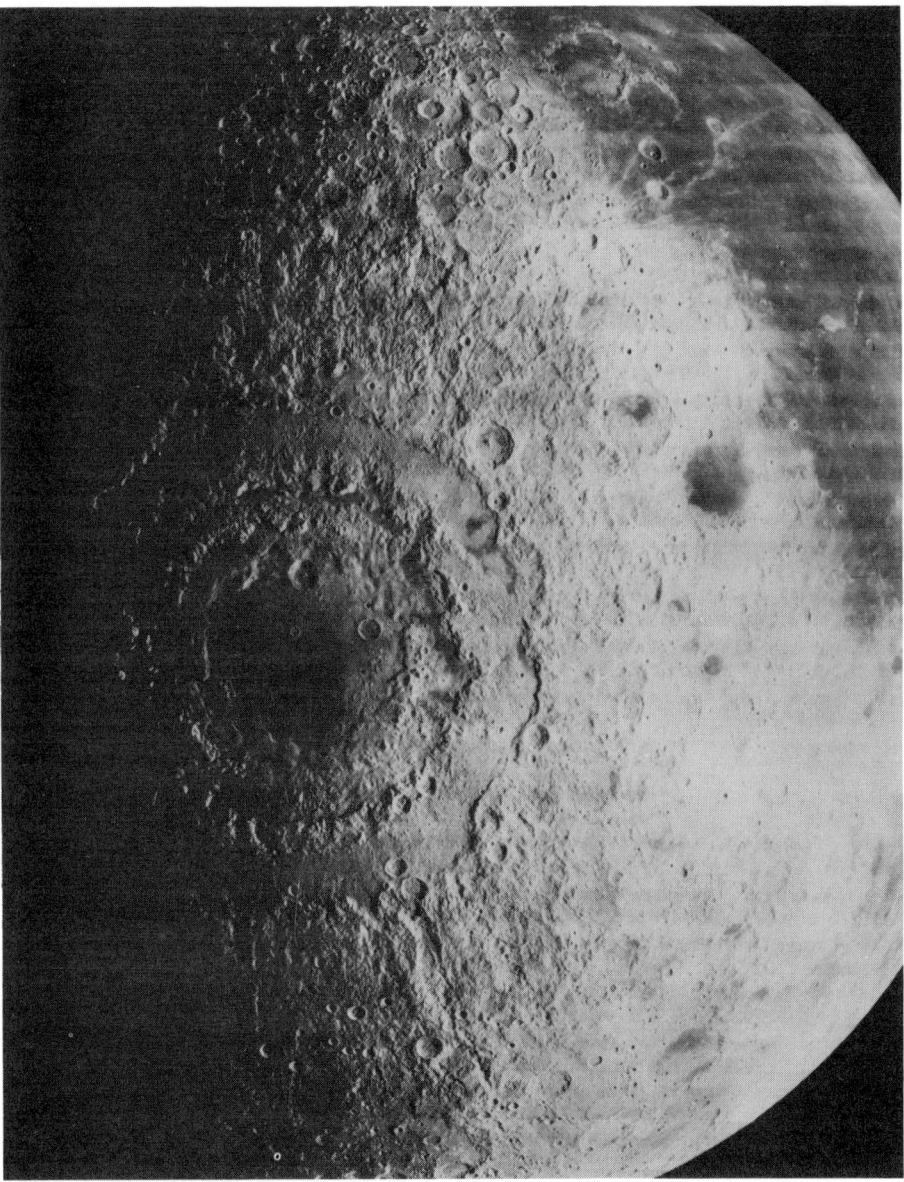

Uncommon views of the far-side of the Moon include this one of Mare Orientale (Eastern Sea) at 20^0 south, 95^0 west. 300 km in diameter, its concenric mountain systems include an outer ring (Montes Cordillera) 1500 km in diameter and 900 km across. Rim deposits exhibit a well-developed radial texture. (Lunar Orbiter IV, M-187).

View of far-side crater Mendeleev [5.6^0 north, 141.5^0 east: 330 km diameter], looking westward along the groundtrack. North is to the right, and crater Schuster [4.5^0 north, 146.5^0 east: 103 km diameter] is in the lower left-hand corner. (Apollo 16 metric camera frame 0472).

Session On Space Physics

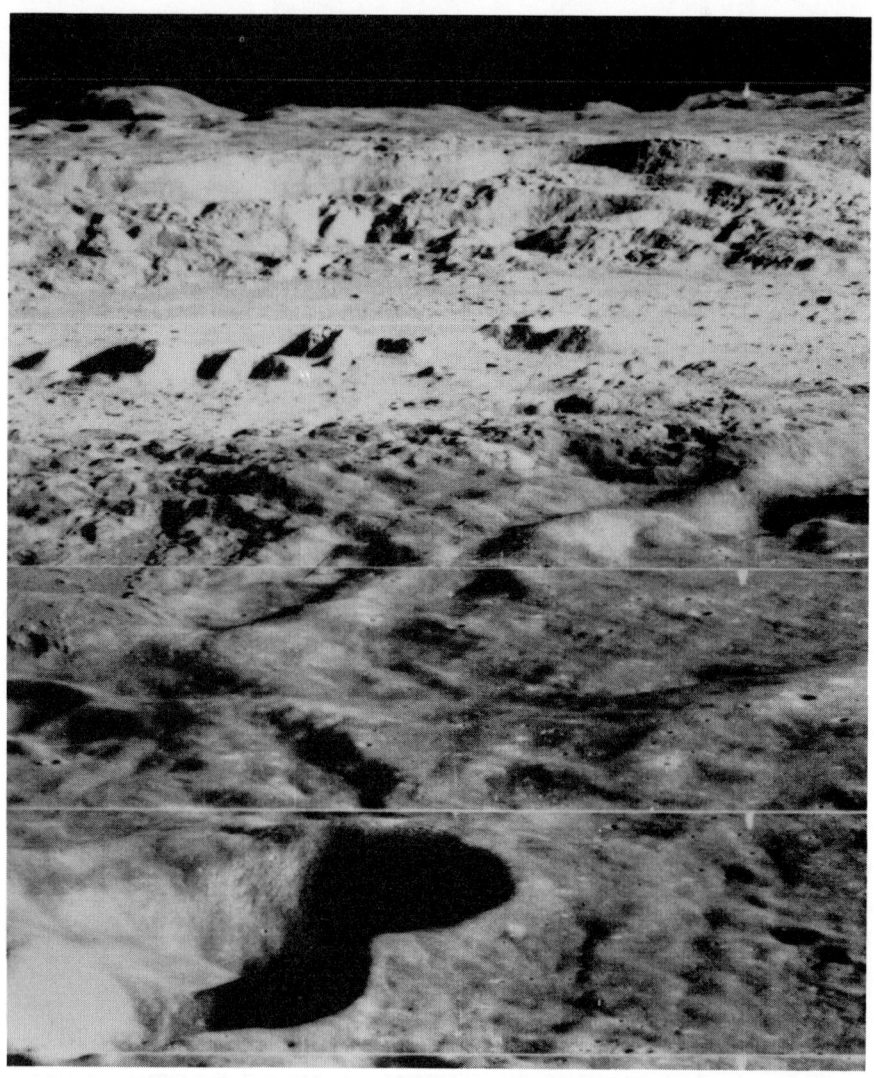

Looking north across Copernicus on the horizon [9.7⁰ north, 20.0⁰ west: 93 km diameter]. To its south and in the foreground is Fauth [6.3⁰ north, 20.1⁰ west: 12 km diameter]. (Lunar Orbiter II-162, H3, H2).

Space Physics from a Lunar Base Project

F. Curtis Michel

Space Physics and Astronomy Department, Rice University,
Houston, Texas 77251
May 20, 1989

A lunar base provides a unique and extremely valuable opportunity for Space Science. Figure 1 shows what I assume to be a baseline lunar base support system. Because it is largely from the platforms provided by the latter that space science observations would be made, I will not discuss at length the manned surface activities, although they will be invaluable in distributing and recovering surface experiments. I will simply take this ability to be a given.

The first element in a lunar base support system would be an Earth orbiting space station node, specialized for such support (i.e., not THE Space Station, which is a general-purpose craft as presently conceived). The obvious facilities required would be fuel storage and transfer, vehicle assembly, and personnel working quarters. To reach this Earth orbiting space station, we would need something like the shuttle or a follow-on to such systems.

To get to the Moon, some sort of transfer vehicle would be required to move fuel, equipment, and personnel to Lunar orbit, which would then return to Earth orbit (returning lunar material after the initial stages of setting up the base). At least two such vehicles would be required, given that one could be disabled in lunar orbit by malfunctions in which case the other would become a rescue/repair vehicle.

Once in Lunar orbit, a second space station would be required once again for exactly the same functions as the one in Earth orbit, except that this one would also support several (about 3) lunar excursion modules to reach the lunar surface and return.

For all practical purposes, this suite of space stations and transfer vehicles would constitute a lunar base regardless of whether a permanently manned surface habitat were maintained.

In addition, a number of unmanned satellites will be required. A triad of equally spaced tracking and communications satellites in lunar equatorial orbit will be required to assure communications with farside activities and provide a continuous data uplink and a command downlink to experiments on the surface. Of course, accurate locations of instruments and personnel on the lunar surface would be essential as well. Notice in this regard that the Moon offers a superior way to "anchor" a "cluster" of satellites. For a cluster in Earth or Solar orbit, the relative positions and directions between the elements of the cluster are continuously changing, which complicates data reduction. For satellites anchored to the Moon, the spacecraft spacing is constant and while the relative directions do change, they do so in a very regular and repeatable way.

Finally we require at least two mapping and survey satellites in low Lunar orbit, one in nearly polar orbit and the other in near equatorial orbit. It would be desirable but not essential

© 1990 American Institute of Physics

that these satellites have propulsion capabilities in order to customize their orbits. The polar orbiter would map the entire Moon at high resolution while the equatorial orbiter would concentrate on site selection for early surface operations.

Trade-Offs and Strategy

We assume here that the overall justification for a lunar base would not be scientific in nature, representing instead international cooperation or national prestige as the situation would have it. Nevertheless, scientific trade-offs are not entirely irrelevant, because ancillary scientific missions will inevitably be optimized to become cost-effective. We can make some crude estimates to illustrate the considerations involved: The outlay per pound for payload into Lunar orbit is about 33 times that into Earth orbit. The outlay per pound for payload onto the Lunar surface is 20 times that into Lunar orbit. The total outlay for setting up the project, *excluding* building suface habitat and costs of long-term maintenance of either, turns out to be about 16% less than the annual budget of the Department of Defense. These numbers reflect current experience (within factors of a few). NASA is in a better position to provide refined estimates if needed. Historically, *manned* projects have been resistant to economies owing to the requirements for safety of flight.

Thus, as an example, the scientific trade-offs for a parabolic radio telescope in Lunar orbit versus one on the Lunar surface are roughly the following: In Lunar orbit one would have a fully steerable instrument easily accessible from the Lunar space station which would be free from Earth interference about 40% of the time that it is in the farside shadowed orbit and probably relatively free of interference (of order 40 dB or more reduced from operation on Earth) the remaining time. In comparison, a surface installation on the farside would be free from interference 100% of the time but would be relatively inaccessible (a Lunar habitat would almost certainly be on the visible side of the Moon for public opinion reasons) and would very likely be a transit instrument (like Arecibo, but more restricted). All in all these trade-offs seem comparable until the comparative costs are examined. This discussion is illustrative, not definitive; it is not my intention to suggest that a radio telescope on the Lunar surface is impractical. It is only if one makes the above *assumptions* that one might arrive at that conclusion. The critical issue is whether such devices can largely be manufactured on the Lunar surface: If one can produce the massive elements on the Moon, it immediately becomes more sensible to do that than transport them to Lunar orbit. I will focus, however, on activities feasible in the early stages when the support system has just been set up and prior to significant surface activities.

Space Science Unknowns

Although a large number of unmanned missions have been flown in Earth orbit, they have proven less than expected owing to the intrinsic variability of the magnetosphere and the solar wind which perturbs it. We have built up a general cartoon approximation for the magnetosphere which describes the general trends seen from satellite observations (Fig. 2) and which is quite detailed but nevertheless is only a first approximation to what the magnetosphere must actually be like. For this reason, first-order uncertainties still exist over even apparently elementary questions such as what causes the aurorae and what is the role of magnetic reconnection in the overall physics of the magnetosphere. Reconnection is a fundamental space plasma process whose details remain poorly known and is therefore an on-going problem of importance. It is now proposed that during magnetic storms, large "plasmoids" break off the geomagnetic tail and are ejected anti-sunward. Unfortunately, deep tail regions are difficult to probe with a satellite because a trajectory that would take it into an interesting regime will be a very eccentric one, and after a few passes the orbit of Earth around the Sun will cause the trajectory to precess out of the region of interest, with the result that for most of the year the satellite will be cruising in regions of near-Earth space not of (original) interest. (A satellite in circular orbit would sample the same region more often, but without giving an radial information, and the precise locations of activity are presently unknown.) Structures reminiscent of plasmoids have been observed but the structures are very complex and have been hard to discern even where two spacecraft have been able to provide simultaneous data. In the same way, the structure of the forcing function – the solar wind – is difficult to monitor because the solar wind is variable both in time and in structure, a fact easily seen in satellite data but impossible to unscramble unequivocably from data from a single satellite. Thus even if we knew how the magnetosphere responded (the first difficulty discussed above), we would not necessarily know what it was responding *to*!

Space Science Opportunities from Lunar Base

Routine operations of a Lunar Base (with or without a habitat) open opportunities to resolve the above basic issue. First, consider the Lunar transfer vehicle. A trajectory that takes it from the Earth to the Moon is largely a radial one (as seen along a fixed Earth-Sun line). Thus it is well suited to sample just the regions of high uncertainty in the geomagnetic tail. From an instrument point of view, it immediately follows that the transfer vehicles be appropriately instrumented (magnetometer, particle flux detectors, radio receivers). From an operational point of view, some latitude in when transfers are (normally) made is essential if the down-tail regions are indeed to be explored. Thus it is important that these decisions be made early on and not at a late date where they become regarded as science-as-a-nuisance to the operational objectives. They need to be part of the mission objectives if at all possible.

Next, consider the tracking and communications satellites. If instrumented with a standard suite of space science instruments (e.g., magnetometer, particle flux detectors, radio receivers), these will give a high quality resolution of spatial versus temporal solar wind properties. Such data will not only yield fundamental information about the nature of the solar wind (presently inferred largely from single satellite measurements) but will provide a quantitative description of the "forcing" function to which the magnetosphere responds. With the three tracking and communications satellites one obtains not only temporal resolution but also resolution in two axes.

The Moon itself responds to the impressed magentic field of the solar wind, and this response allows one to probe the lunar interior in a unique way. Thus magnetometers on the Lunar surface and on the mapping satellites, in conjunction with the global monitoring from the tracking and communications satellites, should permit a rather precise determination of how the moon reacts to changes in the magnetic field which the solar wind bathes it in. Such behavior can be obtained simply by choosing events in which the same magentic field is observed at all three magnetometers (i.e., a reasonably uniform field) and which simultaneously changes to another uniform configuration.

Because the Moon spends a significant time in the geomagnetic tail, a number of active experiments (e.g., propagation of low frequency radio waves) can be performed to trace the magnetic field lines between the Earth and the Moon.

In summary, a lunar base would permit:
- Regular solar wind monitoring and intercorrelation from Lunar Orbiting Communications Satellites.
- Systematic probing of Geomagnetic Tail in otherwise inaccessible regions from Transfer Vehicle, together with multiple-point and long-term observations at 60 R_E.
- Probing of deep Lunar interior with surface magnetometers or on the mapping satellites which can then be intercorrelated with solar wind monitors on the communications satellites.
- Active experiments between the Moon and Earth (each way) which would permit a determination of how the magnetic field lines actually connect (low frequency radio, particle injection, etc.).
- High precision determination of the Lunar gravity distribution (and hence internal structure) from tracking of the mapping satellites.

Without a lunar base, there is very little likelihood that such an extensive system for unraveling the magnetospheric behavior of the Earth would be put into place.

Operational Opportunities

The scientific measurements possible on the Moon are also of direct relevance to the entire mission. It is clear from the trade-off and strategy estimates that one needs to exploit any available lunar resources *at the earliest possible opportunity*. The key resource that might be found on the Moon is water. Because any atmosphere has a residence time of only weeks, water injected into the extremely thin atmosphere will simply be lost. However, if there are intermittent outgassing events from the Lunar interior (which is not excluded and which may have been observed from the Apollo ALSEP experiments), it is imperative that they be localized and possibly tapped. Such a source would permit liquid oxygen/hydrogen propellant to be manufactured on-site at essentially no cost compared to hauling propellants from the Earth (which, in addition, have significantly lower specific impulse because they will typically NOT be cryogenic for a number of reasons: safety of flight, high mass overhead of in-flight refrigeration, etc.).

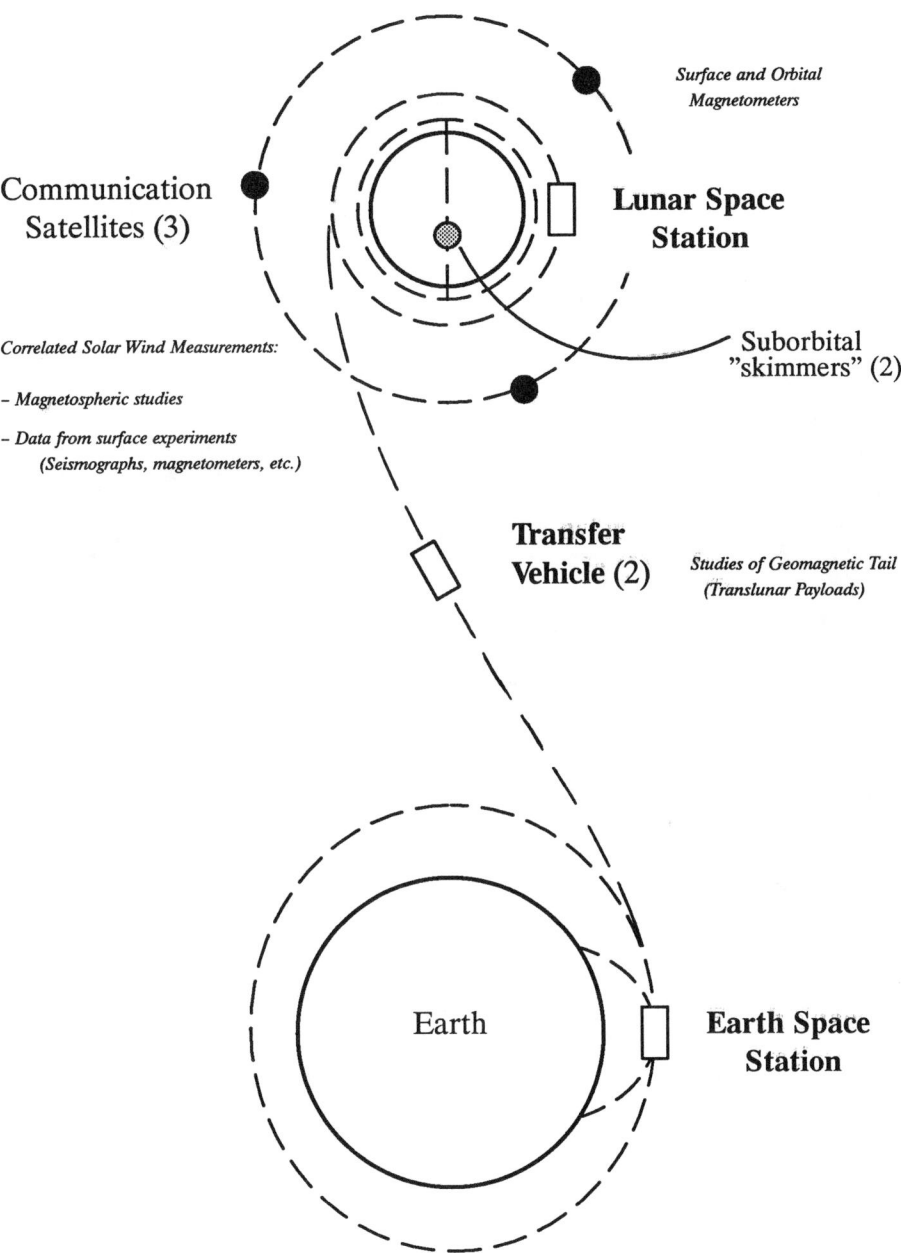

Figure 1. A baseline Lunar Base support system.

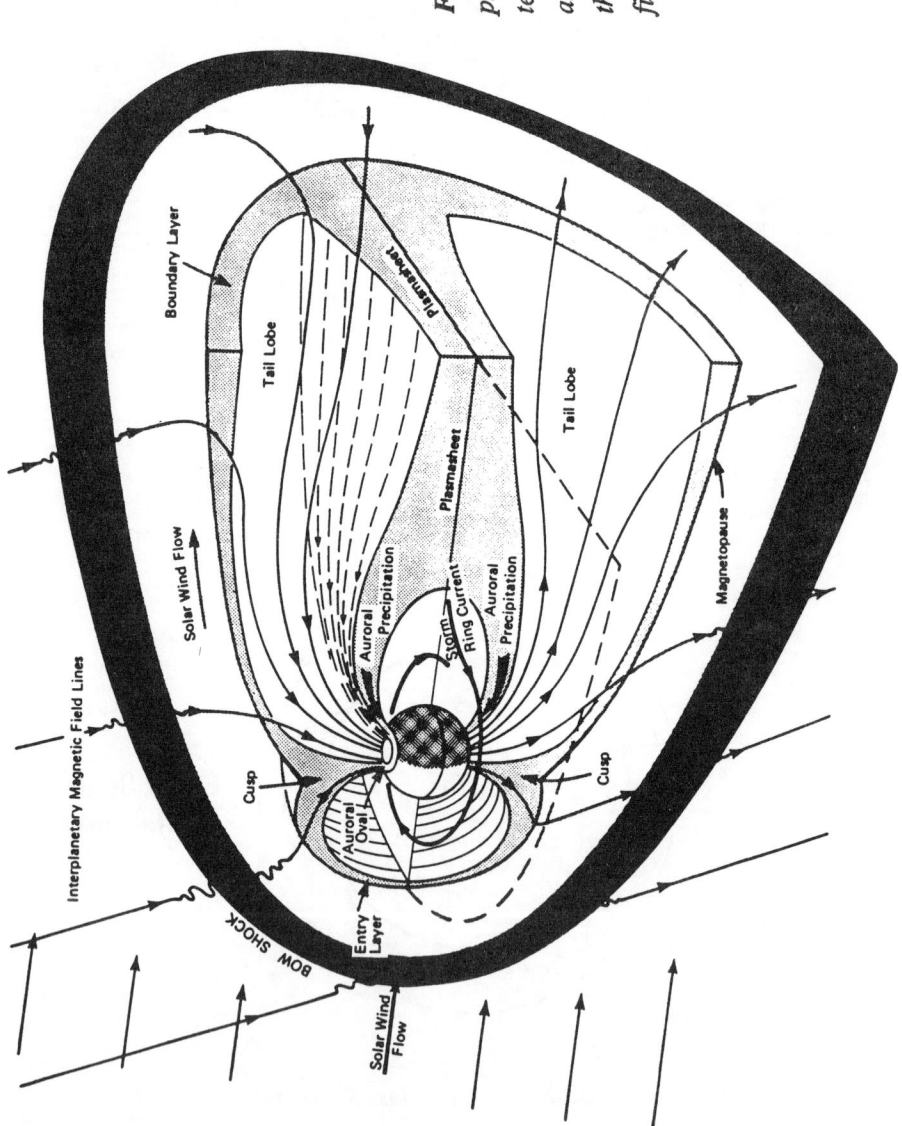

Figure 2. Sketch of the principal regions of the terrestrial magnetosphere and its connections to the solar wind magnetic field (not to scale).

The Moon and the Magnetosphere and Prospects for Neutral Particle Imaging

John W. Freeman, Jr.
Rice University, Houston,Tx 77251

Abstract

Despite nearly thirty years of study by earth-orbiting satellites, major questions remain concerning the dynamics of the earth's magnetosphere. This is the unavoidable result of attempting to understand a large complex and continuously changing system with a few isolated, single-point measurements. A solution to this problem may lie in imaging the magnetosphere with energetic neutral atoms generated by charge-exchange reactions between the magnetospheric plasma and exospheric hydrogen atoms. A lunar base would make an excellent location from which to carry out such an experiment.

Lunar Plasma Regions

As the moon makes its monthly journey around the earth, it is exposed to four plasma environments. Three of these are in the magnetospheric tail of the earth and the fourth is the solar wind itself. Each of these environments has a unique set of charged particle characteristics; fluxes, temperatures and densities. The principal plasma components of each of these regions are electrons and protons, however, in some cases heavy ions are present in significant numbers and their charge state provides a clue to their solar or atmospheric origin. Figure 1 shows these regions and summarizes the characteristics of each region as determined partly by the Apollo ALSEP experiments.

The hottest region is the Plasmasheet, which lies at the center of the magnetospheric tail, and which is characterized by electrons and protons with temperatures ranging up to 10 keV.[1,2] Electron and ion densities in the plasmasheet are quite low, however, and while large scale plasma motions exist, they are irregular and episodic in nature. Figure 2 shows two sample plasmasheet ion spectra.

Outside the plasmasheet lies the magnetosheath. The magnetosheath is heated and decelerated solar wind plasma. It is somewhat cooler than the plasmasheet,[3] but has a bulk motion antisunward that approaches solar wind velocities. The density of the magnetosheath is higher than that of the plasmasheet.

Above and below the plasmasheet, lies a region called the tail lobes. The thickness of the plasmasheet is less than the latitudinal excursions of the lunar orbit hence the moon is often carried into the lobe plasma. This plasma, thought to originate from the plasma mantle, high above the poles, is very cold and tenuous but moves with a steady flow velocity of about 100 km/s away from the earth.[2] Figure 2

© 1990 American Institute of Physics

also shows a lobe plasma ion spectrum. This plasma contains ions heavier that hydrogen, thought to originate in the earth's ionosphere.

The solar wind itself is a region where the flow energy exceeds the particle thermal energy. Velocities range from about 250 to over 1000 km/s and ion temperatures range from 0.3 to over 50 eV.[4] Over most of the moon the local magnetic field is sufficiently low that the solar wind reaches the lunar surface unimpeded.[5,6,7]

Magnetospheric Dynamics

The magnetosphere is not static. Plasma flows through the magnetosphere continuously forming a regular convection pattern which generally consists of motion toward the sun. Figure 3 shows this pattern as viewed in the noon-midnight meridian cross-section plane. As plasma from the plasmasheet in the deep tail flows toward the earth and encounters the stronger magnetic field near the earth, it is deflected north and south to form the auroral zones and east and west to form the ring current.

The magnitude of this flow is not steady but rather is modulated under the varying influence of the solar wind pressure and the direction of the interplanetary magnetic field. Long term enhancements of the convection velocity, plasma density, and temperature, with time scales from many hours to days, are called magnetospheric storms. Shorter time scale modulations, of the order of a few hours or less, are often convulsive in nature and each convulsion is referred to as a substorm.

The details of the magnetospheric convection pattern and its associated variations, as they are understood to date, have been inferred primarily from single-point measurements from a relatively small number of isolated satellites. Such measurements are fraught with a large number of difficulties including ambiguities such as the inability to distinguish spatial variations from temporal variations. Moreover, it is almost impossible to piece together the big picture on a single storm. What is needed is a series of snapshots of the plasma motion of the entire magnetosphere at once. It might be possible to do just that through neutral particle imaging.

Neutral Particle Imaging of the Magnetosphere

In addition to the tenuous plasma, the magnetosphere is filled with relatively cold exospheric neutral hydrogen atoms. The energetic, convecting ions can undergo electron transfer processes with the cold neutrals giving rise to fast neutral hydrogen atoms and cold ions. The resulting fast neutral hydrogen atoms leave the magnetosphere on straight-line trajectories, unimpeded by the earth's magnetic field. Viewed from a distance, they will give a picture of the entire magnetosphere since their source regions will represent the regions of enhanced plasma activity. A fast neutral detector on the moon could, therefore, image the magnetosphere and monitor the full dynamics of a magnetic storm. In this fashion, it would be possible to watch the development of a storm throughout the entire magnetosphere at the same time. Figure 4 illustrates the concept.

Indeed, Roelof et al.[8] have already reported the observation of energetic neutral atoms (ENAs) with energy of 50 KeV from the magnetospheric ring current using solid-state detectors aboard the IMP 7/8 and ISEE 1 satellites and they suggest this as a technique for imaging the rest of the magnetosphere. We propose to do just that using the moon as the base for the imaging detector and lowering the energy of the observed neutrals.

Feasibility

We can examine briefly the feasibility of this proposal. Using values typical of solar maximum, for a neutral hydrogen density of 10 atoms/cm^3 and an ion flux of 10^8/cm^2-s, we obtain a fast ion production rate of order 10^{-6} neutrals/cm^3-s at the source region. With a source volume of 2x2x5 earth radii, about right for a region of enhanced plasma near the plasmasheet boundary (see figure 4), we obtain a flux of fast neutrals at the orbit of the moon of about 1 atom / cm^2-s. This is a low flux, however, the magnetosphere changes on a time scale of hours and integration times of the order of 10 minutes should allow the dynamics of a substorm to be resolved.

For the imaging detector, we might consider a pin-hole camera with a CCD array or microchannel plate for the sensor as shown in figure 5. A CCD array would permit the detection of lower energy protons than the solid-state detectors used by Roelof et al.[8] The image size of the magnetosphere in such a device would be about 10 degrees. A stripping foil might be required to convert the neutrals to ions before impinging on the CCDs and also to exclude light.

The moon is a logical place from which to conduct imaging of the earth in other wavelengths as well. For example, atmospheric, ionospheric and, in particular, auroral phenomena could be imaged in the UV, X-ray, radio and infra-red wavelenghts.

Advantages of the moon for magnetospheric and/or atmospheric imaging include the following:

• the distance between the moon and the earth is about right for a good sized image. For example, all parts of the inner magnetosphere can be seen in an image size about 10 degrees in diameter;

• the phase-lock nature of the moon provides a stable platform with a continuous view back toward the earth;

• the direction or cross-sectional view is favorable particularly when the moon is at first or third quarter; in other words when the sun is at right-angles to the earth-sun line;

• cosmic ray background can be reduced by placing the detector in a hole below the lunar surface.

Summary

Imaging of the earth's magnetosphere by neutral atoms generated in charge-exchange processes appears to be a practical and useful experiment to consider for a nearside lunar base.

The results of such an experiment could lead to a very substantial increase in our knowledge of the dynamics of the earth's magnetosphere.

Features which make this concept advantageous also apply to imaging of the earth's atmosphere and ionosphere in the UV, x-ray, infra-red and radio wavelengths as well. Plans for a lunar base should include further evaluation of various forms of imaging af the earth and the magnetosphere from the moon.

References:

1. T. W. Hill, J. Geophys. Res., **12**, 379 (1974).

2. David A.Hardy, John W. Freeman and H. Kent Hills, Magnetospheric Particles and Fields, Ed. B M. McCormac, D. Reidel Publishing Co. (1976).

3. G.D. Sanders, J.W. Freeman and L.J. Maher, J. Geophys. Res., **86**, 2475 (1981).

4. Ramon E. Lopez and John W. Freeman, J. Geophys. Res., **91**, 1701 (1986).

5. G.L. Siscoe and B. Goldstein, J. Geophys. Res., 78, 6741 (1973).

6. B.J. O'Brien and D.L. Reasoner, NASA Spec. Publ. 272, 193 (1971).

7. C.W. Snyder and D.R. Clay and M. Neugebauer, NASA Spec.Publ. 235, (1970).

8. E.C. Roelof, D.G. Mitchell and D.J. Williams, J. Geophys. Res., 90, 10,991 (1985).

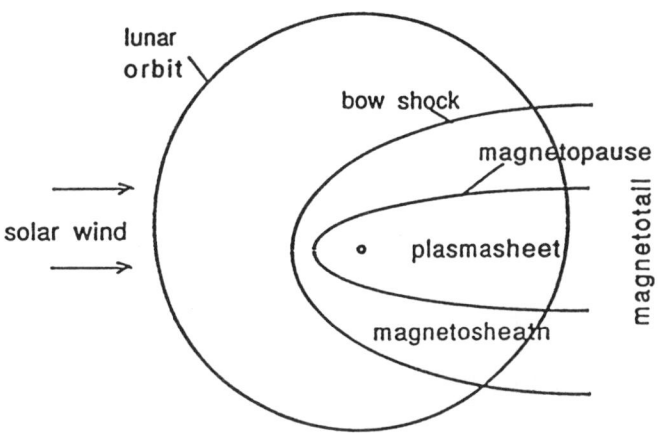

Solar Wind:
  Protons: V=400-1000 Km/s
    kT=0.3-50 eV
    n=1-15/cm^3
  Electrons: kT=1-100 eV
    n=1-15/cm^3
  Misc. Heavy ions with multiple ionization states

Tail Lobe:
  Protons etc.: V=100 km/s
    kT=1-10 eV
    n=0.1-1 /cm^3
  Electrons: ?

Magnetosheath:
  Protons: V=100-500 km/s
    kT=50-100 eV
    n=10/cm^3
  Electrons: kT=50-100 eV
    n=10/cm^3

Plasmasheet:
  Protons: kT=500-5000 eV
    n=0.25-1 /cm^3
  Electrons: kT=1000-10,000 eV
    n=0.25-1/cm^3
  Misc. Heavy ions singly ionized

Plasma Regions in Lunar Orbit

Figure 1

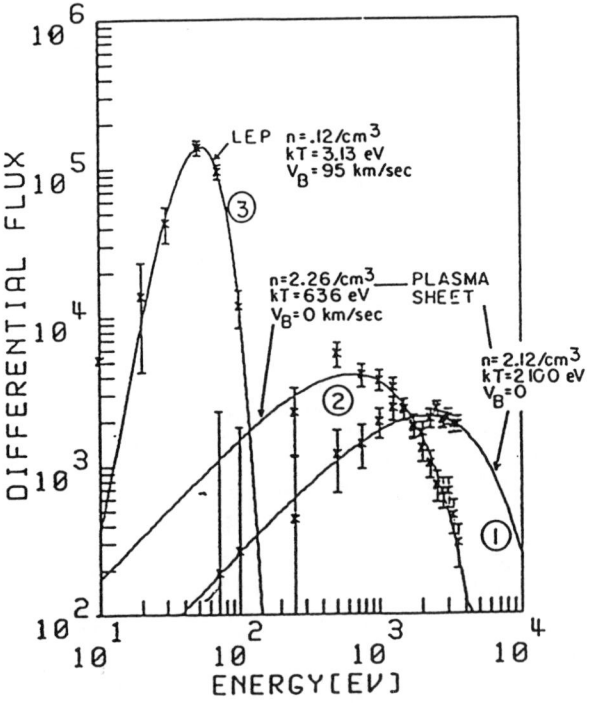

Figure 2. Differential flux spectra for the Lobe Plasma (LEP) and the Plasm ions. The flux is in ions / cm^{-2} s^{-1} sr^{-1} eV^{-1}.

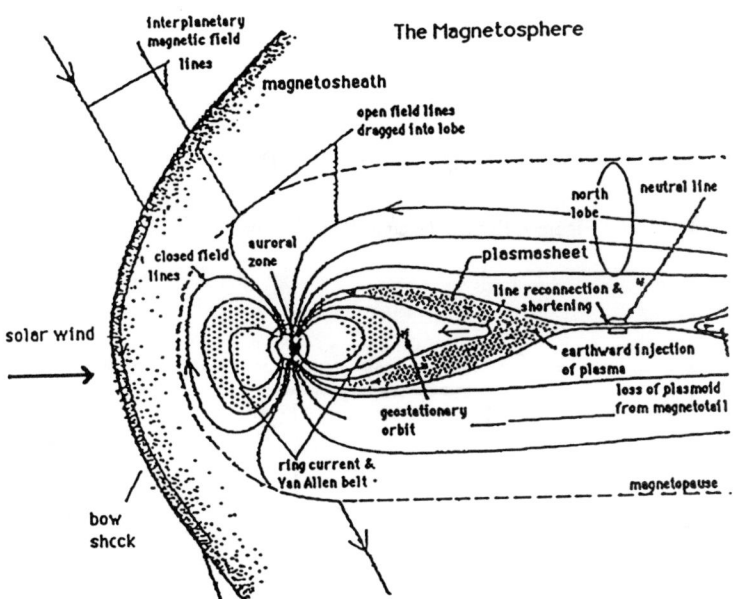

Figure 3. The magnetosphere in noon-midnight cross-section.

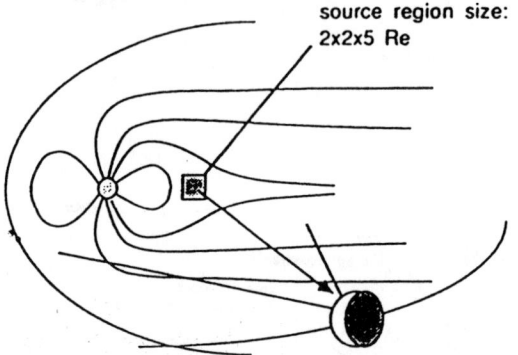

Figure 4. Fast neutral atom source region.

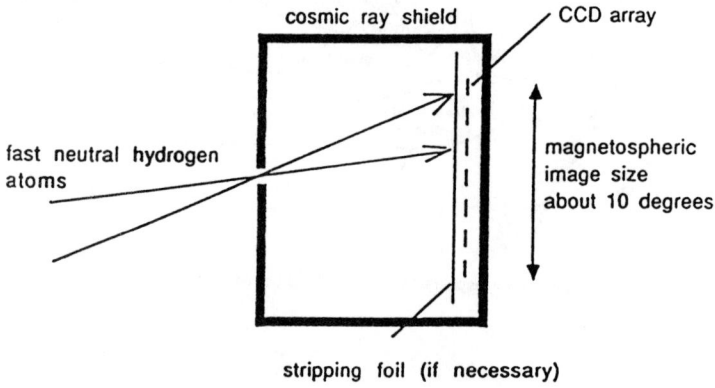

Figure 5. Fast neutral imaging detector

Space Plasma Physics Research at a Lunar Base

Richard Vondrak
Space Sciences Laboratory
Lockheed Palo Alto Research Laboratory
3251 Hanover Street
Palo Alto, CA 94304

Introduction

The lunar surface is an ideal location for several types of space plasma physics experiments. Observations of space plasmas can be made on the moon due to the absence of a substantial atmosphere or magnetic field that would shield the lunar surface. The lunar orbit traverses several distinct regions of space, spending each month about twenty days in the solar wind, four days in the magnetosheath, and four days in the magnetospheric tail. Furthermore, the moon provides a unique location and perspective for viewing the earth and the earth's magnetosphere. Images of the earth, the sun, and other solar system objects can be made over a wide range of wavelengths because of the absence of atmospheric absorption on the moon.

A variety of techniques can be used: direct in-situ measurements, remote sensing observations, and active experiments with controlled releases for diagnostics and for perturbations. Table 1 lists some representative space plasma physics experiments that can be made at a lunar base.

Table 1. Space plasma physics Research At a Lunar Base

- In-situ PLASMA MEASUREMENTS
    - Magnetospheric Phenomena
    - Solar-wind
    - Lunar Ionosphere

- REMOTE-SENSING OBSERVATIONS
    - Magnetospheric Structure
    - Terrestrial Ionosphere and Aurora
    - Solar Phenomena
    - Comet and Planets

- ACTIVE EXPERIMENTS
    - Gas Releases
    - Radiowave Experiments

© 1990 American Institute of Physics

Space plasma physics research in connection with a lunar base can provide answers to many outstanding questions such as,

- o Do plasmoids form in the geomagnetic tail?
- o How does the polar aurora in conjugate regions respond to variations in the solar wind?
- o What is the isotopic composition of the sun? How does it vary in the present and ancient sun?
- o What is the composition and variation of the lunar ionosphere and photoelectron layer?
- o What is the configuration of the Jovian magnetosphere?

This paper briefly describes several kinds of plasma research that can be accomplished as part of a lunar base program and shows some examples of previous observations.

Terrestrial Ionosphere and Magnetosphere

Instrumentation at a lunar base can be used to study the earth's space plasma environment. In-situ plasma and field measurements can be made of the magnetosheath and magnetospheric tail. The general configuration of the earth's magnetic field in space is shown in Figure 1. The flow of plasma from the sun, referred to as the solar wind, distorts the earth's magnetic field into a comet-like structure, known as the magnetosphere. Each month the moon passes through the long cylindrical magnetotail (see Figure 1). The magnetotail is not a static structure. In its interaction with the solar wind the magnetosphere gains energy and plasma, some of which is dissipated in the earth's upper atmosphere in the aurora. An outstanding objective of space plasma physics research has been to identify the process by which magnetic field energy in the magnetotail is converted into the kinetic energy of the auroral plasma. Evidence from limited satellite observations indicates that during an auroral substorm part of the earth's magnetotail is pinched off into a large closed body of magnetized plasma, called a plasmoid. The dynamics of such structures is not understood, primarily due to the paucity of satellite observations. A network of plasma sensors on the lunar surface could resolve the outstanding questions regarding the occurrence and characteristics of plasmoids in the magnetosphere. These observations would be greatly enhanced by plasma measurements made on lunar transport vehicles in cislunar space.

Imagers on the moon can be used to study the earth's atmosphere and the aurora. At present such measurements are made from the earth's surface or from low earth orbit with a very limited field-of-view. Imagers at geosynchronous orbit are unable to view well the polar atmosphere and aurora. At present the only way to image the entire polar aurora is with a sensor on a spacecraft in eccentric polar orbit. Such a sensor can view only one polar region. The moon is far enough away from the earth so as to obtain a synoptic view of an entire hemisphere and to view the atmosphere in both polar regions simultaneously.

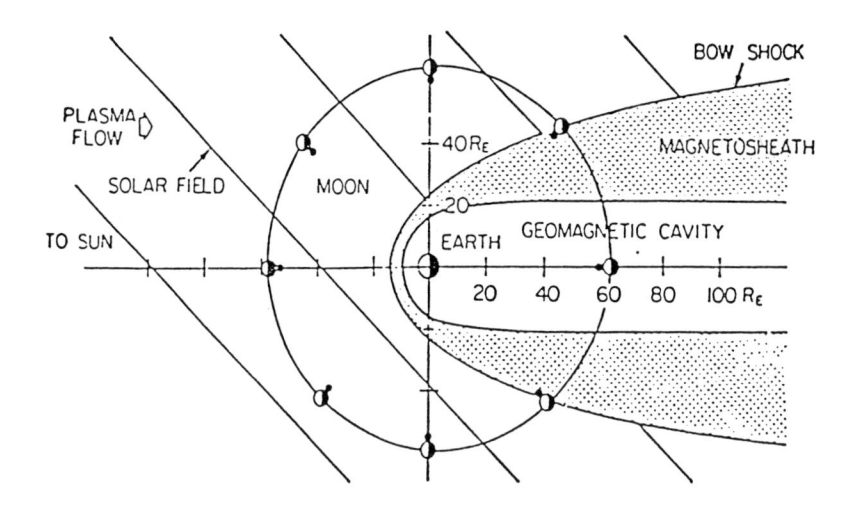

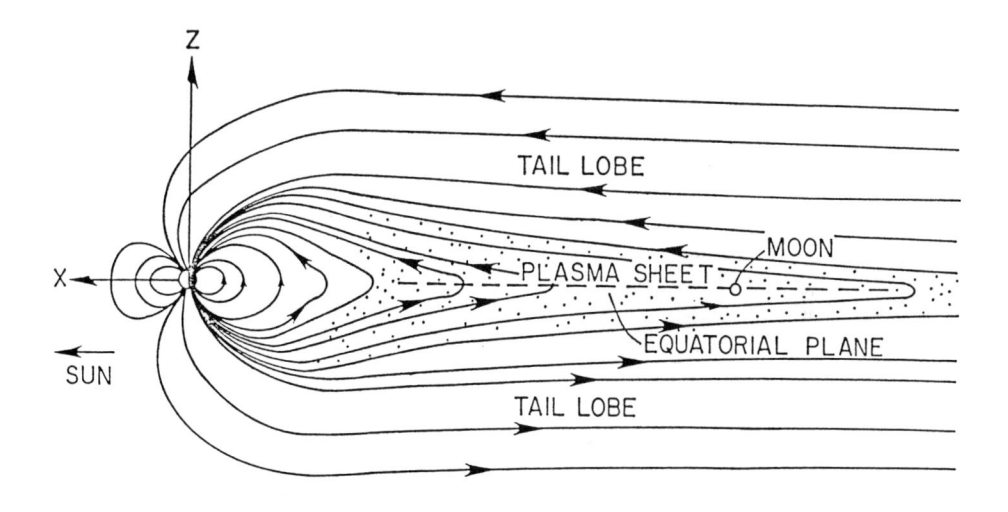

Figure 1. Cross-sections of the earth's magnetosphere in the ecliptic plane (above) and in the meridional plane (below).

Imaging of the earths's plasma environment from the lunar surface was accomplished by a far ultraviolet camera/spectrograph deployed during the Apollo 16 mission (Carruthers and Page, 1976). This instrument was based on an electrographic Schmidt camera having a 20° field-of-view and an angular resolution of 2 arc min. Images were made of the aurora and airglow in the wavelength range of 1250 to 1600 Å, which includes emissions of atomic oxygen and molecular nitrogen. Figure 2 is an image of the earth that shows the dayglow, the polar aurora, and the tropical airglow belts. Also shown is a diagram illustrating the orientation of the earth as seen from the Apollo 16 landing site and the approximate location of the far ultraviolet features. These Apollo images were the first ever made of the aurora in the far ultraviolet and the first that showed the aurora in both polar regions simultaneously.

An exciting new possibility is imaging of the entire magnetosphere including the high-altitude auroral acceleration regions and the magnetospheric tail. The magnetic field configuration shown in Figure 1 has been deduced generally from numerous single-point observations by isolated satellites. A major goal of future missions, such as the CLUSTER and Global Geospace Science missions during the International Solar-Terrestrial Physics (ISTP) program is to use multi-satellite measurements to define better they dynamics of the magnetosphere. However, to identify the instantaneous configuration of the magnetosphere and the dynamics of magnetospheric boundaries it will be necessary to use imaging systems. As shown in Figure 3 (Space Science Board, 1988), magnetospheric imaging from the moon has been identified as a primary objective for post-ISTP research. Imaging of magnetospheric regions can be accomplished by observing magnetospheric ions in resonantly scattered sunlight (Chiu et al., 1986; Swift et al., 1989). Table 2

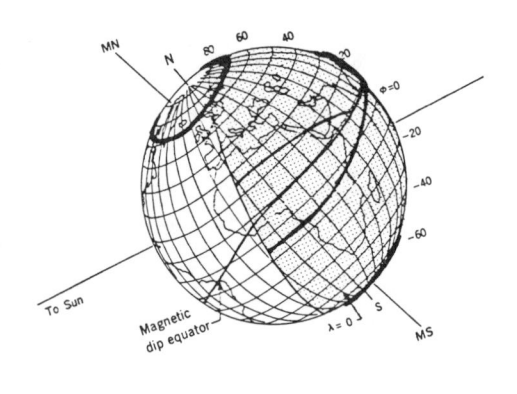

Figure 2 An image of the earth's airglow and aurora made with the far ultraviolet camera/spectrograph deployed on the lunar surface during the Apollo 16 mission. The orientation of the earth is shown for reference (Carruthers and Page, 1976).

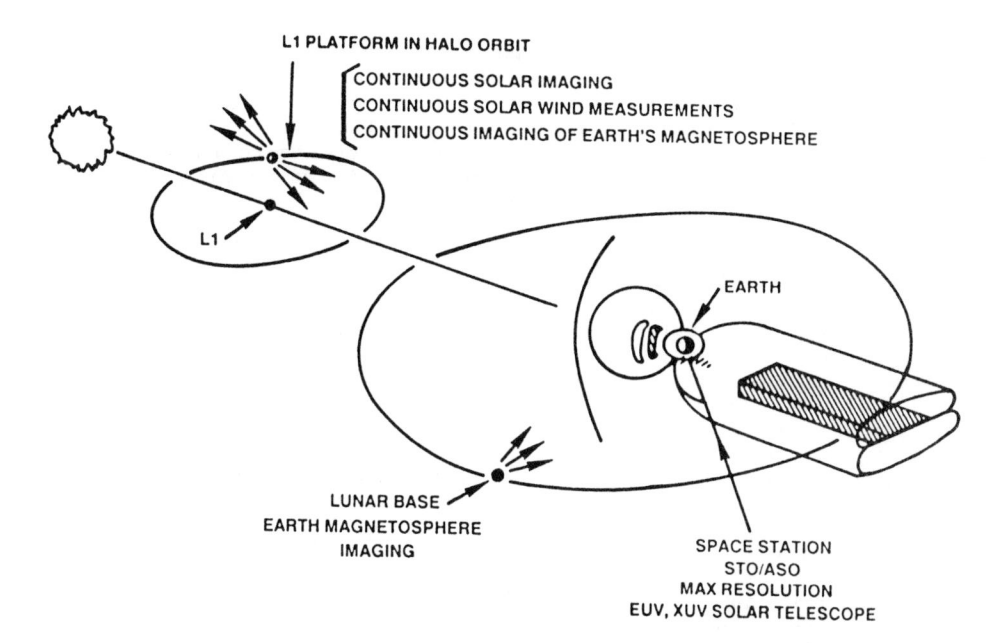

Figure 3. Location of instruments for imaging the terrestrial magnetosphere, as recommended by the Space Science Board, 1988.

Table 2. Solar resonance imaging of the magnetosphere

Line (Å)	Ion	Region
304	He+	Plasmasphere
834	O+	Auroral outflow, Plasma sheet
1216	H	Geocorona
1304	O	Exosphere

lists some of the wavelengths that might be used. Other techniques to image the magnetosphere from the moon could use neutral particle cameras to observe energetic neutrals produced by charge exchange in the inner magnetosphere. Another possibility is to obtain three-dimensional topographic images of the magnetosphere by monitoring coherent radiowave signals from satellites in earth orbit (Chiu et al., 1989).

Solar and Heliospheric Physics

At the lunar surface the sun can be observed each month for 14 days. Continuous viewing of the sun for such an extended period would allow improvement in helioseismological study of the structure and dynamical behavior of the solar interior. Continuous long-duration observations of the sun are difficult on the earth and require coordination of worldwide observatories or the use of polar facilities that are frequently obscured by weather. Other solar phenomena, such as solar flares and coronal mass transients, can be observed at many wavelengths because of the absence of an absorbing lunar atmosphere.

Plasma and field sensors on the lunar surface can measure directly the composition and dynamics of the solar wind and solar flare particles. A network of such sensors on the lunar surface can be used to identify small-scale structures in the solar wind.

An outstanding question in solar physics is the isotopic composition of the solar corona and its variation in time. The solar wind ions possess enough energy to penetrate several hundred angstroms into soil grains lying on the lunar surface. Isotopic analysis of these implanted gases suggests that the N15/N14 ratio has increased in the sun during the past few billion years (Kerridge, 1989). Such a change is inexplicable within our present understanding of solar processes. The lunar surface may provide a record of solar history and cosmic ray variations analogous to the record of the terrestrial climate and atmospheric variations that has been extracted from the Greenland icecap.

Planetary and Cometary Physics

The moon is a unique object in the solar system with respect to solar wind interactions. All other large bodies have a thick atmosphere (e.g. Venus, Mars, Comets) or a large-scale magnetic field (e.g. Mercury, Earth, Jupiter) that prevents the solar wind from penetrating to the planetary surface. Although surveyed during the Apollo program, the plasma interactions between the moon and the space environment are still not fully understood. Unresolved questions remain regarding the characteristics and dynamics of the lunar ionosphere, the lunar photoelectron layer, and the lunar surface electric field.

Planets and other bodies in the solar system can be observed with the imagers used to study the earth's magnetosphere in aurora. In particular, the magnetospheres of Jupiter and of comets could be measured, although with a reduced spatial resolution.

Active Experiments

An important method of space plasma physics research uses the controlled injection of material for perturbation and diagnostic tracer experiments. Injection at a controlled rate can be made with fluorescent chemicals, tracer ions, electron beams, or radiowaves. Such experiments test our understanding of the processes that energize and transport plasmas in space.

Figure 4 shows one proposed active experiment (Hones, 1985) to observe the acceleration and motions of a plasmoid down the magnetospheric tail. A satellite would release several kilograms of calcium ions that would be visible from the moon. The motion of the cloud would be followed as a tracer of plasmoid dynamics during an auroral substorm.

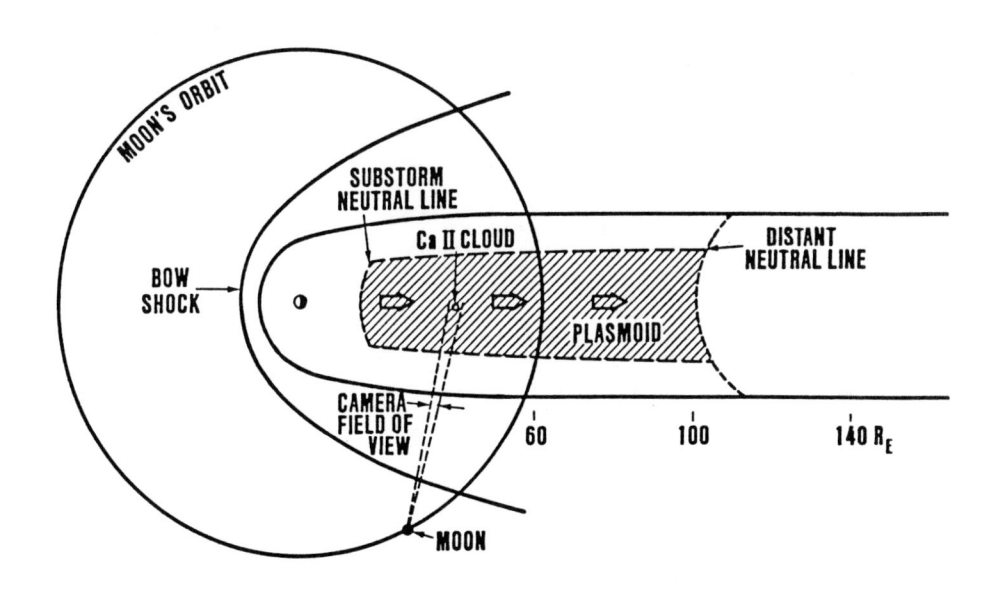

Figure 4. Conceptual experiment using a calcium ion cloud, viewed from the moon, to trace the development and motions of a plasmoid in the geomagnetic tail (Hones, 1986).

The transportation activities required for deployment and support of a lunar base will result in substantial chemical releases in the form of rocket exhaust or gases released from the surface. Measurements of the plasma phenomena associated with such activities provide an opportunity to study the interaction of neutral gases with space plasmas. For example, on April 15, 1970 the Apollo 13 S-IVB stage impacted the nighttime lunar surface approximately 140 km west of the Apollo 12 ALSEP site and 410 km west of the dawn terminator (see Figure 5). Beginning 20 s after impact the Suprathermal Ion Detector Experiment and the Solar Wind Spectrometer observed a large flux of positive ions (maximum flux ~ 3 x 10^8 ions/cm2s sr) and electrons (Lindeman et al., 1974). Two separate streams of ions were observed: a horizontal flux of predominantly heavy ions (>10 amu), which probably were material vaporized from the S-IVB stage. An examinations of the data shows that collisions between neutral molecules and hot electrons (50 eV) were probably an important ionizations mechanism in the impact-produced neutral gas cloud. These electrons, which were detected by the Solar Wind Spectrometer, are thought to have been energized in a shock front or some form of intense interaction regions between the cloud and the solar wind. Thus, strong ionization and acceleration are seen under conditions approaching a collisionless state. This observation was the first experimental evidence that the phenomenon known as "the critical velocity hypothesis" operates in the space environment.

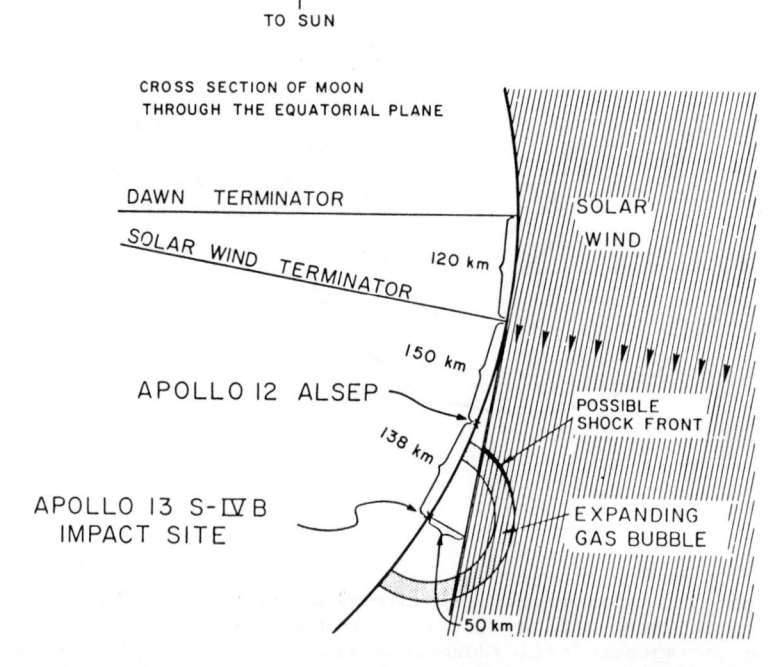

Figure 5. An equatorial cross-sectional view of the moon showing the location of the Apollo 13 S-IV B impact with respect to the Apollo 12 ALSEP site and the dawn and solar wind terminators (Lindeman et al., 19784).

Conclusions

Significant advances in our understanding of space plasma physics can result from research accomplished as part of a lunar base program. The moon is a unique location for many significant observations.

The experimental sensors used for space plasma physics measurements are generally modest, weighing typically 20 kg and requiring about 10 watts of power, and no new technology development would be required for most observations. Instruments can be placed at landing sites, in lunar orbit, or on lunar transfer vehicles operating between the earth and the moon.

Measurements of the plasma environment at the lunar surface will be of significant benefit to other aspects of the lunar base program, such as the search for resources and the monitoring of environmental alteration. Because the lunar ionosphere and atmosphere are subject to alteration by human activities, they need to be continuously monitored as part of a lunar base program. Sensors placed on the lunar surface to monitor the lunar ionosphere and atmosphere could detect any natural gas releases and could identify the location of any endogenous source of volatiles. The discovery of such volatiles would greatly facilitate the establishment of a lunar base.

References

Carruthers, G. and T. Page, "Apollo 16 far ultraviolet imagery of the polar auroras, tropical airglow belts, and general airglow", J. Geophys. R., 81, 483-496, 1976.

Chiu, Y. T., R. M. Robinson, G. R. Swenson, S. Chakrobarti, and D. S. Evans, "Concept and verification of imaging the outflow of ionospheric ions into the magnetosphere", Nature, 322, 441, 1986.

Chiu, Y. T., G. T. Davidson, and R. M. Robinson, "Radio wave tomagraphic imagery of magnetospheres," to be presented at Fall Annual Meeting of the AGU, December, 1989.

Hones, E. W., "Studying space plasmas from a lunar base", Los Alamos National Laboratory Report LA-UR-85-3994, 1985.

Kerridge, J. F., "What has caused the secular increase in solar Nitrogen-15?", Science, 245, 480, 1989.

Lindeman, R. A., R. R. Vondrak, J. W. Freeman, and C. W. Snyder, "The interaction between an impact-produced neutralgas cloud and the solar wind at the lunar surface", J. Geophys. Res., 79, 2287, 1974.

Space Science Board, "Space science in the twenty-first century: Imperatives for the decades 1995-2015", National Academy Press, 1988.

Swift, D. W., R. W. Smith, and S.-I. Akasofu, "Imaging the earth's magnetosphere", Planet. Space Sci., in press, 1989.

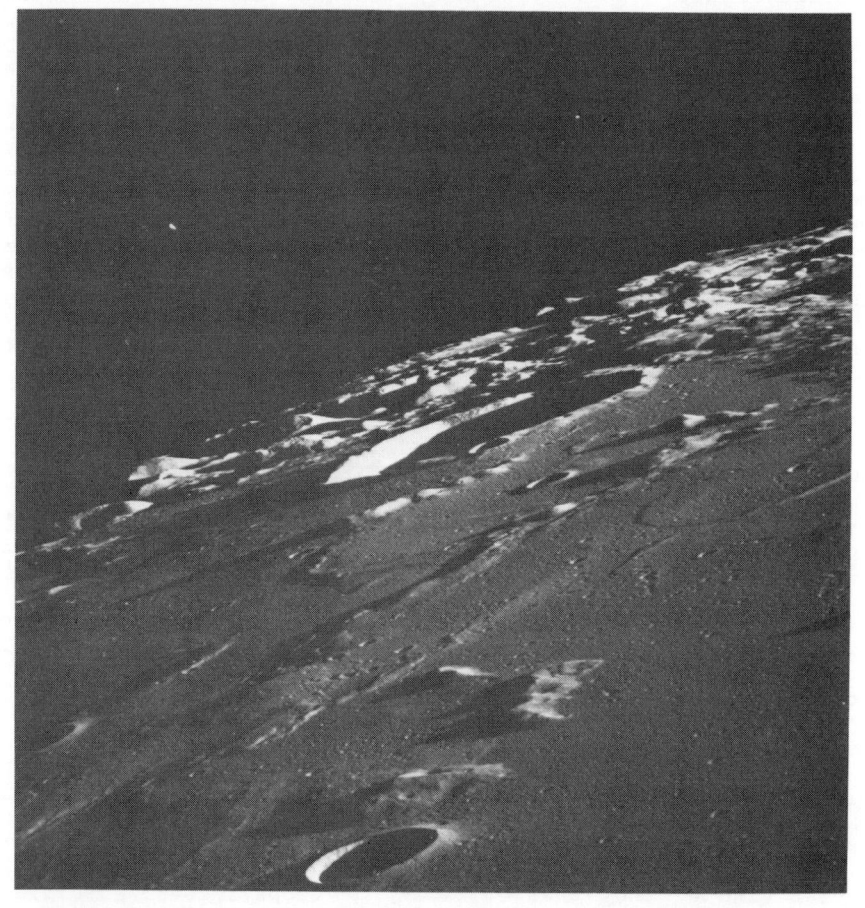

View looking west from spacecraft as it passed northwest of Mare Tranquillitatis (Sea of Tranquility). (AS8-13-2347).

Session On Cosmic Ray Physics

Tsiolkovsky represents a very interesting far-side candidate as a lunar base site. It is as far away from the regolith γ-ray radiation and radioactivity (antineutrinos) of Oceanus Procellarum (Ocean of Storms) as one can get. North is top center. (Apollo 15 metric camera frame 0889).

The Atmosphere as Particle Detector

Todor Stanev

*Bartol Research Institute, University of Delaware,
Newark, Delaware, 19716*

Abstract

The possibility of using an inflatable, gas-filled balloon as a TeV γ-ray detector on the Moon is considered. By taking an atmosphere of Xenon gas there, or by extracting it on the Moon, a layman's detector design is presented. In spite of it shortcomings, the exercise illustrates several of the novel features offered by particle physics on the Moon.

1. Introduction

TeV and higher energy γ-ray astronomy is a new experimental branch of high-energy astrophysics, which concentrates on the acceleration of charged particles by astrophysical systems. The electrons produce TeV γ-rays through inverse Compton scattering, while the accelerated cosmic ray beams interact inelastically with the matter or the radiation fields at the source and produce secondaries, including π^0, which decay into γ-rays. Unique environments are necessary to accelerate cosmic rays to such high energies and the observed γ-ray fluxes of individual sources[1,2] are surprisingly high. At 1 TeV the typical fluxes are $\sim 10^{-11}$ cm^{-2} s^{-1} which requires luminosity

$$L\gamma \sim 10^{35} \text{ erg s}^{-1} \times (D/Kpc)^2 \quad ,$$

where D is the distance to the source.

The sources of TeV γ-rays include rapidly rotating isolated neutron stars (the Crab pulsar) or binary X-ray sources, of which Cygnus X-3, Hercules X-1, and Vela X-1 are the best examples. Most signals are sporadic with outbursts on time-scales from several minutes to several days. Young supernova remnants, such as the recent SN1987A, are also expected potential sources[3]. There is an observation[4] of a 2-day outburst of SN1987A in TeV γ-rays, which may be an indication of particle acceleration at the remnant[5]. The strong variation of the source luminosity requires uninterrupted observations, which are very difficult from the Earth. Space

© 1990 American Institute of Physics

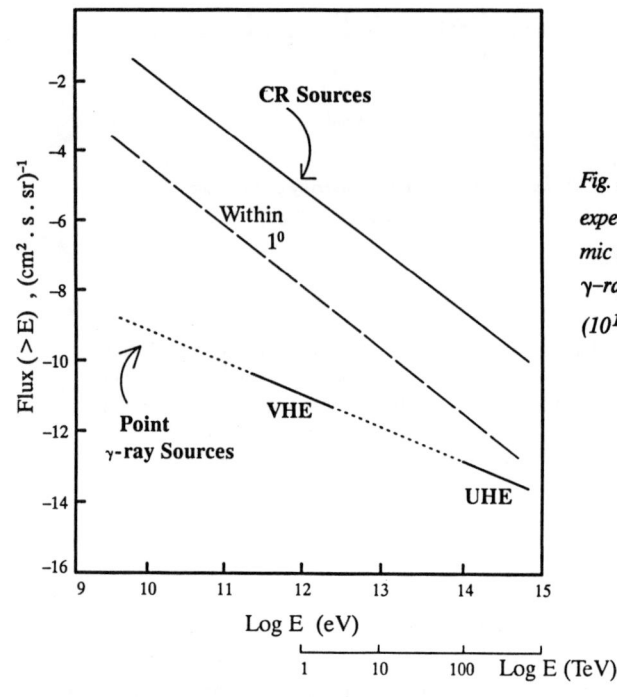

Fig. 1. Comparison of expected fluxes, for cosmic rays and for point γ–ray sources at TeV (10^{12} eV) energies.

Station-based detectors are not likely to be efficient because TeV γ-ray fluxes are too low and the detection of particles of such energies requires massive detectors.

2. TeV Gamma–Ray Astronomy on the Moon

At Earth the detection of VHE and UHE γ-rays is only possible because of the atmosphere. Earth's atmosphere is a poor guy's calorimeter. It is cheap and a very decent one: At 30^0 angle, it is about 30 radiation lengths and 30 interaction lengths deep. γ-rays interact in the atmosphere and develop cascades that deposit enough energy for detection. The problem is that even in a small angle around a potential source (as seen in Fig. 1) the flux is dominated by isotropic cosmic ray nuclei. These nuclei also interact in the atmosphere and produce enough secondary γ-rays to swamp the astrophysical γ-ray fluxes unless the source emission is unusually strong.

One could think of atmospheres with a better composition from the point of view of the TeV γ-ray astronomy. These would not be especially popular at Earth, but one could think

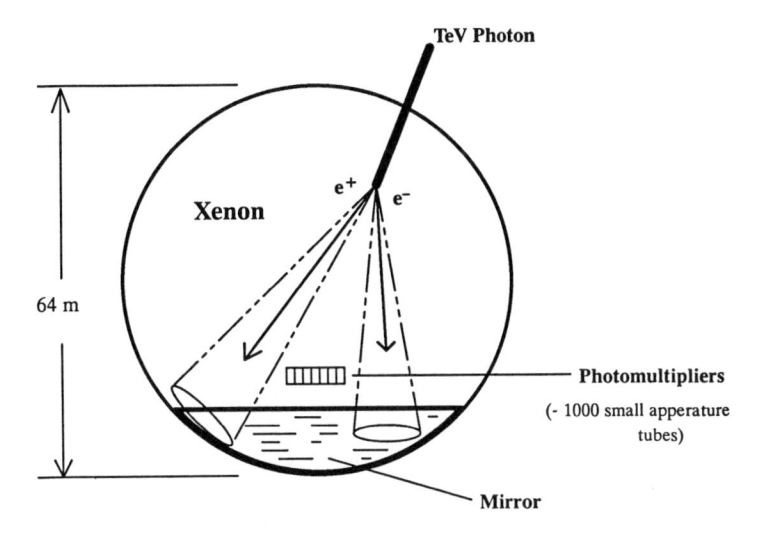

Fig. 2. A Xenon-filled TeV γ-ray detector, consisting of an inflatable balloon 64 m in diameter.

of creating a small piece of specially-designed (and well-contained) atmosphere at the Moon and use it for detection of astrophysical TeV γ-rays. If we use a heavy gas with high ratio of interaction to radiation length our detector will naturally select γ-rays and thus suppress the cosmic ray background.

The concept of a lunar TeV γ-ray detector is shown in Fig. 2, which illustrates a layman's detector design for a Xe-filled mylar balloon as a TeV γ-ray lunar telescope. It is a cross-breed of the IMB (Irvine-Michigan-Brookhaven) proton decay detector with a Cherenkov telescope. When TeV photons produce pairs in the Xe gas, the resulting electron-positron pairs will generate Cherenkov radiation, detected with photomultipliers at the mirror's focus. The field of view could be very large (45⁰ cone or 1.8 steradian). Imaging consists of reconstructing the γ-ray energy from the amplitude of the signal and its direction from the timing of the elec-

Table I. Counting Rates (s^{-1})

$E_{Threshold}$	Cosmic Rays	In 1⁰ Cone	γ-ray Source	Signal-to-Noise Ratio
10 GeV	5x10⁴	46	0.1	2.2^{-3}
100 GeV	846	0.8	0.01	1.2^{-2}
1000 GeV	15	0.014	0.001	7.0^{-2}

tron-positron pair. The reconstruction of the γ-ray arrival direction should be possible to a small fraction of a degree. An area of 10^4 m² produces 1 count/sec above 1 GeV from a 'standard' astrophysical source. (1 radiation length of Xe is 8.48 g/cm² and the density assumed is $\rho_{Xe} = 1.3 \times 10^{-3}$ g/cm³.) The background suppression factor for Xe is 20.

Table I summarizes the expected counting rates. At 100 GeV, for example, the expected rate is 36 events/hour.

3. Conclusions

To obtain Table I's event rates, the volume of Xe is 1.4×10^{11} cm³ and the mass of Xe is 180 metric tons. This is certainly too heavy to be transported to a lunar base. The size can be obviously scaled down with both the density or the linear dimensions of the balloon. The decrease of the density is a better solution, because it will keep the spatial resolution of the detector high. For example, 3 metric tons of Xenon gas will still yield 9 events/hour.

On the other hand the mylar balloon will also be a perfect micro-meteorite detector, with the precious Xenon leaking out or simply bursting in the near-vacuum at the Moon. The Xe balloon is only an idealized detector with optimum solid angle and background suppression characteristics. The idea suffers from the common problem of most particle detectors currently proposed for the lunar base: It is not practical.

It would be much more practical if the gas were extracted or produced *in situ* – on the Moon, rather than transported there at great expense. It is possible to use the un-used rocket fuel in the landing modules as target substance (K. Lande, *private communication*). The problems with the balloon could be circumvented by use of an existing or specially-tunnelled cavity in a lunar crater or rille. The proposed crew habitation quarters, for example, will be inflated and then covered with a layer of regolith for micro-meteorite protection. A similar clever technique may prove feasible whereby the enormous benefits of a simple TeV γ-ray telescope can come to fruition at a mature lunar base.

The author enjoyed the encouragement and help of Dr. T. L. Wilson in preparation of this manuscript. A discussion with Professor K. Lande is highly appreciated. This work is in part funded by NASA and the NSF.

References

1. T. C. Weekes, Physics Reports **160**, 1 (1987).
2. D. E. Nagle, T. K. Gaisser, R. J. Protheroe, Ann. Rev. Nucl. Part. Sci. **38**, 609 (1988).
3. T. K. Gaisser, A. K. Harding, and T. Stanev, Ap. J. **345**, 423 (1989).
4. I. A. Bond *et al.*, Phys. Rev. Lett. **61**, 2292 (1988).
5. V. S. Berezinsky and T. Stanev, Phys. Rev. Lett. **63** (1989), to be published.

LOW ENERGY COSMIC RAY STUDIES FROM A LUNAR BASE

Mark E. Wiedenbeck

Enrico Fermi Institute and Department of Physics
University of Chicago, Chicago, Illinois 60637 USA

ABSTRACT

Studies of cosmic ray nuclei with energies $\lesssim 7\,\text{GeV/nucleon}$ in low Earth orbit are hampered by the geomagnetic field. Even in high inclination orbits these effects can be significant. The lunar surface (or lunar orbit) provides an attractive site for carrying out low energy cosmic ray studies which require large detectors. The rationale and requirements for this type of experiment are described.

ASTROPHYSICAL STUDIES USING CHARGED PARTICLES

Studies of energetic charged particles provide a unique source of information on astrophysical processes on the Sun, in the interplanetary medium, and in our Galaxy. Charged particle trajectories are randomized as the result of interaction with magnetic fields and one generally can not identify the site of origin, as one does when imaging photon sources. However, these particles are a direct sample of matter from a remote site and comprise all of the stable nuclides in the periodic table (as well as some of the long-lived radioactive ones). The relative abundances of the various nuclides provide a source of information on nucleosynthesis, chemical evolution, charge and mass fractionation, and particle acceleration and transport which is not easily obtained by other means.

The energetic particles observed in the solar system have a variety of different origins. At the highest energies, above $\sim 50\,\text{MeV/nucleon}$, one observes the galactic cosmic rays, which originate outside the solar system, mostly in our own Galaxy. At somewhat lower energies, ~ 0.1 to $50\,\text{MeV/nucleon}$, one observes particles accelerated in solar flares and the so called "anomalous cosmic ray component", thought to result from local interstellar neutral atoms which enter the solar system and are then ionized and accelerated in the outer heliosphere.

From the composition of these particle populations one can obtain fundamental astrophysical information. The cosmic rays can provide relative abundances of a wide range of nuclides in the galactic "source" material from which these particles are accelerated. In fact, the only significant limitation preventing the determination of source abundances for *all* stable nuclides is the fact that the galactic cosmic rays contain a significant "secondary" component due to fragmentation of heavier nuclei during interstellar propagation. However, even considering only those nuclides with small secondary contributions, the number of galactic isotope ratios which can be determined from cosmic ray measurements exceeds the number presently accessible by other means (e.g., millimeter wave studies of interstellar molecules).

If one adopts the widely held view that the anomalous cosmic rays are derived from the neutral component of the interstellar gas in the immediate vicinity of the solar system, then one can use the measured composition of this particle population to investigate isotopic abundances of selected nuclides (those with high ionization potentials, such as the rare gases) in the interstellar gas. Comparison of this modern composition of the interstellar gas in the solar neighborhood with the composition of the Sun and that of other galactic material will make it possible to investigate the effects of chemical evolution of interstellar matter.

The nuclei observed in solar flares are accelerated from the solar corona. While large flare-to-flare variations in composition are common, it is observed that these variations are strongly correlated with the charge state of the particles. Thus using measured charge state distributions one is able to extract reliable coronal abundances. In addition, the systematics of the processes by which the composition of the corona is derived from that of the photosphere are sufficiently well established that one can also derive high quality photospheric abundances from the flare measurements. Elemental abundances derived in this way are in good agreement with those obtained by spectroscopic means, in cases when both methods are thought to yield accurate results. However, with solar flare particle measurements one can obtain the abundances of elements inaccessible by spectroscopic means, and can also measure isotopic composition, thereby greatly extending our knowledge of solar photospheric abundances. This is particularly significant because although "solar system" abundances are widely used as a test of nucleosynthesis theories and as a benchmark against which other composition measurements can be compared, present compilations of solar system abundances are largely based on meteoritic elemental abundances together with terrestrial isotope ratios. While these abundances are important in their own right, it is not clear that they should be representative of the primordial solar nebula, since the vast majority of the solar system material resides in the Sun.

Recent reviews of our knowledge of each of these particle populations can be found in reference 1. In discussing the advantages of a lunar base for measurements of low energy charged particles, I will focus on the galactic cosmic rays, with special emphasis on the requirements for carrying out *isotopic* composition studies. Similar conclusions are reached when one considers composition studies of the anomalous cosmic ray component or of solar flare accelerated particles.

LIMITATIONS ON MEASUREMENTS IN LOW EARTH ORBIT
Energy Region for Optimum Event Rate.

Panel (a) of Figure 1 is shows schematically the integral energy spectra of selected cosmic ray nuclides in space near the Earth. The spectra have similar shapes, indicating that the relative abundances of different nuclides are approximately independent of energy. The integral spectra are reasonably flat out to $\sim 1\,\mathrm{GeV/nucleon}$, and then fall rapidly, approaching a power law $J(>E) \propto E^{-1.7}$ at high energies. On the right hand side are indicated typical geometrical apertures required for instruments designed to study particles at energies above a particular point on one of the plotted curves. (Actually, instruments capable of isotopic identification operate over less than a decade of energy, but since the spectra fall steeply at high energies, the use of the integral spectra will not result in a gross overestimate of the expected count rates.) Here I have assumed that an experiment will accumulate data for one year, and have imposed a minimal requirement that at least 100 particles be collected to obtain useful results for a particular element. Instruments having $\sim 10\,\mathrm{cm}^2$ ster geometrical factor are typical of present day high quality isotope spectrometers for space missions. They typically operate at energies $\sim 50\text{--}500\,\mathrm{MeV/nucleon}$. Although with several year exposures, these instruments are capable of measuring all of the abundant nuclides with atomic numbers $Z \lesssim 30$, they are too small to measure rarer species, including all nuclides with $Z > 30$, and obtain only very limited statistics for lighter, low abundance nuclides. For studies of rare species one concludes that: 1) there is a distinct advantage in making measurements at low energies where the yield will be greatest; and 2) even at low energies, instruments with geometrical factors significantly greater than $10\,\mathrm{cm}^2$ ster are needed in order to obtain statistically significant samples.

There has been a considerable amount of work on developing high resolution particle detectors with large geometrical factors, since such instruments can be used to measure the more abundant cosmic ray species from high altitude balloons[2-6]. In this case the large area is used to make up for the short exposure time (typically 1 day for a balloon experiment vs. several years for a satellite experiment). These balloon instruments have geometrical factors ranging from a few hundred cm^2 ster to $\sim 1\,\mathrm{m}^2$ ster. With a several year exposure in space they could collect up to 1000 times as many events as the present, smaller space instruments.

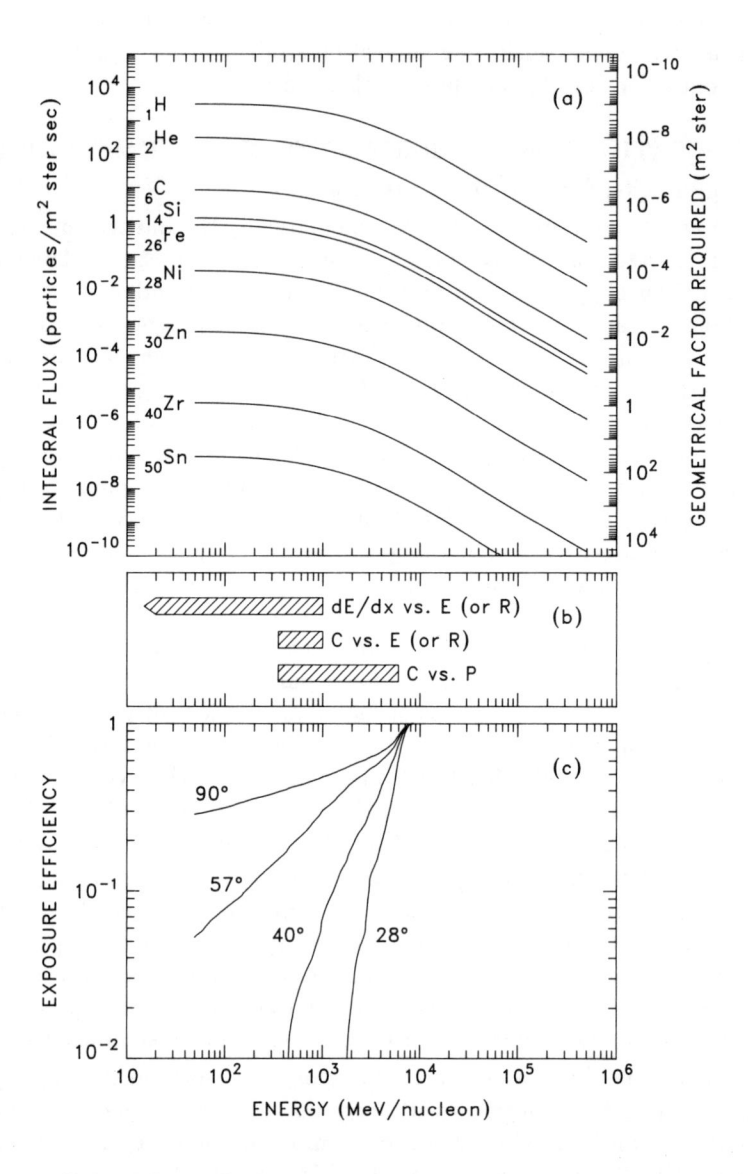

Fig. 1. Comparison of cosmic ray integral spectra (panel a); intervals over which various techniques can be used to measure isotopic composition (panel b); and exposure efficiency for experiments carried out in low Earth orbit, inside the Earth's magnetosphere (panel c). Spectra for H, He, and C are based on observations, and the others are scaled from C using measured low energy abundance ratios. The geometrical factors indicated in (a) assume collection of 100 particles in a 1 yr exposure.

Energy Region for Optimum Sensor Performance.

Operation at low energies is not only dictated by the higher fluxes obtained but, more importantly, by the fact that present detector systems achieve their best resolution at low energies. Instruments which are presently being used or developed for cosmic ray isotope studies rely on the measurement of two parameters to derive particle masses. One of these must depend on particle mass, M (and possibly on velocity, βc, and charge, Z, as well), while the other generally depends only on the particle velocity and charge. In isotope experiments carried out to date, the mass dependent parameter has been total kinetic energy, $E \propto M(\gamma - 1)$ (where $\gamma \equiv 1/\sqrt{1-\beta^2}$ is the usual Lorentz factor), the particle's range, $R \propto \left(M/Z^2\right)(\gamma-1)^{1.7}$ (approximately), or its magnetic rigidity, $P \propto (M/Z)\beta\gamma$. Obtaining sufficiently precise measurements of E or R has required that the nuclei be brought to rest in the detector. Thus instruments which use this approach are effectively limited to energies below $\sim 1\,\text{GeV/nucleon}$, since at higher energies the particle ranges become significantly longer than the nuclear interaction lengths for the particles in the detector material, and most of the cosmic rays fragment before stopping. One can work at somewhat higher energies when P is used as the as the mass dependent parameter. Here the approach is to measure particle deflections in a known magnetic field. While this technique could, in principle, be applied up to very high energies, practical limitations on the size and strength of the magnetic field presently restrict isotope studies to energies $\lesssim 10\,\text{GeV/nucleon}$. For example, the *Astromag* superconducting magnet facility[7] being considered for the Space Station would provide a field strength of a few Tesla over a volume $\lesssim 1\,\text{m}^3$.

The energy range over which precise isotope measurements are possible is also presently limited by the sensors available for measuring particle velocities. Measurements of specific ionization (dE/dx) can be used down to energies of a few MeV/nucleon (depending on Z), where electron attachment reduces the sensitivity of dE/dx to particle velocity. At these very low energies, time-of-flight measurements are useful, but they can not be extended to high energies without very long flight paths. Cerenkov counters (C) can provide very precise velocity measurements just above the threshold for Cerenkov light emission. The useful range goes from just above 320 MeV/nucleon (the threshold for a counter with refractive index of 1.5) to almost 10 GeV/nucleon where the accuracy starts to be limited by the low photoelectron statistics available with counters of reasonable thickness. This upper limit is not "firm", but could be increased somewhat with the development of improved integrating or ring imaging Cerenkov counters.

Panel (b) of Figure 1 indicates the approximate energy ranges over which good isotopic identification is possible with present sensor technology.

Effect of the Geomagnetic Field.

The deflection of charged particles in the Earth's magnetic field leads to the exclusion of nuclei with low rigidity from equatorial regions where the field has a significant horizontal component. Thus at any point on the Earth one can define a "vertical cutoff rigidity", P_c, such that particles with $P < P_c$ can not arrive at the Earth travelling in the vertical direction, while those with $P > P_c$ can. (The definition of P_c is actually somewhat more complicated due to the existence of penumbral bands.) Panel (a) in Figure 2 shows the range of P_c values encountered as a function of geographical latitude, λ.[8] For a given latitude there exists a range of P_c values since the symmetry axis of the dipole field does not coincide with that of the Earth. Panel (b) shows the relative amounts of time spent at different values of λ for spacecraft in circular orbits of selected inclinations.

If one considers particles of a particular rigidity, P, the efficiency for these particles to reach a detector in a circular low Earth orbit of a particular inclination is just the fraction of the area under the appropriate curve in Figure 2b occurring at λ values for which $P > P_c$, as found from Figure 2a. Panel (c) of Figure 1 shows this transmission efficiency as a function of energy per nucleon, assuming nuclei with mass-to-charge ratio $A/Z = 2$ (see the right hand scale in Fig. 2a). Comparing these curves with the spectra (Fig. 1a) and with the energy ranges accessible with present instrumentation (Fig. 1b) one sees that the statistical accuracy of isotopic composition studies will be seriously degraded in low Earth orbit. In the 28° Space Station orbit, techniques requiring the measurement of total energy or range are essentially ruled out, and even in polar orbit the transmission efficiency below 500 MeV/nucleon is only $\sim$ 40%.

For balloon-borne experiments, it is possible to choose a high latitude launch site where the cutoff lies below the energies of interest. But as discussed above, the short exposure time of such experiments limits their objectives to the more abundant species. This situation could be improved somewhat with the development of the capability for long duration balloon flights of payloads in the south polar vortex. However, for many of the measurements of interest, balloon experiments are limited by the energy loss and fragmentation of cosmic rays in the residual atmosphere, and do not provide a viable alternative to space experiments.

Spacecraft experiments to study the more abundant species at low

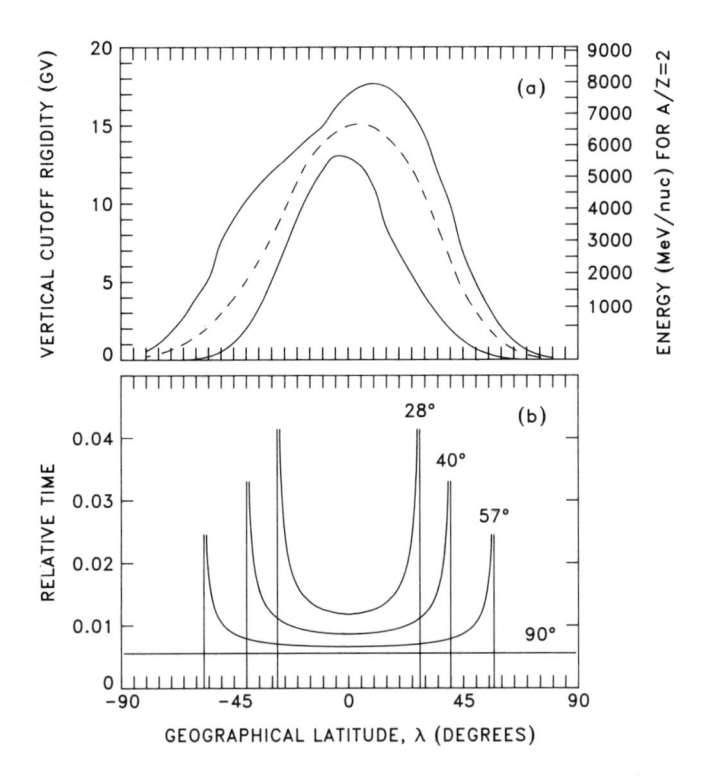

Fig. 2. (a) Minimum and maximum vertical cutoff rigidities (solid lines) and average cutoff (dashed line) as a function of latitude, from ref. 8. The right hand scale shows the energies to which these rigidities correspond for nuclei with $A/Z = 2$. (b) Relative times spent at different latitudes by spacecraft in circular orbits with the indicated inclinations. The relative times are given as a fraction of the orbital period per degree of latitude, and are plotted to within 1° of the inclination value.

energies have been successfully carried out through the use of special orbits. Halo orbits about the L1 Lagrangian point, 1.5×10^6 km from the Earth towards the Sun, have proven especially valuable since there the spacecraft can spend 100% of the time outside the Earth's magnetosphere. Other alternatives are highly eccentric orbits with apogee at many Earth radii, so that the only a minor fraction of the orbital period is spent in regions where the cutoff is significant. At present these orbits are impractical for very large, Shuttle-launched instruments. These difficulties are compounded for experiments which require periodic servicing, such as the replenishment of cryogen for a superconducting magnet.

ADVANTAGES OF A LUNAR BASE

The geomagnetic cutoff at a particular magnetic latitude scales as $1/r^2$ where r is the distance from the center of the Earth. For the Moon, $r = 60R_E$, so in the absence of interplanetary fields the 18 GV maximum cutoff in low Earth orbit would be reduced to ~ 0.005 GV at the Moon. In fact, the lunar orbit lies well outside the Earth's bow shock, which marks the boundary between the terrestrial and interplanetary fields. Thus the Moon provides an excellent platform for the investigation of low energy cosmic rays.

While the primary advantage of carrying out such experiments there is the Moon's favorable orbit outside the magnetosphere, there may also be significant secondary advantages as a result of the infrastructure which will inevitably accompany any serious development of a manned lunar facility. The difficulties in resupplying consumables such as liquid helium cryogen for a superconducting magnet or specialized gases for ionization detectors should be comparable to those on the Space Station, other than the extra lift capability needed to get them to the Moon. If facilities are eventually developed for extracting volatiles from the lunar surface, experiments requiring the resupply of these gases could benefit.

A lunar base would provide an additional benefit for experiments involving large magnets. In low Earth orbit such magnets are practical only if they have zero net dipole moment, since the torque which tends to align the dipole with the ambient field is significantly greater than the torques used for maintaining the spacecraft attitude (e.g., gravity gradient torques). The requirement of no net dipole moment precludes many favorable field configurations, such as that produced by a pair of Helmholtz coils. (Elaborate arrangements of pairs of identical magnets with opposite orientations will permit cancellation of the dipole moments, but are impractical because of the complexity of the required mechanical and cryogenics subsystems.) Since the ambient magnetic field on the Moon is negligible, and since in any case the Moon provides a stable platform which will be unaffected by the torques that would be encountered, magnetic spectrometers for operation on the Moon will not be required to have negligible dipole moment. This greater freedom of design should permit improved field configurations and enhance experiment performance.

CONCLUSIONS

A lunar base could provide a highly advantageous site for carrying out investigations of the composition of low energy charged particle populations—galactic cosmic rays with energies < 10 GeV/nucleon, the anomalous cosmic ray component, and particles accelerated in solar flares. The main advantage is simply that the ambient magnetic field is negligible and does not hinder the access of low energy charged particles. The experiments that will be required to investigate the composition of the rarest species (e.g., cosmic ray elements with $Z > 30$) will have to be large and will benefit from the infrastructure which will be developed to transport, install, and maintain apparatus on the Moon.

ACKNOWLEDGEMENTS

I would like to thank R. A. Mewaldt for providing calculations of the transmission efficiency through the geomagnetic field (Fig. 1c). This work was supported, in part, by NASA grant NGL 14-001-005.

REFERENCES

1. Waddington, C. Jake (ed.), *Cosmic Abundances of Matter*, (American Institute of Physics, N.Y.), 1989.
2. W. R. Webber, J. C. Kish and D. A. Schrier, *Proc. 19th Internat. Cosmic Ray Conf.* (La Jolla) **2**, 88 (1985).
3. J. A. Esposito, B. S. Acharya, V. K. Balasubrahmanyan, B. G. Mauger, J. F. Ormes, R. E. Streitmatter, W. Heinrich, M. Henkel, M. Simon and H. O. Tittel, *Proc. 19th Internat. Cosmic Ray Conf.* (La Jolla), **3**, 278 (1985).
4. E. R. Christian, J. E. Grove, R. A. Mewaldt, S. M. Schindler, T. Zukowski, J. C. Kish, and W. R. Webber, *Proc. 20th Internat. Cosmic Ray Conf.* (Moscow), **2**, 382 (1987).
5. J. J. Connell, J. W. Epstein, M. H. Israel, J. Klarmann, W. R. Binns, and J. C. Kish, *Proc. 20th Internat. Cosmic Ray Conf.* (Moscow), **2**, 398 (1987).
6. R. A. Leske, P. Meyer, D. Ruffolo, C. Smith, and M. E. Wiedenbeck, *Nucl. Instr. Meth. in Phys. Res.*, **A277**, 627 (1989).
7. J. F. Ormes, M. H. Israel, R. Mewaldt, and M. Wiedenbeck, *Proc. 20th Internat. Cosmic Ray Conf.* (Moscow), **2**, 378 (1987).
8. M. A. Shea and D. F. Smart, *Tables of Asymptotic Directions and Vertical Cutoff Rigidities for a Five Degree by Fifteen Degree World Grid as Calculated Using the International Geomagnetic Reference Field for Epoch 1975.0*, Air Force Cambridge Research Laboratories Report AFCRL-TR-75-0185, 1975.

Measurement of Ultra-Heavy Cosmic Rays at a Lunar Base

M.H. Salamon [1], P.B. Price [2], and G. Tarle [3]

[1] Physics Department, University of Utah, Salt Lake City, UT 84112
[2] Physics Department, University of California at Berkeley, Berkeley, CA 94720
[3] Physics Department, University of Michigan, Ann Arbor, MI 48109

A wealth of information regarding cosmic ray synthesis and propagation is contained in the ultra-heavy ($Z > 60$) cosmic ray abundances; to extract this information, however, requires a detector capable of acquiring large statistics for these rare particles, as well as a charge resolution adequate to separate neighboring charge peaks at very large Z. A large, passive surface array of nuclear-track-detecting glass plates would meet these requirements. These glass plates could be periodically processed and analyzed for tracks at a lunar base, then melted/annealed for reuse in a continuously recycled detector array.

Ultra-Heavy Cosmic Rays:
An accurate measurement of the elemental abundances of the ultra-heavy ($Z > 60$) cosmic rays would contribute substantially to our understanding of the origin, acceleration, and propagation of cosmic rays within our Galaxy. In particular, a high statistics measurement of the abundances of the actinide elements in the cosmic rays, along with the possible detection of transuranics, would provide definitive answers to questions that have remained outstanding since the discovery of the cosmic ray heavy-nucleus component in 1948 [1].

One of the foremost questions regards the source of the nuclear cosmic rays. Data from the HEAO-3 [2], HEAO-C [3], and Ariel 6 [4] satellites show that for $Z < 60$ the galactic cosmic ray (GCR) elemental distribution is consistent with the assumption of a source whose elemental abundances are the same as that of our solar system (SS). Observed deviations from unity of these (normalized) GCR/SS abundance ratios can be attributed to ionization selection effects, viz., the probability of initial acceleration (hence selection) of an ion in a source is dependent upon its first ionization potential (FIP). The striking similarity of the SEP(solar energetic particle)/SS elemental abundance ratios to the GCR/SS ratios suggests [5] that nascent GCR's originate in stellar chromospheres, where temperatures are such that the element's FIP determines ionization fraction and ejection to the stellar corona.

These ions then reach GCR energies via first-order Fermi acceration within shock waves that are produced by neighboring supernovae or stellar winds. A collorary of this picture is that SS abundances (as determined by meteorites and solar photospheric measurements) are essentially the same as those of the present local interstellar medium (ISM).

The above picture, however, does not disallow variations in isotopic and elemental composition in various source regions due to ongoing chemical evolution within the ISM, and these are in fact seen. Isotopic GCR data from IMP [6] and ISEE [7] satellites show significant deviations in certain isotopic ratios (such as $^{22}Ne/^{20}Ne$), indicating that additional nuclear processing has occurred for some fraction of the GCR. The data for GCR's of Z > 60, from HEAO-C and Ariel 6, suggest that there may be a significant enhancement of r-process [8] elements in this component [3], so that we may be directly observing the products of (possibly recent) explosive nucleosynthesis.

This raises the possibility that a fraction of the ultra-heavy GCR's may be freshly synthesized nuclei from recent supernovae, accelerated to cosmic ray energies by the supernova shocks. Confirmation of this would give us strong evidence for the role of shocks in nuclear GCR acceleration, and would provide an observational window to nucleosynthetic environments perhaps different from our own. A high-statistics measurement of the abundances of the GCR actinides would provide a definitive answer to this question. Figure 1 (taken from Ref. 9) shows the abundances of the actinide elements from Z=90 to 96 as a function of time after explosive nucleosynthesis. Since the propagation lifetime of GCR actinides should be less than 10^7 yr [10], it is clear that the GCR actinide composition from freshly synthesized material would be quite distinct from that of old (> 10^9 yr) ISM material. Due to their differing lifetimes, the U/Th ratio, equal to 0.60 in SS material [11], would be greater than unity in fresh r-process material. In addition, the observation of comparable fluxes of transuranic elements, such as Cm whose halflife is 1.6×10^7 yr, would be an unambiguous signature of fresh r-process enhancement of the GCR's.

To make these observations, however, requires very large detector area and collection time, as well as unprecedented charge resolution to obtain clean charge separation in the actinide region. To put this in perspective, we note that a grand total of 4 actinides were detected by the HEAO-C and Ariel 6 instruments; the (superior) charge resolution of the HEAO-C instrument was considerably worse than 1 charge unit in the actinide region, making impossible the identification of individual actinides. To achieve an accuracy of 10% in the U/Th ratio, over 400 U-Th ions must be detected (with a charge resolution of 0.5 charge units or better);

it is evident that satellites with electronic payloads cannot easily be scaled up in size to meet these collection requirements.

A detector that is capable of cleanly separating even- and odd-Z elements in the Pt-Pb/sub-Pt-Pb region would also do much to improve our understanding of the CR propagation pathlength distribution (PLD) for pathlengths on the order of 1 g/cm^2 and less, since the nuclear interaction length of these nuclei in the interstellar medium is on the order of 1 g/cm^2. The simplest model of CR propagation, the leaky box with a negative exponential PLD, fails to account for the observed abundances of CR secondaries (spallation products of heavier cosmic-ray parents) in the Li-B and sub-Fe regions. To obtain agreement with these data, an energy-dependent truncation of the smallest pathlengths (< 1 g/cm^2) in the PLD must be made [12]. This supports the hypothesis of a nested leaky box model [13], in which newly accelerated CR's must travel through significant grammage before leaving their site of acceleration. Thus the 0-1 g/cm^2 region of the PLD can very likely yield clues to the structure of the astrophysical sites within which CR's are born. Element-resolved abundance measurements of the Pt-Pb/sub-Pt-Pb region are the best way to constrain this important region of the PLD; again, high statistics are required to obtain the necessary accuracy in the secondary/primary ratios, which argues for detector sizes far larger than those previously flown. We note that for these measurements it is imperative that there be as little upstream mass as possible; even a fraction of a gram per square-centimeter will cause significant fragmentation of the ultra-heavy CR flux, contaminating the astrogenic secondary CR flux. Thus balloons, whose payloads typically have float altitudes of several g/cm^2, are unsuitable for this type of measurement, leaving orbiting facilities, and now a lunar base, as ideal deployment sites.

Finally, we note that a high-statistics measurement of the Z > 60 CR elemental abundances may yield unexpected surprises. If superheavy (Z > 100) elements are in fact produced in supernovae and have halflives in excess of 10^7 years, then in the event that a fraction of the nuclear CR's are comprised of fresh r-process material, a detector such as that described below might provide the first evidence for the existence of an island of nuclear stability beyond the transuranics. Speculative particles such as quark nuggets [14] may be searched for as well.

A Lunar-Based Heavy Nucleus Collector (LBHNC):

The recent discovery [15] of the extraordinary properties of certain phosphate glasses as nuclear track detectors [16] has made possible the construction of an extremely simple and powerful detector of ultra-heavy cosmic rays of Z > 60. When an ionizing projectile passes through a generic nuclear-track-detecting medium, it creates a latent track, which is a region of damage (typically Angstroms in radius) about the projectile's trajectory. Subsequent to its exposure, the material is chemically etched in a corrosive liquid, which preferentially removes material from the damaged region, creating a visible track or cone whose geometric parameters, determined by microscopy, determine the ratio of projectile charge to velocity [16]. A detector composed of several thin layers of this material measures this ratio in successive layers as the ionizing particle is slowing; the variation of track etch rate with depth in the material uniquely identifies the charge of the particle and its incident energy.

Conceptually, an ultra-heavy CR detector comprised of nuclear-track-detecting material, such as sheets of certain types of glass or plastic, is very simple. A given detector module would consist of a sufficient number of thin sheets of (e.g.) glass stacked together to ensure that a particle's charge be unambiguously identified by the method described above; nuclear interactions cause a loss of signal, so that it would be important to keep the glass sheets as thin as possible. To minimize temporal variations in temperature the modules would be thermally isolated with thermal standoffs and multilayer insulation. Total collection area could be increased as desired by the addition of modules to the experiment site. No power would be required for such a detector, as it would be completely passive during its acquisition of latent tracks. Thus the achievement of a total detector area on the order of 10^2 m^2 would be quite straightforward; with a collection time of a few years, given the absence of a significant magnetic field on the Moon, the number of observed CR actinides would be increased by over 2 orders of magnitude above the present world's total (assuming solar system abundances for the CR actinides).

One nuclear-track-detecting glass in particular, BP-1 [17], has remarkable immunity to the problems that must be faced by any long-duration experiment composed of nuclear track detectors that is deployed in space. Those problems include the fading of latent tracks over exposure periods of months or years at even fairly low temperatures [18], the variation of the sensitivity of the detecting medium with temperature [19] and oxygen pressure [20], and the growth of latent track reactivity with time [21]. BP-1 glass, produced by Schott, suffers no fading of latent tracks up to the test limit of 5 months at 50 °C [22], ensuring that no detectable fading would occur at a lunar base site even over a several year period. The temperature dependence of the glass sensitivity is also remarkably small at -20 °C (the designed operating temperature of the passive detector) with a fractional shift in the subsequent track etch rate of $\sim 1 \times 10^{-3}$ per degree Centigrade of temperature at the time of track creation. In addition, unlike many other track-detecting media, the response of BP-1 in vacuum is the same as in air, so that no pressure vessels (with their unavoidable mass overburden) need be employed. There also is no measurable variation of latent track reactivity with time in BP-1 glass.

The charge resolution of BP-1 glass is unequalled in the high-Z regime. Figure 2 shows data on the fragmentation of 1 GeV/u Au (Z=79) ions produced by Lawrence Berkeley Laboratory's Bevalac. A total of 5 sheets (10 surfaces) of glass sufficed to yield a charge resolution of 0.06 charge units (note that the ordinate is logarithmic)! The underlying physical reason [23] for such extraordinary charge resolution is that a latent track in glass is created only by electrons that receive low momentum transfers from the ionizing projectile (and thus deposit their energy close to the particle's trajectory); their number is very large, so that the statistical variation of damage within the track is negligible. This is in constrast to the energy deposited far away from the track by high-energy electrons (delta-rays); the number of high-energy electrons is far smaller, and thus statistical fluctuations are of importance; detectors which are sensitive to the energy deposited by these electrons (glass is not) will therefore suffer larger fluctuation in response.

The actual charge resolution of the LBHNC detector is a complicated function of the intrinsic resolution given above and the ability to uniquely fit response curves of individual elements to the measured track etch rate vs. stack depth of any given event. Figure 3 shows the expected charge resolution versus CR energy as determined by a Monte Carlo calculation for a module composed of 14 sheets of 0.2 cm thick BP-1 glass. It is seen that a majority of the ultra-heavy CR events will have their charge determined to an accuracy of < 0.3 charge units.

A lunar site is arguably the ideal location for an HNC detector. The absence of a magnetic field means that lower energy ultra-heavy CR's are not screened from the detector, as they would be for a detector in a low-inclination orbit about the Earth. Assuming solar system abundances, the integral flux of CR actinides of energy > 0.85 GeV/u (ensuring adequate range within the glass) would be ~ 1 U/m^2-yr and 2 Th/m^2-yr; a 100 m^2 detector of mass ~8000 kg would therefore collect close to 1000 actinides in a 3 yr period. This would be more than adequate to determine the U/Th ratio to < 10%; such a detector would also be sensitive to transuranics even if the fresh r-process component of the GCR's comprised only a few percent of total CR flux.

After its 3-year exposure, the glass could be returned to Earth for chemical processing and automated track microscopy and analysis. Alternatively, if a chemical etching facility and an automated track imaging system were to be established at the lunar base, the glass culd be harvested, analyzed, then either annealed or remelted at the lunar base, then placed back onto the lunar surface for additional exposure; in effect, the LBHNC detector could be indefinitely recycled at no additional payload cost.

References:

[1] P.S. Freier et al., Phys. Rev. $\underline{74}$, 213 (1948).
[2] P. Goret et al., Proc. 18th ICRC (Bangalore) $\underline{9}$, 139 (1983), and references therein.
[3] W.R. Binns et al., to appear in Ap. J., Nov.15 ,1989.
[4] P.H. Fowler et al., Ap. J. $\underline{314}$, 739 (1987).
[5] J.P. Meyer, Ap. J. Suppl. $\underline{57}$, 173 (1985).
[6] M. Garcia-Munoz, J. Simpson, and J.P. Wefel, Ap. J. Lett. $\underline{232}$, L95 (1979).
[7] M.E. Wiedenbeck and D.E. Greiner, Phys. Rev. Lett. $\underline{46}$, 682 (1981).
[8] E.M. Burbidge, G.R. Burbidge, W.A. Fowler, and F. Hoyle, Rev. Mod. Phys. $\underline{29}$, 547 (1957).
[9] J.B. Blake and D.N. Schramm, Astrophys. and Sp. Sci. $\underline{30}$, 275 (1974).
[10] T.G. Guzik et al., Proc. 19th ICRC (La Jolla) $\underline{2}$, 76 (1985).
[11] A.G.W. Cameron, in **Essays in Nuclear Astrophysics**, ed. C.A. Barnes et al., Cambridge University Press (1982).
[12] J.P. Wefel, in **Genesis and Propagation of Cosmic Rays**, p.1, eds. M.M. Shapiro and J.P. Wefel (Reidel, Dordrecht, 1988).
[13] R. Cowsik and L.W. Wilson, Proc. 14th Inter. Cosmic Ray Conf. $\underline{2}$, 659 (Munich, 1975).
[14] E. Witten, Phys. Rev. $\underline{D30}$, 272 (1984),
[15] P.B. Price et al., Nucl. Instr. Meth. $\underline{B21}$, 60 (1987).
[16] R.L. Fleischer, P.B. Price, and R.M. Walker, **Nuclear Tracks in Solids**, University of California Press (1975).
[17] Shicheng Wang et al., Nucl. Instr. Meth. (in press, 1989).
[18] M.H. Salamon, P.B. Price, and J. Drach, Nucl. Instr. Meth. $\underline{B17}$, 173 (1986).
[19] A. Thompson et al., Proc. 11th Inter. Conf. on Solid State Nuclear Track Detectors (Bristol, England, 1981).
[20] J. Drach et al., Nucl. Instr. Meth. $\underline{B23}$, 367 (1987).
[21] J. Drach and P.B. Price, Nucl. Instr. Meth. $\underline{B28}$, 275 (1987).
[22] A.J. Westfall et al., to be published.
[23] G. Tarle, S.P. Ahlen, and P.B. Price, Nature $\underline{293}$, 556 (1981).

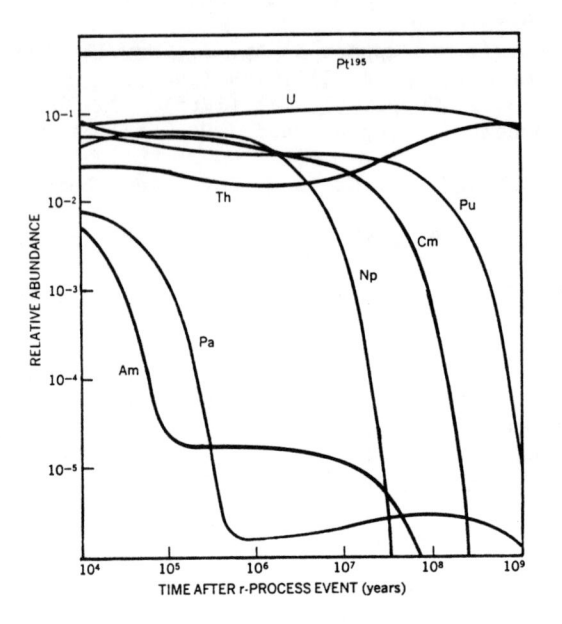

Figure 1

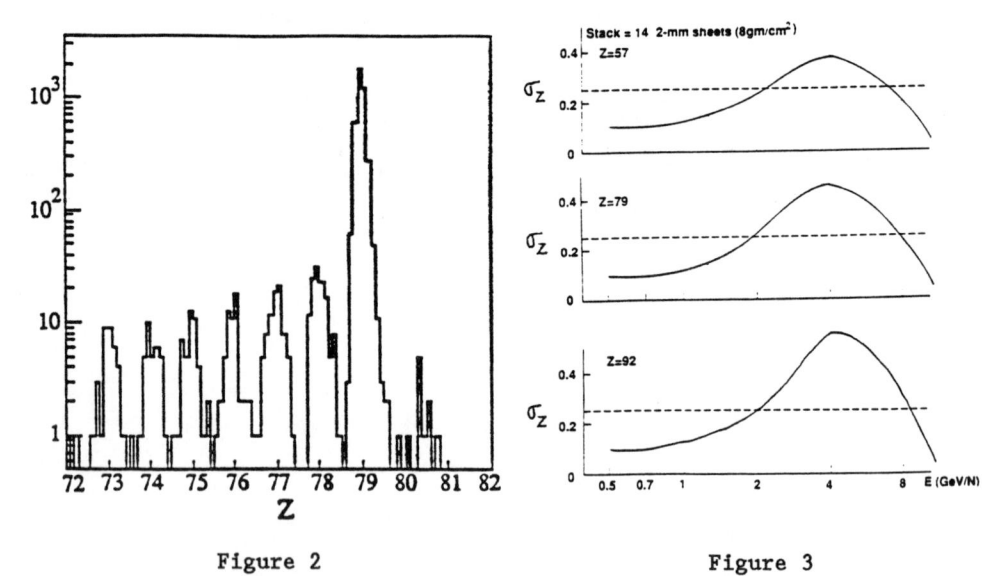

Figure 2

Figure 3

Study of Chemical Composition of Ultra–High Energy Cosmic Radiation From a Lunar Base

Gautam D. Badhwar

National Aeronautics and Space Administration, Lyndon B. Johnson Space Center, Houston, Texas 77058

Abstract

There is a need for direct charge and energy measurements of cosmic rays above 100 TeV. A method for performing such a study, which cannot be done directly on Earth or from satellite, is proposed for a lunar base.

Cosmic ray charged particles of energy above 100 TeV (10^{14} eV) have a gyro radius in the interstellar magnetic field that is comparable to the size of our galaxy. These particles, therefore, do not suffer much scattering and if these particles are of galactic origin their observation allows one to view the source. On the other hand, if these particles are of extragalactic origin then one may not see this anisotropy. Thus a study of the cosmic radiation composition above 100 TeV may reveal the source(s) of such radiation.

The chemical compostion and energy of these ultra–high energy particles is currently studied by the use of Extensive Air Showers (EAS). As the name implies, these are cascade showers of electrons, muons, and nucleons, produced by the interaction of the primary cosmic particle in the upper layers of the atmosphere. Because of the low density of the atmosphere, the secondary particles are spread out in lateral extent for many kilometers. This makes the detector farms difficult and expensive to run. In addition, only a part of the EAS is sampled by the detectors.

The richness of the EAS in muons is taken as evidence of the presence of heavy nuclei, a totally indirect method. The current meager experimental information indicates that the relative abundance of heavy nuclei decreases as energy increases and vanishes at energies greater than 10^{17} eV. Much of the evidence at energies greater than 10^{17} eV comes from interpreting multicore EAS as being due to heavy primaries, again a totally indirect method. There is thus a need for a direct charge and energy determination for energies above 100 TeV.

© 1990 American Institute of Physics

The Moon, with no appreciable atmosphere, provides a unique vantage point and offers a chance to study this part of the cosmic radiation that cannot be done effectively from the Earth or from free-flying platforms.

A method is proposed here for studying the cosmic radiation above approximately 100 TeV from a lunar base. This method provides a direct determination of the charge of the primary particle before it intereacts with the lunar soil overburden (needed for radiation protection anyway). It also produces cascades that are very small in lateral extent (due to the higher density of lunar rock compared to the Earth's atmospheric density) and can be more easily studied. Since the flux of these particles is extremely small, these detector systems can be left operating to gather data with minimal attention and thus provide information about the anisotropy. A lunar base thus provides an opportunity to perform this basic astrophysics experiment that cannot be done in any other realistic way.

Session On Neutrino Physics

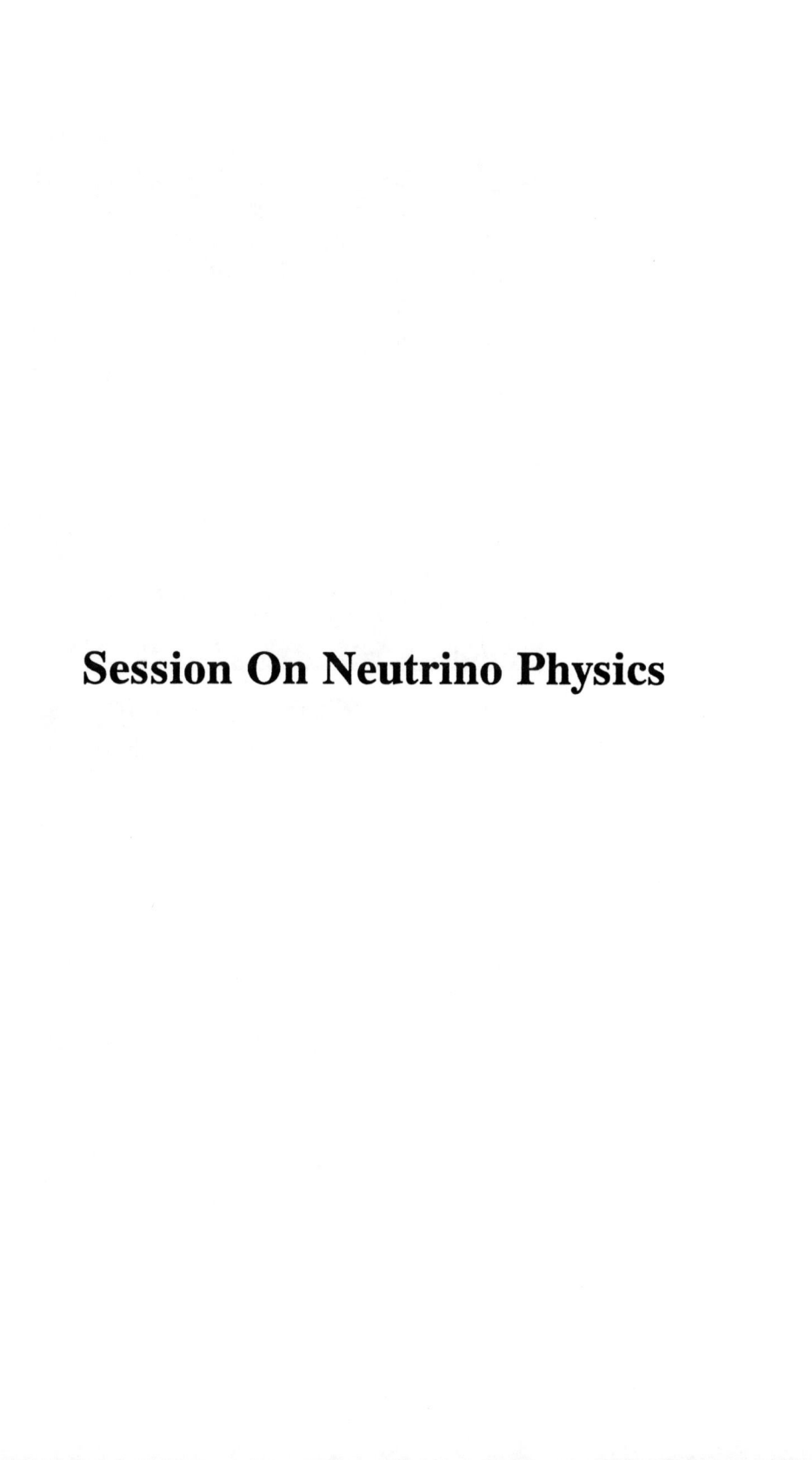

Oblique view southwest across Tsiolkovsky, 265 km from the central peak of high-albedo material. The dark deposit on the crater floor would serve as a smooth area for a landing, launch, and habitation site to support the physics experiments discussed in the workshop. (AS15-91-12383).

East end of Tsiolkovsky's central peak. A plumb line through the area of high radioactivity on the Earthside in Oceanus Procellarum, through the center of the Moon, passes through this crater. This makes the crater as far as possible from known regolith radiation noise. Orbital γ-ray spectrometer measurements indicate its regolith is quiet.
(AS15-96-13017).

Medium and High–Energy Neutrino Physics From a Lunar Base

Thomas L. Wilson

National Aeronautics and Space Administration, Lyndon B. Johnson Space Center, Houston, Texas 77058

Abstract

The prospect of neutrino physics at a lunar base is analyzed from the point of view of neutrino astronomy for medium and high energy. This is done by introducing "particle astronomy" as a new discipline of physics. In addition, the entire Moon is considered as a neutrino detector. Due to the drop in flux of atmospheric neutrinos from the Earth, a high-energy ($1-10^3$ TeV) "window" into our Galactic center is proven to exist at the lunar surface which is obscured for Earth-based observatories. In Sect. 8 it is further demonstrated that all of the Earth's neutrinos can be eliminated for astrophysical sources, an important new result. Another issue addressed is long-baseline particle physics in the Earth-Moon laboratory (Sect. 10), which cannot be accomplished for any other planetary initiative such as Earth-Mars, nor from satellite or the Earth alone. This includes neutrino oscillation studies (Sect. 11) as well as neutrino exploration of the Earth-Moon and antineutrino radionuclide imaging of the Moon's interior at planetary densities (Sect. 7). The important new vista of a lunar base for neutrino (antineutrino) astronomy is proposed as a scientific justification for returning to the Moon. A robust and bold new frontier of fundamental physics is advocated, using the Moon as an astrophysics testbed.

1. Introduction

Neutrinos, unlike electromagnetic and gravitational radiation, are fundamental probes of astrophysical objects such as our Galactic center, stars, supernovae, and other astronomical sources which photons and gravitational radiation cannot penetrate. However, they suffer the experimental difficulty of low event rates at medium energies due to small cross-sections, and these small events are often overwhelmed for Earth-based observations by a neutrino background produced in the Earth's atmosphere.

From the surface of the Moon, on the other hand, some aspects of neutrino physics take on a surprisingly new perspective: Terrestrial neutrinos exist there, but can be removed from the data by virtue of a compact cut theorem (Sect. 8) in "particle" astronomy.[1] Since the particle is a neutrino, the observer can see directly back along the particle's geodesic to its origin in space–time.[2] We will explore the scientific advantage of this neutrino (ν) property from a lunar base, and we shall demonstrate that a "window" in the Universe exists from 1–10^3 TeV for neutrino astronomy there. The compact cut theorem will further open up the entire neutrino sky for lunar base particle astronomy. This new vista into the inner structure of our own Galactic center justifies the construction of a modest if not substantial ν detector on the Moon.

It will be argued (Sects. 4,7) that by means of antineutrino measurements, the internal radionuclide abundances of the Moon can be investigated, demonstrating that neutrino (antineutrino) astronomy can play a role in determining the structure, origin, and thermal evolution of the Moon. The fledgling science of antineutrino astronomy is only beginning to make its debut on Earth, and its time has come for further consideration. By performing measurements which are inaccessible to thermal convection experiments and seismicity analysis, neutrino (antineutrino) detectors can serve a dual role.

An additional point is made that a number of the scientific motivations put forward to justify the construction of a National Underground Science Facility[3] (NUSF) are met by a man–tended lunar base.[4] If we are to confront the expanding frontiers of high–energy physics in a realistic fashion, it is clear that a fundamental high–energy physics facility must be a substantial part of any serious lunar base initiative [Appendix]. As we continue to complicate the Earth's natural radiation environment (e.g. with reactor antineutrinos[5]), the issue of an international science facility on the Moon is a pressing one.

The case will be presented here that Galactic neutrino astronomy is one of the few[6,7] experiments which uniquely require the Moon to be performed well. It cannot be done by current estimates on Earth, or from Earth orbit.[8] Because the object is to get away from the Earth's atmospheric neutrinos, the Moon presents itself as a very interesting candidate for this type of fundamental physics.

Emphasis will be on a flux analysis only. This will identify the spectral "windows" at a lunar base as a function of energy, from which a scientific justification is proven and established. The detectors themselves[9] will be left for the work of others.

2. Beyond the Earth's Atmosphere

Considering that the Earth itself is virtually transparent to low– and medium–energy neutrinos, it is one of the ironies of physics that our atmosphere (which is much less dense than the Earth) should be such a copious source of neutrinos when our own planet is not. But the Earth is bombarded by cosmic rays which collide with dust, gas, and particles of its atmosphere creating unstable mesons (μ's, π's, and K's) that decay and produce this noisy background flux of neutrinos.[10] The key point is that the unstable secondaries have time to decay (e.g. $\mu^- \to e^- + \bar{\nu}_\mu + \nu_e$)[11] due to a large mean free path. When and if the cosmic rays reach the

Earth's crust, on the other hand, the unstable mesons interact with matter before they can decay and produce neutrinos ν (antineutrinos $\bar{\nu}$). The production of cosmic ray ν's($\bar{\nu}$'s) by the Earth is thus suppressed.

The net result is a spherical shell about the Earth which serves as the origin of a diffuse background of atmospheric neutrinos. This situation is depicted in Figure 1, where the Earth-based neutrino astronomer (A) is imagined to exist inside such a shell emitting a flux ϕ_A of neutrino radiation. This diffuse background has been studied by a number of individuals[12-15] and has been the concern of serveral groups[16-20] attempting to define the diffuse and point sources in the neutrino sky. The atmosphere's differential flux spectrum (xE^3, a representation due to Lagage[20] although introduced earlier by Volkova[13]) is depicted in Fig. 2a and its integral flux in Fig. 3a, both for an Earth-based observer. The result for the diffuse atmosphere is the dashed hump spectrum[21] peaking around 10 GeV in Fig. 2a, and the bold-face flux curve in Fig. 3a..

For an astronomer at the Earth's surface, the flux ϕ_A is not isotropic and possesses horizontal and vertical components (e.g., see Ref. 13 and 16). But a flux is a scalar quantity (such as a number of events). A simple relativistic argument permits us to determine what happens when the astronomer leaves the Earth's shell in Fig. 1 and moves to the Moon. If astronomer A is at the center of mass of the Earth[22], ϕ_A is isotropic and consists only of the vertical component at one Earth radius (R_e). A Lorentz (Poincare) transformation first to the Earth's surface ($1R_e$) and then to the center of mass of the Moon 60 Earth radii (60 R_e) away simply drops the flux ϕ_A off as $(60)^{-2} = 2.8 \times 10^{-4}$ while simultaneously making it a compact source (no longer diffuse) of 1.9^0 angular subtent in the neutrino sky.[23] The flux spectra (Figs. 2 and 3) are logarithmic, whereby the vertical constituent drops by 2.8×10^{-4} (modulo the geometric factor[23]) when A moves to B, the center of mass of the Moon. Then $\phi_B = (60)^{-2} \phi_A$ has become a compact (not a diffuse) source when viewed from a lunar base.[24] The resulting approximation is Fig. 2b and 3b for the Earth's atmospheric neutrinos as seen from the Moon. The seasonal stellar and galactic fluxes are unchanged. This procedure is further illustrated in Fig. 4 and 5 using

The Neutrino Astronomer
(Earth-to-Moon)

$\phi_B = \phi_A (60)^{-2}$
$= 2.8 \times 10^{-4} \phi_A$

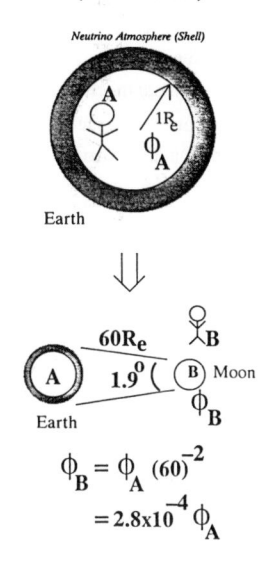

Figure 1. Earth-to-Moon flux transformation.

the conservative estimates of Stecker.[16] The results of Volkova,[13] advantageous because the spectra are clearly tabulated, are in basic agreement with those of Ref. 16-18.

3. Neutrino (Antineutrino) Sources in the Lunar Base Sky

A broad range of neutrino (antineutrino) sources has been investigated for Earth-based observation which serves as the basis for the Earth-Moon flux comparison here. These are depicted for medium energies (10^{-1}-10^5 MeV) in Fig. 2 and for high energy (10^{-1}-10^7 TeV) in Fig. 3. In some cases such as the solar neutrinos[25] and the atmospheric neutrinos[12-17] they have been studied exhaustively, while in other cases such as a supernova (SN) burst at our

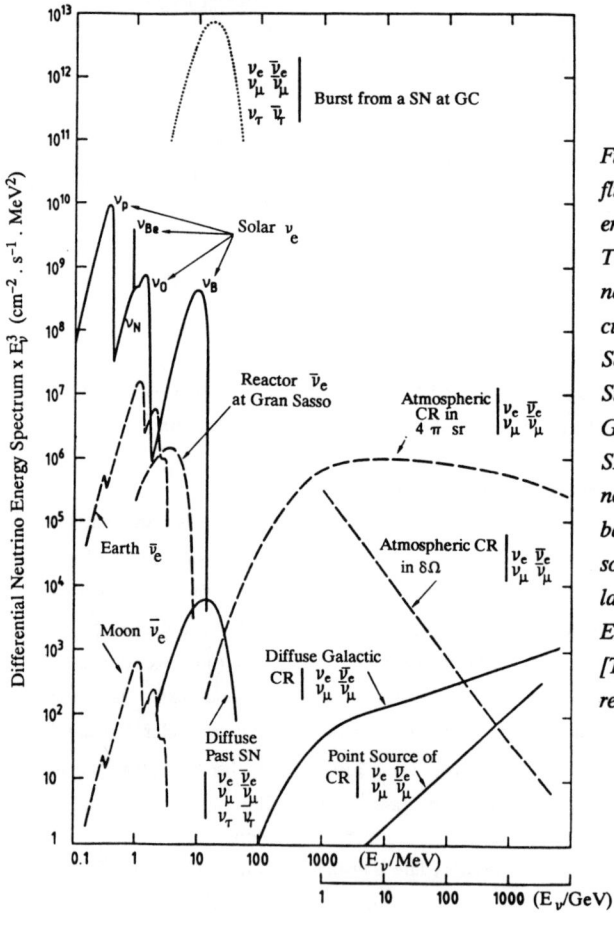

Fig. 2a. Differential $\nu(\bar{\nu})$ flux (xE^3) for medium energies on the Earth. The antineutrino luminosity of the Moon is discussed in the text. Sources are for the Sun[25], SN burst at the GC[26], diffuse past SN[29,31], Earth's antineutrinos[31], Earth-based reactors[5], point source CR[33], diffuse Galactic CR[16], and the Earth's atmosphere[12,13]. [Taken from Ref. 20, and revised.]

Galactic center (GC) impressive theoretical work[26-29] has been done on gravitational collapse to form a neutron star, but only one mechanism[26] has been shown. Supernova 1987A in the Large Magellanic Cloud has produced a flurry of research[30] on Type II events but no radical changes to the two SN curves in the Fig. 2. Supernovae have clearly been occurring throughout much of the existence of the Universe, and many papers have addressed the neutrino luminosity of present[26-28] and past[29,31-32] (relic) supernovae and their remnants. Diffuse past SN spectra are sensitive to assumptions in cosmology, and only the spectrum of Ref. 29 is shown, with a normalization to the cosmological redshift argument of Ref. 31. Point sources such as Cygnus X-3 of cosmic ray (CR) neutrinos are illustrated, and are taken from Ref. 33 by equating the ν flux to the γ-ray flux. Higher ν/γ flux ratio mechanisms, however, have been

Fig. 2b. Differential $\nu(\bar{\nu})$ flux (xE^3) for medium energies on the Moon. The antineutrino luminosities of the Earth and Moon roughly interchange, while the Earth's atmospheric flux[12,13] drops, following the argument of Fig. 1. [Reactor antineutrino luminosities have been deleted to simplify the figure.]

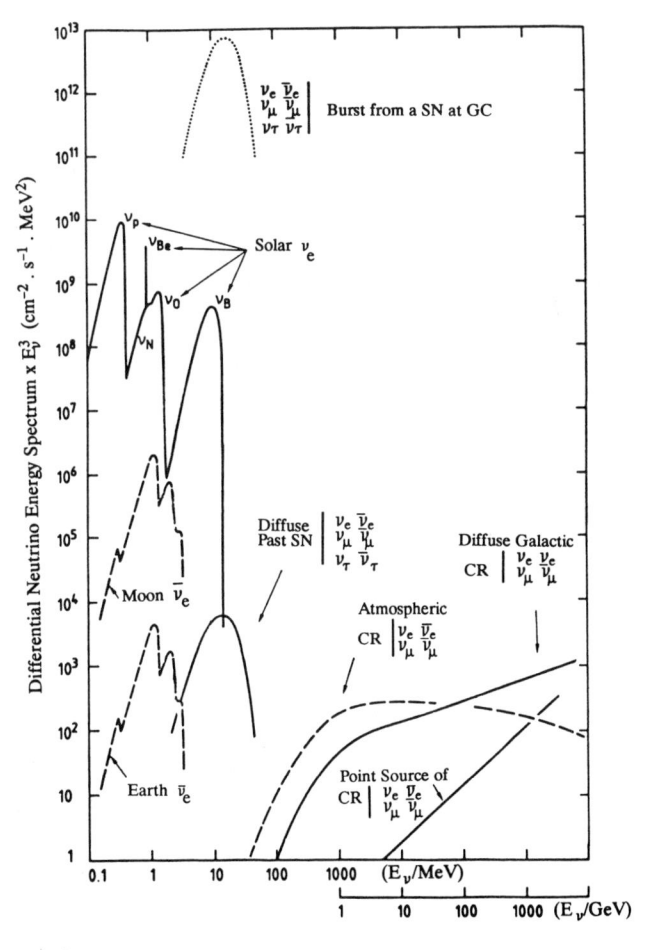

studied.[34-35] The big-bang relics (Weinberg[36-37] relics) are of extremely low energy (~10^{-9} MeV) and are not addressed here. Their spectrum is given in Fig. 2 of Ref. 31.

For high-energy cosmic neutrino physics (> 10^{-1} TeV), diffuse sources in the neutrino sky are depicted in Fig. 3. Typically, astrophysical mechanisms are sought which produce the unstable mesons (π's and K's), and these decay to produce the neutrinos. Shown are the atmospheric neutrinos[13,16-17], the diffuse flux from the inner Galaxy[16] due to cosmic-ray interaction (primarily proton-proton collisions) with interstellar gas, related pulsar production,[17] and finally the photoproduction of pions[17] (π-mesons). The inset (a) of Fig. 3 is for particle astronomers, defining the deep-inelastic scattering cross-section used as that of the standard electroweak model with an intermediate vector boson mass M_w = 80 GeV. The quark-parton distri-

The Neutrino Sky
(As Seen From The Earth)

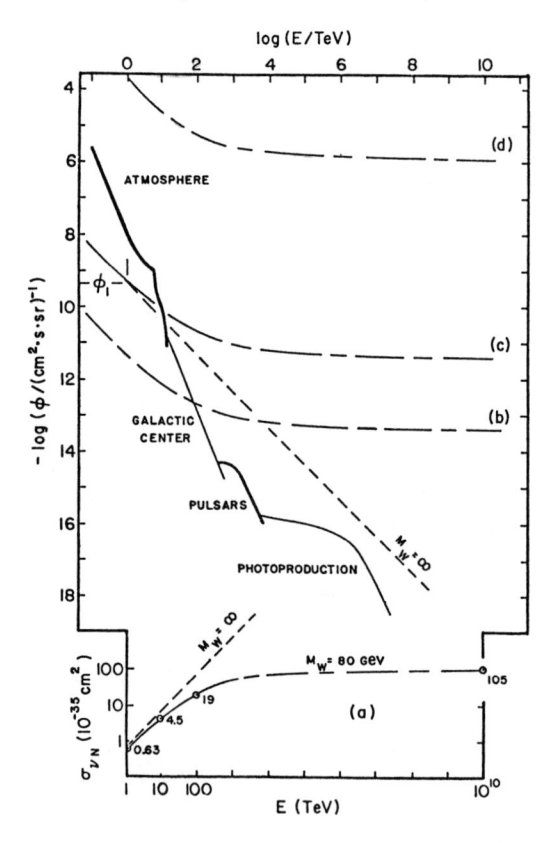

Fig. 3a. Integral $\nu(\bar{\nu})$ flux for high energies on the Earth. This is a conservative view of diffuse sources in the neutrino sky, based upon estimates from Ref. 13, 16, 17. [Taken from Ref. 18.]

bution of Buras and Gaemers[38] was assumed (1-10³ TeV), giving the logarithmic detector sensitivity (1 event/day) obtained by flipping curve (a) upside down.[39] In Fig. 3, curve (b) is for a 10^{11} and (c) is for a 10^9 metric ton detector (of the DUMAND type). Detector (d) is a 2-kton Shuttle external tank in low Earth orbit filled with water.

Figs. 4 and 5 stress the case for lunar base neutrino physics even further. Stecker's analysis[16] of the cosmic neutrino flux is reproduced in Fig. 4a and 5a, for an Earth-based observer. His was a very lucid assessment of the experimental difficulty in measuring neutrinos from the Galactic center, not only because they are enshrouded by the atmospheric neutrinos, but because the prompt atmospheric neutrinos mimic the Galactic ones. (They have the same slope.) Under the Earth-to-Moon flux transformation used here (Fig. 1), the Galactic center flux is

The Neutrino Sky
(As Seen From A Lunar Base)

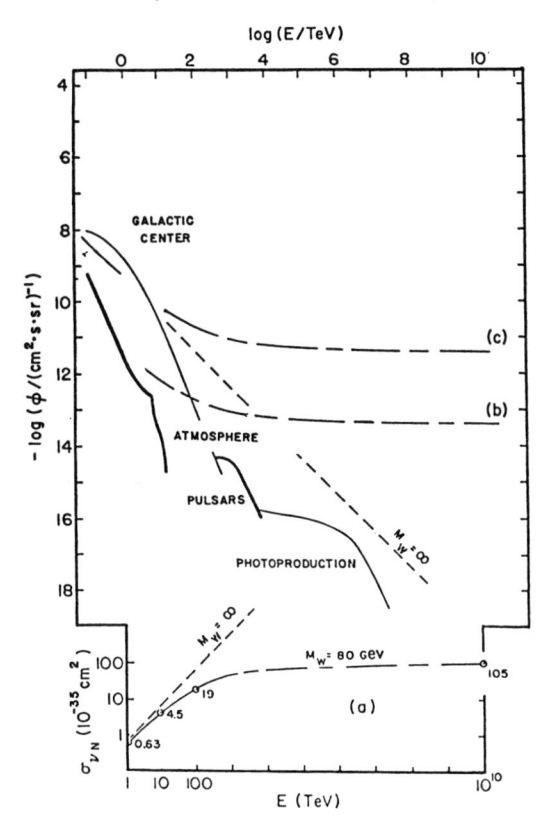

Fig. 3b.   Integral $\nu(\bar{\nu})$ flux for high energies on the Moon. The atmospheric neutrino flux has dropped sufficiently to open up a complete new "window" ($1-10^3$ TeV) into the Galactic center, using the flux transformation of Fig. 1.

Comparative Neutrino Spectra
(Earth–Based Versus Lunar–Based)

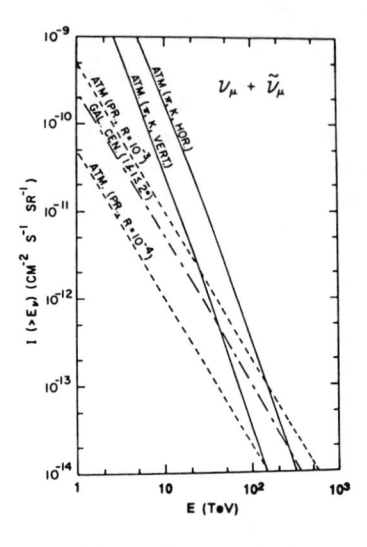

(a) Earth–Based Flux

(b) Lunar–Based Flux

Fig. 4. The integral muon $\nu(\bar{\nu})$ flux in TeV neutrino astronomy. The π and K meson production as well as the direct or prompt ν's are compared with those from the inner galaxy. (a) is the Earth-based and (b) is the lunar-based flux. [Fig. 4a is taken from Ref. 16.] The Galactic center is clearly unobstructed in the lunar-based flux (1–10^3 TeV).

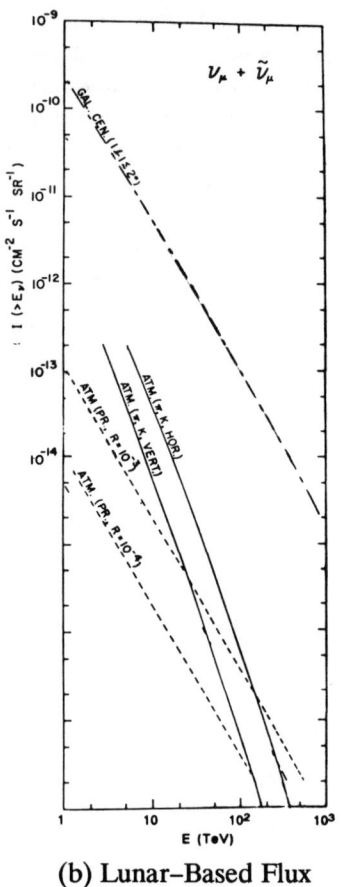

uneffected, while the atmospheric flux drops off by $(60)^{-2}$ as illustrated in Fig. 4b and 5b.

It is clear from the comparative neutrino spectra in Fig. 4b and 5b that a "window" exists in Galactic neutrino astonomy for a lunar-based detector, which is obscured from Earth-based observation. This is true for the energy range 1–10^3 TeV, and has been proven to exist for studies surveyed here. It will be shown in Section 8, however, that the atmospheric ν's ($\bar{\nu}$'s) can be cut from the data altogether.

Comparative Neutrino Spectra
(Earth–Based Versus Lunar–Based)

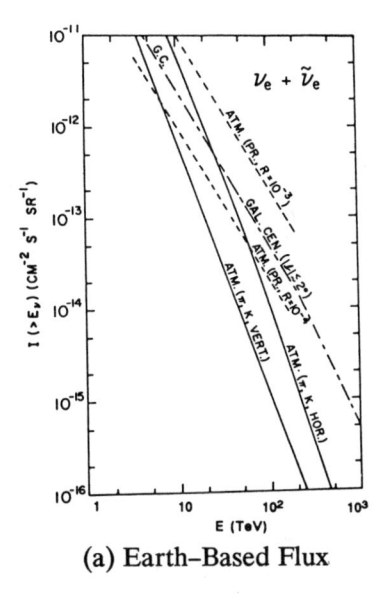

(a) Earth–Based Flux

Fig. 5. The integral electron $\nu(\bar{\nu})$ flux in TeV neutrino astronomy. Similar to Fig. 4. Again, the Galactic center is clearly visible in the neutrino sky of a lunar-based observer (1–10^3 TeV).

(b) Lunar–Based Flux

4. The Antineutrino Luminosity of the Moon

The atmospheric neutrinos (antineutrinos) of the Earth are not the only ones effected by the Earth-to-Moon flux transformation. Another phenomenon occurs, which will be very important to the present discussion. The antineutrino fluxes due to the bulk radioactivity of each planetary object interchange, roughly speaking. That the Moon has an antineutrino luminosity is a conclusion that can be drawn from orbital measurements[40-44] of radionuclides such as uranium (U) and thorium (Th) deposits in the lunar regolith using an orbital gamma–ray spectrometer, as well as numerous studies[45-51] of the radioactivity of lunar surface samples returned to Earth. Not only is the Moon an abundant source of radioactivity, surprisingly enough the relative abundance analyses[46] indicate it has three times the relative U and Th content of the Earth while only one-half the relative ^{40}K abundance (Ref. 46, Table 3).

Even though very little is understood even for the Earth as regards its thermal evolution and history, heat flow studies[52-53] indicate that U–Th–K systematics are critical to planetary composition and structure. Uranium and thorium decays appear to be responsible for most of the heat production within the Earth, although the earliest studies of the antineutrino luminosity of the Earth[54-57] demonstrated that the most abundant antineutrino emitter was ^{40}K.

Krauss, Glashow, and Schramm[31] performed an extremely important investigation into the possibility of studying the Earth's antineutrino luminosity by relating it to geophysics. Their model consisted of a radioactive lithosphere of 1/300 Earth masses (2×10^{25} g) with a conservative shell depth of 30 km, from which was derived an antineutrino flux (Ref. 31, Equations 2 and 3) from standard abundances. They then reconsidered the Marx and Lux analysis (Ref. 54, Fig. 1) and improved it extensively. Their result (Ref. 31, Fig. 2) was later modified by Lagage[20] to give it the "head-and-shoulders" appearance illustrated here in Fig. 2a as Earth $\bar{\nu}_e$.

From chemical composition studies,[45-51] seismic analysis of the Moon's internal composition,[58-60] and modeling of its internal density,[61] one cannot expect the published lunar abundances[46] to be rigorous and accurate when the data is incomplete. First of all, important questions regarding thermal evolution are unanswered which can lead to inconsistencies in abundance modeling, and second, the U–Th–K abundance[46] is not consistent with the Earth model.[31] That is, the lunar U–Th abundance is triple and the ^{40}K only one-half that of Earth. What this does is change the shape of the "head-and-shoulders" spectrum in Fig. 2a when we try to use the Earth as a model and adapt it to the Moon.

It is the ^{40}K abundance which fixes the peak of the antineutrino luminosity flux in Fig. 2, and not the U–Th abundance. This follows quickly from the Marx & Lux,[54] the Krauss, Glashow, & Schramm,[31] and the Lagage[20] analyses. Furthermore, it is impossible to determine the full spectral shape since there is insufficient data at the present time. Consequently, in order to adopt a model for the lunar antineutrino flux, we approximately halve that of Earth according to Ref. 46, Table 3 (half the ^{40}K abundance of Earth), use the Earth model of Ref. 31 with a lunar radius, and neglect lunar variations in U–Th–K systematics. That is, the mass and radius of the Moon relative to Earth are $M_m = 0.01228\ M_e$ and $R_m = 0.27280\ R_e$. The relative abundance of ^{40}K is $A_K = [96\mu g/g]/[170\mu g/g] = 0.565$. Then dimensionally the lunar $\bar{\nu}_e$ differential flux is

$$\phi_{Moon} = [A_K M_m / R^2_m] \phi_{Earth}$$
$$= [0.09323] (1.1 \times 10^7 \text{ cm}^{-2} \cdot \text{s}^{-1} \cdot \text{MeV}^{-1})$$
$$= 1.03 \times 10^6 \text{ cm}^{-2} \text{ s}^{-1} \text{ MeV}^{-1} ,$$

and in terms of the head-and-shoulders flux in Fig. 2,

$$E^3 \phi_{Moon} = [0.09323] E^3 \phi_{Earth}$$
$$= [0.09323] (2.15 \times 10^7 \text{ cm}^{-2} \cdot \text{s}^{-1} \cdot \text{MeV}^2)$$
$$= 2 \times 10^6 \text{ cm}^{-2} \cdot \text{s}^{-1} \cdot \text{MeV}^2 .$$

This does not take into account the mean density differences, $\bar{\rho}_e$ = 5.522 g cm^{-3} for the Earth and $\bar{\rho}_m$ = 3.3437 g cm^{-3} for the Moon. What it does establish is a terrestrial $\bar{\nu}_e$ flux model for the Moon at its reduced radius, mass, and abundance for ^{40}K at 1.25 MeV, which has been identified as the Moon $\bar{\nu}_e$ spectrum in Fig. 2b. The remote Earth (Moon) $\bar{\nu}_e$ flux in Fig. 2b (2a) is obtained from Fig. 2a (2b) by the (60)$^{-2}$ conversion of Fig. 1. That is, they approximately interchange.

5. The Entire Moon as a Neutrino Detector

Probably one of the most powerful ideas which emerged from the Apollo program was that of using the entire Moon as a gravitational radiation detector, by measuring quadrupole excitations with gravimeters. It was suggested by Weber[62] whose gravimeter experiment did succeed in gathering some data.[63–65] Similar suggestions have been made for the Sun.[66–67]

A related proposal[4,68] was made by this author to view the entire Moon as a neutrino detector. After all, it weighs in at 10^{20} metric tons which when considered as a space platform is bigger than anything the Space Agency has ever considered launching into Earth orbit. Its remote isolation from Earth in space vacuum with low magnetic fields makes it ideal as an astrophysics testbed for large-scale arrays of instrumentation in particle (neutrino) astronomy.

This point of view is represented by Fig. 6 in an effort to delve into its meaning. What is the bulk, global response – if any – of the Moon to neutrino radiation? For this question to have scientific meaning, we must imagine a Moon instrumented with some network of detectors D (e.g. calorimeters) as illustrated. A neutrino burst of intensity I$_o$ sees a bulk detector of scattering cross-section σ and opacity κ, a burst which is attenuated along a cylindrical cord length L, 1 cm^2 in area and containing N nucleons.[69]

In order to determine or count the number N of nucleons in the 1 cm^2 column through the target of Fig. 6, at angle ψ from the detector's nadir, we need a density model of the Moon. As fortune would have it, the only reasonable reference[61] proposes not one but 35 or so density models and none of these is tabulated. The wide variation is due to paucity of data, no data at all from the far-side surface, and general controversy as to the origin and thermal evolution of the Moon. Consequently, a five-layer density model of our own is derived here which is consistent with the envelopes of Ref. 61, Fig. 6a, and based upon the same principle conditions of mean density $\overline{\rho}_{av}$ = 3.3437 g cm^{-3}, a crustal mean density $\overline{\rho}_{Crust}$ which varies from 2.8 to 3.05 g cm^{-3}, and a moment of intertia constraint C/MR^2 = 0.3905. This is clearly defined in Fig. 7 with tabulated density values in italics at corresponding shell radii (dashed lines).

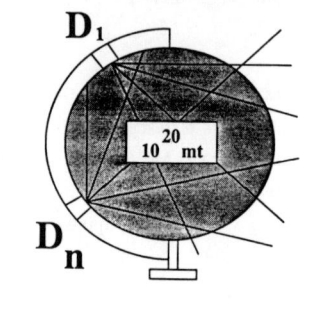

The Moon as a Neutrino Detector

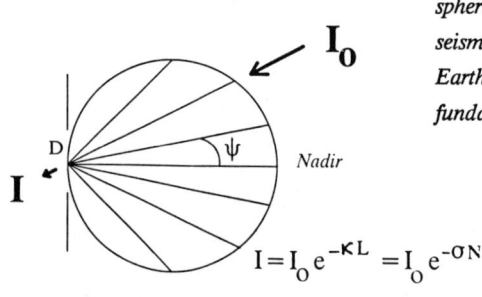

Fig. 6. The bulk Moon as a space platform of 10^{20} metric tons (mt) configured with detectors D. The absence of atmospheric noise[73] and the much quieter seismic background as compared to the Earth, provides a new emerging class of fundamental physics experiments.

$$I = I_o e^{-\kappa L} = I_o e^{-\sigma N}$$

Now we are prepared to discuss what the Moon looks like to a TeV neutrino.[70] To do so, we begin by considering the Earth where this problem has already been worked out,[18] using the Preliminary Reference Earth Model (PREM).[71] That is, one simply plots the percentage attenuation [100 (1 - I/I$_o$)] in polar coordinates (using Ref. 18, Table 1) for the nadir angle ψ. The result for Earth is Fig. 8a. Similarly using the lunar density model in Fig. 7, the neutrino opacity for the Moon is derived and tabulated in Table I. When plotted in polar coordinates, the result for the Moon is Fig. 8b.

The first observation to make is that a bulk, multi-layered spherical mass such as the Earth and Moon appears tear-drop shaped to a high–energy neutrino. Noting that the total cross-section is a sum of elastic and diffusive inelastic terms, a bulk index of refraction n = k/k$_o$ can be defined where k and k$_o$ are the neutrino momenta in matter and *in vacuo* respectively. Since

$n^2 = 1 - 4\pi Nb/k^2$ for a coherent scattering length b, then by approximating the square root of n^2, we get[72]

$$n = 1 + (2\pi N/k^2) f(0) \qquad (1)$$

as the bulk index of refraction in Fig. 8, for a single-species neutrino and a forward scattering amplitude f(0) by virtue of the optical theorem. Medium-energy radiochemical astronomy,

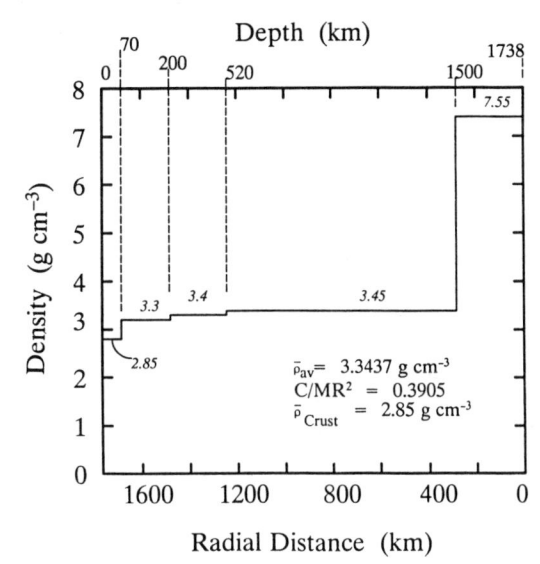

Fig. 7. Five-layer lunar density model used in the determination of the neutrino opacity of the Moon.

Table I. Neutrino Opacity of the Moon

Moon layer (density gcm^{-3})	ψ (deg)	N (x10^{32})	L (x10^{8}cm)	κ (10^{-10} cm^{-1}) 1TeV	10TeV	100TeV	% Attenuation 1TeV	10TeV	100TeV
Center	0.0°	8.30	3.476	0.15	1.075	4.537	0.52	3.67	14.59
Core	7.87°	7.06	3.443	0.129	0.923	3.896	0.44	3.13	12.55
	17.46°	6.79	3.316	0.129	0.922	3.891	0.43	3.01	12.10
	30.00°	6.14	3.010	0.129	0.918	3.815	0.39	2.73	11.01
Mantle	44.50°	4.99	2.479	0.127	0.906	3.824	0.31	2.22	9.04
Upper Mantle	62.24°	3.13	1.619	0.122	0.870	3.673	0.21	1.40	5.77
Crust	73.78°	1.68	0.977	0.108	0.774	3.269	0.11	0.75	3.14

in contrast, treats the Earth as a transparent observing platform orbiting in space, whose sensitivity is isotropic and which can serve as an all-sky monitor (ASM). At higher TeV energies, however, Fig. 8 shows that the bulk behavior is anisotropic, bearing more of a resemblence to an inverted (half- or vertical) dipole radiation pattern in electromagnetism. As a modulation transfer function which is quantum mechanically sound, this antenna-like feature justifies viewing the Moon as a bulk neutrino telescope.

The bulk acoustical properties of the Moon also present interesting possibilities for $\nu(\bar{\nu})$ detection, once again drawing from the comparison with gravitational radiation.[62-67] Seismic and atmospheric noise of Earth[73] impose stringent backgrounds for gravitational radiation interferometers. At a lunar base, not only lower seismic backgrounds but more importantly the complete loss of atmospheric noise (Ref. 73, Fig. 2) might make acoustic detection[74-76] of ν's ($\bar{\nu}$'s) very feasible because of the quieter environment. Enabling research should be pursued for this interesting possibility and its relationships with seismic gravimeter networks and gravitational radiation interferometers, all of which share portions of the same noise spectrum (micrometeoroids and acoustic CR impacts). The network of detectors D in Fig. 6 then might be imagined as an acoustical one.

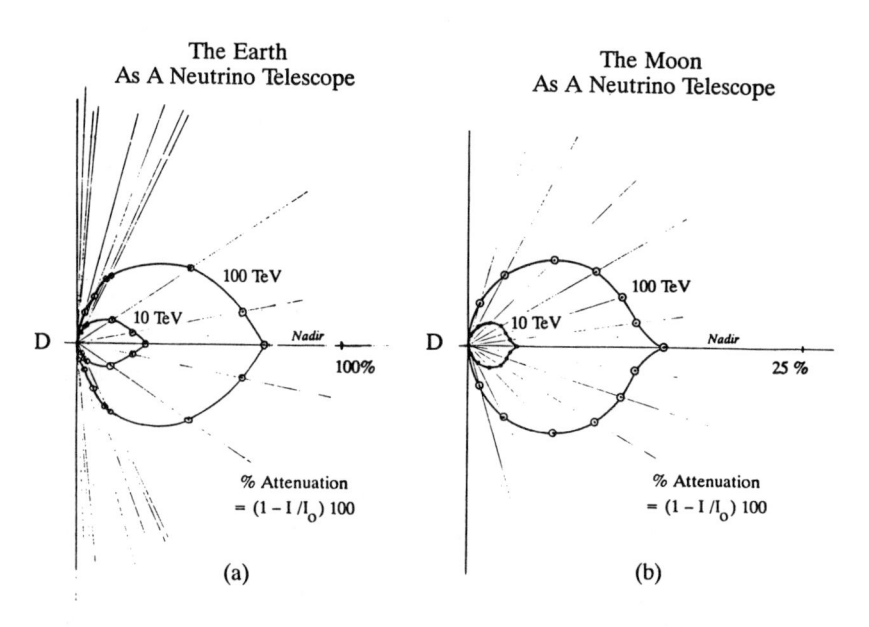

Fig. 8. The bulk inclusive scattering of the Earth and of the Moon due to their neutrino opacity, viewing them as neutrino telescopes. (a) is from Ref. 18, and (b) is from Table I.

6. Particle Astronomy from the Moon: Inverse Scattering Transforms

As a large-scale observational platform spinning in inertial space within the celestial sphere, the Moon can be imagined in Fig. 9a as either an instrumental array of detectors D_i or as a bulk detector with a non-contiguous network of D_i data sampling points (such as discussed in Sect. 5). A diffuse source of neutrinos is depicted, the Galactic Center, and the signal response in space-time can be mapped into momentum space as a Fourier transform $F(u,v,w)$ of which planar slices are shown in Fig. 9b. The transform F is the product of three principle transfer functions plus noise (such as the Earth's atmospheric neutrinos): The neutrino transport through and out of the astrophysical source (S), the transfer or modulation function for the bulk Moon $\mathcal{M}_b(\kappa,\rho)$ which depends upon the opacity κ and the density ρ (using Sect. 5), and finally the detector modulation transfer function $\mathcal{D}$ representing the response and sensitivity of the network D_i. That is, $F = c_d \mathcal{D} \mathcal{M}_b (S + s^*)$ where c_d is a detector scale-factor and s^* is all other extraneous sources of neutrinos.

The Fourier transform $F = F(u,v,w)$ is 3-dimensional and is measurable. Presumably, the modulation transfer function $c_d\mathcal{D}$ for the detector network is also determinable since we construct it ourselves. Thus, we can define $\mathcal{F} = c_d^{-1}\mathcal{D}^{-1} F^*(u,v,w) = \mathcal{M}_b(\kappa) S$ as the experimental transform for a lunar base array, after we have deleted extraneous sources of known noise ($F^* = F - c_d\mathcal{D}\mathcal{M}_b s^*$). Procedurally, particle imaging is the technique by which we isolate $\mathcal{M}_b(\kappa)$ from $\mathcal{F}$ and particle astronomy is the technique by which we isolate S from $\mathcal{F}$.[77]

Fig. 9. Particle (neutrino) astronomy from a lunar base, by solving the inverse scattering problem. (a) The rotating Moon, (b) the measured Fourier transform F, (c)–(d) the cylindrical slice theorem for a stationary detector D which measures only $\bar{F}$ and not F, and (e) the planar slice theorem. (c)–(e) are 2-dimensional cuts of the 3-dimensional transform F. The inverse transforms $\mathcal{M}_b^{-1}$ and S^{-1} determine the structure of the Moon and the Galactic Center respectively. [Taken from Ref. 18.]

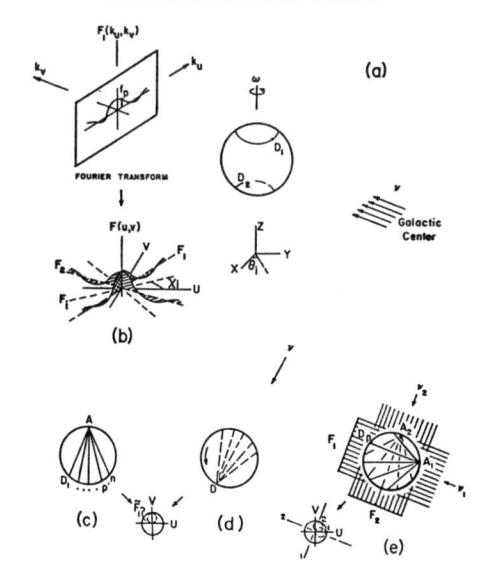

Briefly, Figs. 9a and 9b illustrate the fact that a space–time object (here, the instrumented Moon) as a bulk sphere of mass 10^{20} metric tons with inertial spin ω has an image in (Fourier) momentum space $\mathcal{M}_b(\kappa)$ which is buried in $F(u,v,w)$. It is a 3–dimensional surface in momentum space.[78] If a known radiation source S (such as an accelerator A on the Earth or Moon) is used, then we can study experimentally the momentum–space transform $\mathcal{M}_b(\kappa)$, since

$$\mathcal{M}_b(\kappa) = S^{-1} \mathcal{F}(u,v,w) . \qquad (2)$$

Similarly, when we have determined the transform $\mathcal{M}_b(\kappa)$, we can examine the astrophysical source S, since

$$S(u,v,w) = \mathcal{M}_b^{-1}(\kappa) \mathcal{F}(u,v,w) . \qquad (3)$$

Eqs. (2) and (3) are the inverse scattering transform problems of particle physics.[77] Transform (2) is that of non–invasive imaging (e.g. Ref. 18), and transform (3) is that of large–scale astronomy. The problem is well–posed mathematically when the inverses exist, and it is ill–posed when they do not.

Imaging of the Earth's interior is also conceivable ($\mathcal{M}_b$ now entails both the Earth and Moon), but at 60 R_e the spatial resolution drops off considerably. The Moon as an observatory of the Earth makes more sense for imaging of its magnetosphere whose scale is of the same order of magnitude ($10R_e \times 60R_e$) as the orbit of the detector $\mathfrak{D}$. This is still particle astronomy, but it uses neutral particle emission rather than neutrinos.

7. Tomographic Imaging and Antineutrino Astronomy

Because an antineutrino detector has been proposed for a lunar base,[7] its scientific justification includes imaging of the Moon's radioactive interior, and this will be addressed here. "X–ray" projections using neutrinos as a form of exploration of the Earth's interior have been suggested for some time.[79–81,75,76] Such projections do not determine the density distribution, a point which has already been made,[18,82–84] and such applications of high–energy physics facilities hardly seem worth the effort when they do not even solve the problem.

A ring of detectors[18] D_n or a moveable (mobile) detector[18] installation D is required in order to have a well-posed inverse Radon transform problem in antineutrino astronomy.[85] What the detector D measures at energy E along a cord length l is g,

$$g(\kappa, E) = \ln(I_o/I) = \int_0^L \rho(l) \, dl , \qquad (4)$$

the inverse transform of which is the density $\rho(r)$.[86]

To prove this graphically, Fig. 9d illustrates that the fan-beam configuration in Fig. 6 of a fixed, stationary detector D only measures a single cylindrical cut through the Fourier transform F, shown as a dashed "bubble" $\overline{F}_1$ beneath the detector in momentum space. Only when the detector is rotated around the entire geometric object does the "bubble" $\overline{F}$ eventually intersect all of the transform F from which we arrive at $\mathcal{F}$ for Equations (2) and (3).

The cylindrical slice theorem introduced in Ref. 18 has a direct application to antineutrino astronomy, proposed for geophysics of the Earth[31] and addressed here for the Moon. This is illustrated in Fig. 10, depicting the tomographic transform problem for deriving the density $\rho(x_o, y_o)$ at a point $P(x_o, y_o)$ within the Moon. Seven detectors or detector positions show the relationship, where again a cylindrical slice contour $\overline{F}$ of F is required in order to reconstruct the radioactive intensity at $P(x_o, y_o)$.[87] *All* of the contour in Fig. 10b is required to reconstruct $P(x_o, y_o)$ in Fig. 10a. Obiously, from Fig. 10, *if* we knew F then all we need is one measurement to reconstruct the cylindrical slice in Fig. 10b (since its diameter in momentum space is the same as the radius $r_o = (x_o, y_o)$ at $P(x_o, y_o)$ in configuration space).

The problem is that we do not know F and therefore $\mathcal{F}$ for Equations (2) and (3). Instead, we are trying to measure F. Hence, at least two stationary detectors or one mobile (moveable) detector is required to do antineutrino imaging of the lunar interior. Conceivably, a single detector installation could be operated for long-durations and then re-located.

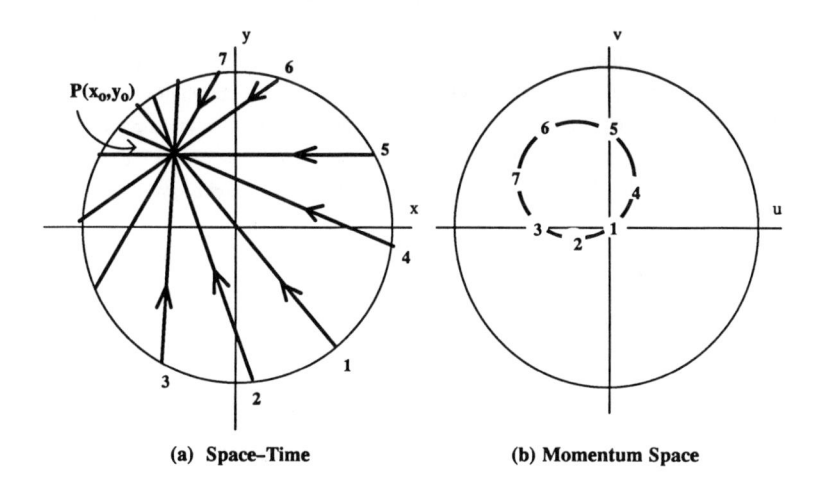

(a) Space–Time

(b) Momentum Space

Fig. 10. Cylindrical slice theorem [Ref. 18], a cut through the 3-dimensional Fourier transform of Fig. 9b illustrating the contour of points in (u,v)–momentum space which are required to reconstruct one point $P(x_o, y_o)$ in real configuration space (x,y).

8. A Compact Cut Theorem for Lunar Base Neutrino Astronomy

Now we come to the simplest and yet perhaps the most significant result of this manuscript. In retrospect, even a child knows that if he/she rotates a small luminescent ball on a long string about a dark room, the walls of the room can still be defined as rigorously as if the ball were not there. The author, however, had to come by this result the hard way – that is, after the diversion into tomographic transforms above.

The single-most adverse problem in neutrino (antineutrino) physics and astronomy can be eliminated from remote astrophysical sources and fundamental particle experiments simply by going to the Moon and building the particle detector there. We can cut the Earth's atmospheric neutrinos entirely from the data of cosmic objects which are sufficiently stationary with respect to the celestial sphere, and from the background of fundamental particle experiments.

This is not true for co-moving or co-rotating particle beam experiments originating from the Earth (Sects. 10–11). Nevertheless, the result follows directly from the Earth–Moon ephemeris and the fact that the Earth at 60 R_e is a compact source of solid angle $\Omega_s = \pi \zeta^2$ = $\pi[(0.95°)(0.01745 \text{ rad/deg})]^2 = \pi[2.75 \times 10^{-4}] = 8.6 \times 10^{-4}$ steradian. The proof will be graphical, using Fig. 10, and will be called the compact cut theorem of inverse scattering theory.

Let Fig. 10a represent the Earth–Moon laboratory rotating within the celestial sphere with the Earth at the origin and the Moon in its orbit at $P(x,y)$. A position $P(x_o,y_o)$ is shown which measures cosmic astrophysical neutrinos from various points on the celestial sphere (the outer circle). There is an osculating cylindrical slice in momentum space shown in Fig. 10b. As the Moon orbits the Earth (the origin), the osculating cylinder rotates about the origin of momentum space. The problematic noise s^* from Earth's atmospheric neutrinos is the compact point ("1") at the origin of Fig. 10b, and can be cut from the Fourier transform F. That is,

$$F^* = F - c_d \mathcal{D} \mathcal{M}_b s^* . \qquad (5)$$

By using a mean background flux from the surrounding momentum space, we "patch" (continuous in first and second derivatives) and smooth over the "hole" left by deleting the Earth s^* from the neutrino sky.[88] The result is a well-posed inverse scattering transform. That is, Equations (2) and (3) are solvable, and the noise from Earth is gone as a result of a compact Earth and Earth–Moon ephemeris. Similarly, one can cut the Solar ν's and the Earth-based reactor $\bar{\nu}$'s as well.

Fig. 10a also illustrates the insurmountable problem of Earth-based neutrino astronomy. In this case, let $P(x,y)$ be viewed as a detector at the Earth's surface, rotating within the celestial sphere (the outer circle), while the entire celestial sphere (e.g. all seven sources) is coated or lined with the diffuse background of atmospheric neutrinos. Then the Fourier transform F in momentum space is saturated with noise s^* from the Earth's atmospheric neutrinos and $\mathcal{F}$ cannot be determined (except at fluxes above that of the atmosphere in Figs. 2–5). Equations (2) and (3) lack causal support because F^* in (5) is indeterminable.

Neutrinos (antineutrinos) from the Earth are present at the lunar surface, but they can be cut by the procedure establilshed here.

9. All–Sky Monitoring of Compact Neutrino Sources

Broadly speaking, Fig. 2 establishes the inter-relationships between various medium-energy compact sources (the Sun, Moon, Earth, nuclear reactors, and supernovae) in the Earth-Moon laboratory proposed here. Similarly, Fig. 3 illustrates the diffuse sources in TeV neutrino astronomy. The fact that some sources (e.g. Solar ν_B) are of much greater flux than others (e.g. the diffuse SN relics) does not preclude a low-flux study from the lunar surface, because a compact source will seasonally move about the celestial sphere and can be removed by the compact cut theorem of Sect. 8. The proposed study of the relic SN spectrum[7] from a lunar base is an example. Medium–energy neutrino (antineutrino) detectors are all-sky monitors, and the procedure for eliminating known extraneous data is clear. They must be designed with sufficient directionality (spatial resolution) that the unwanted, conflicting sources can be cut from and tracked within the data. Within margins, this has just been established as a straightforward procedure in the inverse scattering problem of particle astronomy (Sect. 6–8).

10. Long–Baseline Particle Beam Experiments

From the outset it was stated [Appendix] that the Earth–Moon system provides a unique situation for particle physics which has never been attempted before and which no other realistic space initiative can offer. It was suggested as an argument against an Earth-Mars initiative.

Fig. 11. ν environment in the Earth–Moon laboratory, depicting the basic geometry of the Moon as an orbiting observatory of the Earth. A high-energy (TeV) ν detector D is assumed on the far side of the Moon.

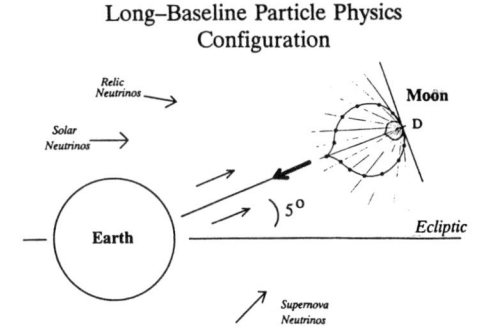

This author proposed[4, Appendix] that should an adequate fundamental particle physics facility be established on the Moon, including a neutrino detector, long-baseline particle beam experiments could be conducted between Earth-based accelerators such as the hadron supercollider or SSC (Superconducting Super Collider[89,90]) and the lunar base.[91]

Fig. 11 shows such a configuration from the point of view of the $\nu\,(\bar{\nu})$ environment. Certainly, this geometry has the distinct feature of providing a "spectroscopic" advantage during a supernova event for gathering both Earth-based and Moon-based data on particle fluxes, arrival times, and coincidences with other contiguous measurements such as X-rays, γ-rays, and possibly gravitational radiation. The dynamic structure of gravitational collapse, neutron star formation, and exotic objects could be better examined while important SN precursory events might be more readily available.

However, the unique possiblity of beaming pulses of neutrinos (antineutrinos) at a detector D on the Moon from a tevatron T on Earth is represented in Fig. 12. An optimal alignment is shown of a DUMAND-type detector on Earth with area $A = 1 \text{ km}^2$ (at 2 R_e) and the detector D on the far side of the Moon (at 61.64 R_e) with beam area $A' = (61.64/2)^2 A = 950 \text{ km}^2$. The

Long–Baseline Particle Physics

$\theta = E_0^s/E_\nu$. For π–mesons,
$A = 1 \text{ km}^2$ when $E_\nu = 1.576$ TeV.
For K–mesons, $A_K = 12.5\, A_\pi$.

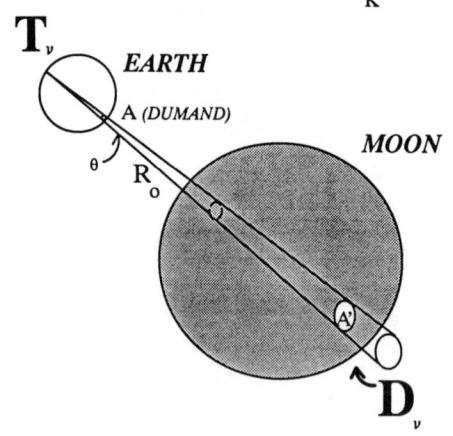

Fig. 12. An Earth-based tevatron T beams ν's to a Moon-based detector D. High-energy ν beams of definite composition and energy (from resultant secondary π and K decay) can be produced by "dumping" a proton beam onto a target. Particle kinematics from Ref. 79 for π and K meson sources are shown, although charm decay has been analyzed.[75] The beam divergence θ is the ratio of meson rest energy to ν beam energy. A beam at 1.576 TeV subtends $A = 1 \text{km}^2$ (10^{10} cm^2) on the opposite side of the Earth (at 2 R_e.) To hold $A' = 1 \text{km}^2$ on the opposite side of the Moon, a ν beam energy of 48.57 TeV is required.

beam forms a conical frustrum of volume $V \simeq LA_d/3$ where A_d is the area of the detector.

The principle new feature or point of view in Fig. 12 is that the decay tunnels and turning magnets (or the "snouts") of the tevatron must be directed downwards into the Earth about the nadir. A number of references have discussed this[76,81,75,79-80] and there appears to be no conceptual difficulty with doing it. However, it would require a modification to the SSC baseline design as well as existing Earth-based acclerator facilities. Future high-energy accelerator designs (including the SSC) should not preclude such a modification.

What Fig. 12 represesents is a frontier in particle physics and particle astronomy, which is virtually unexplored. As a strategy or scheme of things, it conceptually "scales up." That is, a long-baseline (e.g., $L = 2R_e$) experiment would first be established as Phase I of what this author is proposing in Fig. 12,[92] using DUMAND. Phase II would then scale this capability up to an Earth-Moon baseline during or following a full, mature operational phase of the lunar base. Earth-based particle (v) astronomy like DUMAND is part of the enabling research and technology which will support a major initiative in fundamental physics from a lunar base and the next generation of scientific exploration suggested by Fig. 12.

11. Neutrino Oscillation Studies in the Earth-Moon Laboratory

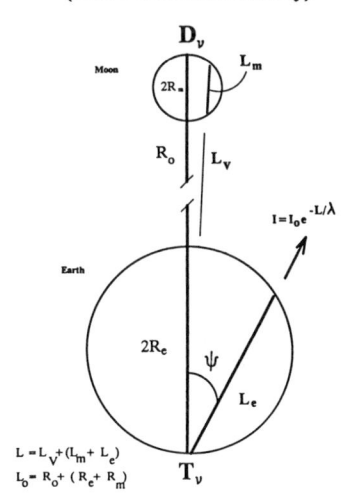

One serious investigation[93] has been made into projecting accelerator beams through the Earth from Fermilab, in order to study neutrino oscillations over baseline cords of the Earth

Fig. 13. Neutrino oscillations can be studied at planetary densities in the Earth-Moon laboratory. Matter (along L_e and L_m) and vacuum (along L_V) oscillations all can be addressed using a pure v eigenstate (pulsed accelerator beam) in tandem with a lunar-based v detector. A relevant feature of oscillation studies is the ratio L/E, where L is the length of the baseline and E is the v beam energy. Mean-free path $\lambda = \kappa^{-1}$ cm.

on the order of magnitude of 10^3 km and detector areas of π km^2. Recent reviews of ν oscillation theory[94] and experiment[95-97] indicate a mounting interest in longer-baselined capability, even though many of the reactor[96] investigations are only of the order of meters.

Fig. 13 addresses quantum physics. Quantum oscillations are one of the fundamental predictions of quantum theory, having been derived from its outset (as *zitterbewegung*[98]) for the electron and subsequently extended to neutrinos. Pure states (in Hilbert space) with mass evolve into mixed states in the presence of interactions,[99] and hence the phenomenon is often called neutrino mixing when applied to neutrinos of different species (or flavors), ever since two varieties were experimentally established.[100] This was first done for neutrinos by Pontecorvo,[101] drawing an analogy with kaon oscillations, and the subject became closely related to the solar neutrino puzzle[25,102] when the work of Wolfenstein[103] was extended by Mikheyev and Smirnov[104,105] (the MSW effect) in an attempt to explain the low ν flux attributed to the Sun.

The phenomenon is equivalent to a precession of the neutrino in flavor space, much like the precession of a particle with a magnetic dipole moment in a magnetic field in space-time or configuration space. As the ν passes through matter such as the Earth and Moon (Fig. 13), it begins to rotate in Hilbert space (not shown) and take on different aspects or colors (flavors).[106] It is a critical proof of the coupling between Hilbert space and space-time predicted by quantum (and not classical) theory.

The single-species refractive index in Eq. (1) becomes a matrix

$$n_{ij} = \delta_{ij} + (2\pi N/k^2) f_{ij}(0) \quad , \tag{6}$$

when two or more flavors of ν_i (i,j = e,ν,τ... lepton families) are allowed to exist in the Hilbert space; that is, $\nu_i = \mathcal{U}_{ij}\nu_j$ for a unitary mixing matrix $\mathcal{U}_{ij}$. The diagonal terms guarantee that different flavors of different mass propagate with different velocities, and the off-diagonal terms appear in matter (and magnetic fields[107,108]) to introduce the mixing. The original model of the bulk Moon (Fig. 8 and 9) becomes birefringent in a 2-species (e and μ) case, and the probability $\mathcal{Q}$ of survival[109] is

$$\mathcal{Q} = \sin^2 2\theta \, \sin^2 1.27 \, \Delta m^2 \, L/E \quad , \tag{7a}$$

where $\Delta m^2 = |m_2^2 - m_1^2|$ is the mass parameter (in eV^2), $\sin^2 2\theta$ is the mixing strength, E is the neutrino energy (in MeV), and L is the oscillation length (in meters). The probability of conversion (disappearance) of the ν from one flavor to another is

$$\mathcal{P} = 1 - \mathcal{Q} \quad , \tag{7b}$$

since the total probability has unit norm and $\mathcal{P} + \mathcal{Q} = 1$. The individual ν_i masses (m_1, m_2, ...) are not measurable by this method of study, and to date only limits on Δm^2 have been established. The sensitivity of an experiment depends upon L/E (and this parameter has been illustrated in Ref. 96, Fig. 1).

Numberous studies[110-116] have considered the effect of matter at planetary densities (the Earth alone in Fig. 13) on resonant enhancement of neutrino mixing, using solar and accelerator ν's. What is clear is that there is great difficulty in using solar ν's because these arrive as mixed eigenstates which are solar-model dependent. The distinct advantage of accelerator experiments (artificially generated neutrinos) is that the neutrinos begin as pure eigenstates for which the resonance enhancements appear to be more pronounced and the experimental assumptions are more certain, particularly when the accelerator beam is pulsed.

Oscillation experiments using artificially generated neutrinos in the Earth-Moon laboratory (Fig. 13) offer the advantage of greater oscillation length as well as the higher density of two core regions instead of the Earth alone. Although this is not a justification for putting a ν detector on the Moon, the existence of a neutrino physics facility at a mature lunar base would offer some fascinating possibilities in this discipline of physics. Oscillation studies at TeV energies could be performed, but there appears at present to be no theoretical reason for doing it.[117]

12. Summary and Conclusions

We have examined the neutrino (antineutrino) fluxes in the Moon's environment, and demonstrated that new vistas in neutrino astronomy exist at a lunar base. The Earth's atmospheric neutrinos are present there, but we have proven that they can be cut from the data, thereby opening up the entire neutrino sky of astrophysics to lunar-based observation. What we have not done is rigorously determine new sources and particle backgrounds which may be coming out of the Moon itself, such as the Moon's muon luminosity as well as new sources of neutrinos due to charm decay, and how these could compromise the Moon's otherwise clear view of the neutrino Universe. The Moon has been addressed as a bulk neutrino detector, and prospects of long-baseline particle physics experiments in the Earth-Moon laboratory have been discussed. Exploration of Earth-Moon density can be done and imaging of the Moon's radioactive interior is feasible, although it would require re-location of the $\bar{\nu}$ detector. Fascinating possibilities in fundamental neutrino physics have been considered, such as qunatum oscillations induced by the Earth and Moon.

In summary, the prospect of a permanent lunar base opens a number of new frontiers in physics and astrophysics which are simply not available to the Earth-based observer. This fact, in conjunction with the discoveries in physics and astronomy during the past several decades which have transformed our view of the Universe, presents itself as a serious scientific justification for returning to the Moon. Some of these opportunities are offered by the Moon alone, and are not available to any other realistic space initiative.

In search of scientific answers to fundamental questions in physics and astrophysics, the strategy of the space age for three decades now has been to extend our reach beyond the Earth by using small spacecraft platforms instrumented with scientific payloads. What the Moon offers instead is a huge, large-scale platform (10^{20} metric tons) in Earth orbit surpassing anything imaginable that we could construct there. It has no appreciable atmosphere or magnetic field, it is of extremely low vacuum and low gravity, and it is 60 Earth radii away from the noise of Earth. Its far side is immune from Earth noise and radiation save for the atmospheric neutrinos, and antineutrinos produced by our nuclear power plants, both of which we have now proven can be cut from the background sky.

The Moon is an extremely interesting scientific asset in its present natural state, as a remote quiet international science facility and astrophysics testbed. We can conclude that it offers a challenging new frontier in fundamental physics for the next generation of scientific exploration which is now emerging.

Acknowledgement

The author would like to thank P. O. Lagage for sharing reprints of Refs. 19 and 20 in a timely fashion, thus making the flux analysis more complete.

References and Footnotes

1. "Particle" astronomy is defined as the measurement of a particle event in a suitable detector, and then determining its (astrophysical) origin and energetics using certain assumptions from particle physics and inverse scattering transforms.
2. If the neutrino has a mass, this statement is complicated by quantum oscillations. These will be addressed later in the discussion of long-baseline particle experiments.
3. A. K. Mann, in *Science Underground, Los Alamos 1982*, eds. M. M. Nieto *et al.*, American Institute of Physics Conference Proceeding **96**, 16 and 445 (AIP, N.Y., 1983).
4. T. L. Wilson, "Astrophysics from a Lunar Base: New Experiments in the Earth-Moon Laboratory," in Proc. Symposium on the Next Supernova: Astrophysics, Particle Physics, and Detectors, Santa Monica, California (February 21-22, 1989).
5. P. O. Lagage, Nature **316**, 420 (1985).
6. Proton decay is another. See J.C. Pati, A. Salam, and B.V. Sreekantan, "The Moon as the Searching Ground for Proton Decay," International J. Mod. Phys. **A1**, No. 1, 147-153 (1986). Also see D. Cline and S. Rudaz in these conference proceedings.

References and Footnotes

7. Relic neutrinos of past supernovae may be another. See A.K. Mann, these conference proceedings.
8. Low Earth orbit (LEO) is even worse than the Earth's surface. If one attempts to do the unimaginable and place a large neutrino detector in orbit (e.g. a Shuttle external tank filled with water or liquid scintillator, as detector (d) in Fig. 3a), one loses all of the advantages of taking the detector underground (as in Reference 3) at less expense and with superior results.
9. The endemic background noise of the bulk Moon itself (e.g. its muon and charm luminosity) as a production source of ν's is not treated here, except for modeling its antineutrinos (Sect. 4). The "windows" (vistas) in Figures 2, 3, 4, and 5 will identify the energy ranges and type (e.g. if radiochemical or other) of detector design.
10. The "μ meson" is now commonly referred to as a muon, to distinguish it from the other mesons π and K. In the quark–lepton theory of elementary particles, the muon is a lepton while the pion (π) and kaon (K) are the strongly interacting bosons known as mesons.
11. Because the electron (e^-) spectrum is a continuum, two neutrinos must be produced to conserve energy–momentum.
12. T. K. Gaisser and T. Stanev, "Neutrino Astronomy and the Atmospheric Background," 19th International Cosmic Ray Conference (ICRC), La Jolla [NASA Conf. Publication 2376, Washington, DC] **8**, 156–159 (1985). See Figure 1.
13. L.V. Volkova, Sov. J. Nucl. Phys. **31**, 784 (1980). Slight modifications to this work are given by K. Mitsui and Y. Minorikawa, 19th ICRC **8**, 144 (1985), Figures 1 and 2.
14. A. Dar, Phys. Rev. Lett. **51**, 227 (1983); and in *Fourth Workshop on Grand Unification*, eds. H. A. Weldon, P. Langacker, and P. J. Steinhardt, p. 101 (Birkhäuser, Boston, 1983).
15. H. Lee and S.A. Bludman, Phys. Rev. **D37**, 122 (1988) and references therein.
16. F.W. Stecker, Ap. J. **228**, 919 (1979). For an experimental analysis of this work, see R. Svoboda *et al.*, Ap. J. **315**, 420 (1987).
17. S.H. Margolis, D.N. Schramm, R. Silberberg, Ap. J. **221**, 990 (1978).
18. T. L. Wilson, Nature **309**, 38 (1984). The inclusive scattering cross-section in this reference has since been corroborated by M. H. Reno and C. Quigg, Phys. Rev. **D37**, 657 (1988).
19. P. O. Lagage, Thesis, "Acceleration and Propagation of Cosmic Rays. Production, Oscillations, and Detection of Neutrinos," Service d'Astrophysique, Centre d'Etudes Nucleaires (CEN) de Saclay, France (June 1987).
20. P. O. Lagage, "Astrophysical and Terrestrial Neutrinos in Supernova Detectors," in *Neutrinos and the Present–day Universe*, ed. T. Montmerle and M. Spiro, Centre d'Etudes Nucleaires (CEN) de Saclay, Gif-sur-Yvette Cedex, France (September 9–13, 1985).
21. The atmospheric flux rolls off at the high and low energy ends as we would expect from

References and Footnotes

two simple arguments. At high energy using a Lorentz transformation, we can boost ourselves into the rest frame of the unstable mesons and see that they live longer due to relativistic time dilation. The longer they survive, the greater their chance of atmospheric collision before decaying into neutrinos, and therefore fewer and fewer neutrinos are produced at higher energies.[12,13] At low energy, the problematic cosmic rays are more easily trapped in the Earth's radiation belts without collision which gives rise to a geomagnetic cutoff[14,15] as their energy and cross-section decreases, and therefore fewer and fewer neutrinos are produced at lower energies.

22. As an observer moves to the center of the Earth, modulo some geometric factors, the zenith flux drops while the nadir flux rises – their sum remaining constant.

23. Only the neutrinos radiated within the conical frustum (with half angle ζ) formed by the Moon and a point source in the Earth's atmosphere are involved. This introduces a geometrical factor $\Omega_s = 2 \times 10^{-5}$ steradian, where the Moon subtends the solid angle 2ζ with respect to the CR ν source at the Earth, and where $0.509^0 < 2\zeta < 0.527^0$.

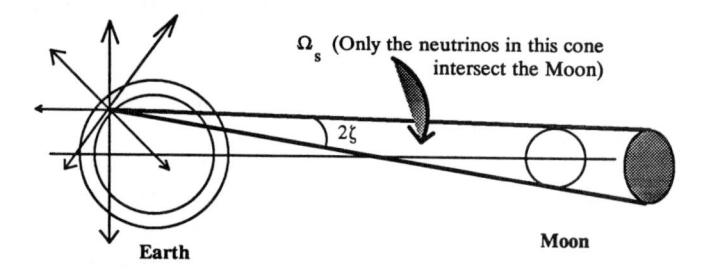

Ω_s (Only the neutrinos in this cone intersect the Moon)

Earth   Moon

24. That the center of the Moon has been chosen for the lunar base flux of Fig. 2b, 3b, 4 and 5 is done to simplify the argument. A lunar base detector can be placed at the surface on the far side, getting as far away from the atmospheric source as possible, and improving matters further. However, the uncertainties due to differences in estimates[12-16] wash out the effect of one additional lunar radius. This also compensates for geometric error introduced in taking the Earth-based surface estimates[12-16] to the center of the Earth in Fig. 1. The geometric solid–angle factor[23] is not taken into account here, since the entire problem is abated by the compact cut theorem in Sect. 8.

25. J.N. Bahcall and R.K. Ulrich, Rev. Mod. Phys. **60**, 297 (1988); also J.N. Bahcall, W.F. Huebner, S.H. Lukow, P.D. Parker, and R.K. Ulrich, Rev. Mod. Phys. **54**, 767 (1982).

26. D. Z. Nadezhin and I. V. Otroshenko, Sov. Astron. **24**, 47 (1980).

27. H. A. Bethe, A. Yahil, and G. E. Brown, Ap. J. **262**, L7 (1982).

References and Footnotes

28. A. Burrows, Ap. J. **283**, 848 (1984).
29. G. S. Bisnovatyi-Kogan and Z. F. Seidov, Ann. N.Y. Acad. Sci. **422**, 319 (1984).
30. See the current review by W. Hillebrandt and P. Hoflich and references therein, "The SN 1987a in the LMC," preprint MPA 458 (May 1989), to appear in *Progress in Physics*.
31. L. M. Krauss, S. L. Glashow, D. N. Schramm, Nature **310**, 191 (1984).
32. W. Zhang, et al., Phys. Rev. Lett. **61**, 385 (1988).
33. M. Samorski and W. Stamm, Ap. J. **268**, L17 (1981).
34. F. W. Stecker, A. K. Harding, and J. J. Barnard, Nature **316**, 418 (1985).
35. H. Lee and S. A. Bludman, Ap. J. **290**, 28 (1985).
36. S. Weinberg, Nuovo Cim. **25**, 15 (1962); Phys. Rev. **128**, 1457 (1962).
37. P. Bandyopadhyay, P. R. Chaudhuri, and S. K. Saha, Phys. Rev. **D1**, 377 (1970). R. Royer, Phys. Rev. **174**, 1719 (1968).
38. A. Buras and K. Gaemers, Nucl. Phys. **B132**, 249 (1978). See also D. W. Duke and J. F. Owens, Phys. Rev. **D30**, 49 (1984).
39. Neutrino cross-sections beyond 10^3 TeV are unknown. The cross-section shown (10^3-10^{10} TeV) is an extension of the logarithmic term in Ref. 38 to emphasize that Bjorken scaling ($M_W = \infty$) is invalid because it violates unitarity. The scaling goes as $\log E$ and not as E. See Ref. 18.
40. A. E. Metzger, J. I. Trombka, L. E. Peterson, R. C. Reedy, and J. R. Arnold, Science **179**, 800 (1973). A notable deficiency in this experiment was that the radioactive thermoelectric generators or RTGs (a known γ-ray source) in the manned-landers were not used to normalize the data, lying as they did in or near the orbital plane of the spectrometer.
41. P. Gorenstein and P. Bjorkholm, Science **179**, 792 (1973).
42. H. C. Urey et al., Phil. Trans. R. Soc. London **A285**, 600 (1977).
43. Proc. Fourth Lunar Science Conference, Vol. **1**, Plate II (1973) [Geochimica et Cosmochimica Acta, Suppl. **4**].
44. S. R. Taylor, *Lunar Science, A Post-Apollo Perspective*, Fig. 5.14, Pp 242-243 (Pergamon, N. Y., 1975).
45. M. Tatsumoto and J. N. Rosholt, Science **167**, 461 (1970); M. Tasumoto et al., Proc. Fifth Lunar Conf., Vol. **2**, 1487 (1974) [Geochimica et Cosmochimica Acta, Suppl. **5**].
46. E. Anders, Phil. Trans. R. Soc. London **A285**, 23 (1977).
47. H. C. Urey and J. A. O'Keefe, Phil. Trans. R. Soc. London **A285**, 569 (1977).
48. V. Rama Murthy, Phi. Trans. R. Soc. London **A285**, 127 (1977).
49. H. Wanke et al., Phi. Trans. R. Soc. London **A285**, 41 (1977).
50. K. L. Rasmussen and P. H. Warren, Nature **313**, 121 (1985); J. Geophys. Res. **92**, No. 35, 3453 (1987).

References and Footnotes

51. Yu. A. Surkov, G. A. Fedoseyev, O. P. Sobornov, and L. S. Tarasov, Proc. Fourth Lunar Science Conf., Vol. **2**, 1437 (1973) [Geochimica et Cosmochimica Acta, Suppl. **4**].
52. C. Jaupart, J. G. Slater, and G. Simmons, Earth and Plan. Sci. Lett. **52**, 328 (1981).
53. J. G. Slater, C. Jaupart, and D. Galson, Rev. Geophys. and Space Phys. **18**, 269 (1980).
54. G. Marx and I. Lux, Acta Phys. Acad. Sci. Hungaricae **28**, 63 (1970).
55. C. Avilez, G. Marx, and B. Fuentes, Phys. Rev. **D23**, 1116 (1981).
56. G. Marx, Czech. J. Phys. **B19**, 1471 (1969).
57. G. Eder, Nucl. Phys. **78**, 657 (1966).
58. Y. Nakamura and G. Latham, J. Geophys. Res. **74**, 3771 (1969).
59. Y. Nakamura, G. Latham, and D. Lammlein, Geophys. Res. Lett. **1**, 137 (1974).
60. Y. Nakamura, J. Geophys. Res. **88**, 677 (1983).
61. L. L. Hood and J. H. Jones, J. Geophys. Res. **92**, No. B4, P. E396 (1987) [Proc. 17th Lunar and Planetary Science Conf., Part 2].
62. J. Weber, Sci. Amer. **224**, No. 5, 22 (1971).
63. J. Weber, in *General Relativity and Gravitation*, ed. A. Held, Vol. **2**, 435 (Plenum, N.Y., 1979), Sect. 18.
64. R. L. Tobias, "The Lunar Surface Gravimeter and the Search for Gravitational Radiation," PhD thesis, University of Maryland (April 27, 1978); J. Weber, personal communication.
65. J. Weber, these workshop proceedings.
66. B. W. Carroll, P. N. McDermott, S. N. Shore, and C. E. Wendell, Nature **308**, 165 (1984).
67. S. P. Boughn and J. R. Kuhn, Nature **308**, 164 (1984); Ap. J. **286**, 387 (1984).
68. This idea also occurred during a conversation with V. J. Stenger while organizing this workshop.
69. The ν opacity $\kappa = \sigma \overline{N}$ is of dimension cm^{-1}. By taking the total mass M along a 1 cm^2 cylindrical column of length L, a mean density of $\overline{\rho} = M/L = M/2a\cos\psi$, for a planetary radius of a, defines a number density $\overline{N} = \overline{m}^{-1}_n \overline{\rho}$ cm^{-3}. Here $\overline{m}^{-1}_n = 6.029030 \times 10^{23}$ g^{-1} is the inverse of the average mass per nucleon, a procedure derived in Ref. 18. Alternatively one can use Avogadro's number $N_A = 6.022169 \times 10^{23}$ mol^{-1} with $\overline{N} = N_A \overline{\rho}$, which is an approximation. In either case, $\kappa L = \sigma \overline{N} L = \sigma N$.
70. A TeV antineutrino cross-section is approximately 1/3 of that in Fig. 3 for 1–10^3 TeV.
71. A. M. Dziewonski and D. L. Anderson, Phys. Earth Planet. Inter. **25**, 297 (1981).
72. Multiple scattering theory in wave mechanics is treated in many places. For the neutron, not the neutrino, quantum optics is discussed by V. F. Sears, Phys. Rep. **82C**, 1 (1982), and Canad. J. Phys. **56**, 1261 (1978); also I. I. Gurevich and L. V. Tarasov, *Low-Energy Neutron Optics* (North-Holland, Amsterdam, 1968). Gurevich and Tarasov first stressed that low–

References and Footnotes

energy coherent scattering (e.g. the relic neutrinos in Ref. 36-37, 31) goes as the atomic number squared, or A^2, and not as A.

73. P. R. Saulson, Phys. Rev. **D30**, 732 (1984).
74. J. G. Learned, Phys. Rev. **D19**, 3293 (1979); also see these workshop proceedings.
75. L. V. Volkova, Nuovo Cim. **8C**, 552 (1985).
76. A. DeRújula, S. L. Glashow, R. R. Wilson, and G. Charpak, Phys. Rep. **99**, No. 6, 341 (1983).
77. In all cases, one must finally return to space-time or configuration space via the inverse Fourier or inverse Radon (Ref. 18) transform. This inverse mapping back out of momentum space then defines the structure of the object under study.
78. This surface is akin to the Fermi surface of solid state physics.
79. L. V. Volkova and G. T. Zatsepin, Acad. Sci. U.S.S.R., Bull. Phys. Ser. **38**, 151 (1974).
80. L. V. Volkova, Sov. J. Nucl. Phys. **45**, 666 (1987).
81. I. P. Nedyalkov, Rep. Bulg. Acad. Sci. **34**, 1495 (1981); private communication (July 30, 1984).
82. T. L. Wilson, Summer School in Nuclear Medicine, University of Kentucky, Lexington (July 25, 1977), where neutrino tomography (Fig. 1 of Ref. 18) was first introduced.
83. T. L. Wilson, MIT Summer School in Transverse Section Imaging and Positron Emission Tomography, Cambridge, Massachusetts (July 23-28, 1978), G. E. Brownell, Convenor, where Fig. 1 of Ref. 18 was again introduced.
84. T. L. Wilson, in *Particles & Fields – 1982*, eds. W. E. Caswell and G. A. Snow, American Institute of Physics Conf. Proc. **98** (AIP, N. Y., 1983).
85. In radionuclide imaging (nuclear medicine and antineutrino astronomy) only the detector must be mobile (moveable) because the radiation source S (Eq. 3) is what is actually being scanned. When the radiation source S is external to the object being scanned (e.g. a particle accelerator A in Fig. 9c) *both* the source S and the detector $\mathcal{D}$ must move relative to the object under study ($\mathcal{M}b$). Otherwise Eq. (2) constitutes an ill-posed inverse Radon transform problem [which was the point of Ref. 18]. Early proposals[79-81,75,76] are not neutrino tomography but rather neutrino photography for this reason.
86. Tomographic tricks other than the Radon transform technique include (a) matrix inversion, whereby Fig. 9e is imagined to be n equations in n unknowns and then inverted; and (b) taking the Radon-Nikodym derivative of Eq. (4), which for circular symmetry is

$$\rho(r) = -d/dr \left| (r/\pi) \int_r^\infty [g(l)/(l^2-r^2)^{1/2}]\, dl \right| = -dI(r)/dr \quad ,$$

whose solution was first obtained by Abel in 1825.

References and Footnotes

87. The cylindrical cut in k-space (Fig. 10b) is the locus of normals to the cords l in configuration space (Fig. 10a).
88. Define a cut about the origin of k-space with radius ϵ > radius of the compact extraneous source s^* (Earth). Create a "patch" $p = c_d \mathcal{D} \mathcal{M} s^*$ determined by observing the compact source seasonally and averaging, or by filling in $c_d \mathcal{D} \mathcal{M} (S-s^*)$ such that $F(\epsilon)$ has continuous first and second derivatives in k-space.
89. E. Eichten, I. Hinchliffe, K. Lane, and C. Quigg, Rev. Mod. Phys. **56**, 579 (1984).
90. J. D. Jackson, M. Tigner, and S. Wojcicki, Sci. Amer. **254**, No. 3, 66 (March 1986). See references on P. 120.
91. One advantage for an accelerator on the Moon would be elimination of need for high-vacuum tunnels (with high life-cycle maintenance costs), a situation very similar to the elimination of vacuum pipes for gravitational radiation interferometers at a lunar base. There appear to be few other scientific justifications, however, for proposing this at the present time.
92. We know that Fig. 12 is optimal but unrealistic in the present era of physics. An optimal configuration would place the SSC near Johannesburg, South Africa and not Waxahachie, Texas in order to support DUMAND. Similarly, DUMAND would have to be located near the St. Paul and Amsterdam Islands in the Indian Ocean rather than at Hawaii to support long-baseline ($L = 2R_e$) experiments with the SSC and Fermilab. A mobile tevatron has been proposed[76] and a moveable (re-locatable) DUMAND installation might be considered for the St. Paul-Amsterdam Islands in the next century. Because demographic and national political factors influence the global site selection of high-energy physics facilities, long-baseline experiments about the nadir will have to be confined to present facilities and their modification. Windows of opportunity based upon Earth-Moon ephemeris and geometry can be worked out for Fig. 12 in each case.
93. A. K. Mann and H. Primakoff, Phys. Rev. **D15**, 655 (1977); A. K. Mann, in *Long-Distance Neutrino Detection - 1978*, eds. A. W. Sáenz and H. Überall, American Institute of Physics Conf. Proc. **52**, 101 (AIP, N. Y., 1979).
94. S. M. Bilenky and S. T. Petkov, Rev. Mod. Phys. **59**, 671 (1987); **60**, 575 (1988). See also S. M. Bilenky and B. Pontecorvo, Phys. Rep. **41**, 225 (1978).
95. V. Flaminio and B. Saitta, Rivista Nuov. Cim. **10**, 1 (1987).
96. F. Boehm, Nucl. Instr. Meth. in Phys. Res. **A264**, 114 (1988).
97. C. Bari, *et al.*, Nucl. Instr. Meth. in Phys. Res. **A264**, 5 (1988).
98. E. Schrödinger, Sitzungsber. preuss. Akad. Wiss., Physik.-math. Kl. **24**, 418 (1930).
99. This happens with fermions in the Dirac theory and bosons in the Duffin-Kemmer theory of elementary particles.

References and Footnotes

100. L. M. Lederman, M. Schwartz, and J. Steinberger, Rev. Mod. Phys. **61** (1989), to be published.
101. B. Pontecorvo, Sov. Phys. JETP **26**, 984 (1968).
102. Also see J. N. Bahcall, R. Davis, Jr., and L. Wolfenstein, Nature **334**, 487 (1988).
103. L. Wolfenstein, Phys. Rev. **D17**, 2369 (1978).
104. S. P. Mikheyev and A. Yu. Smirnov, Nuovo Cim. **9C**, 17 (1986).
105. H. Bethe, Phys. Rev. Lett. **56**. 1305 (1986).
106. The bare ν's can be imagined as "dressed" initially in three pure colors of clothing, red, white, and blue. With time of interaction, the clothing becomes plaid and eventually changes colors. A detector designed to capture red-shirted ν's will miss those that have changed coloredshirts to blue and white.
107. L. B. Okun, "Neutrino Magnetic Moment and SN 1987A," in Proc. Symposium on the Next Supernova: Astrophysics, Particle Physics, and Detectors, Santa Monica, California (February 21-22, 1989).
108. J. Cooperstein and J. Lattimer, Phys. Rev. Lett. **61**, 23 (1988).
109. "Survival" means retaining the pure, initial eigenstate (e.g. in Ref. 106, of red-shirted ν's keeping on their red shirts).
110. G. Auriemma, M. Felcini, P. Lipari, and J. L. Stone, Phys. Rev. **D37**, 665 (1988).
111. A. J. Baltz and J. Weneser, Phys. Rev. **D35**, 528 (1987).
112. E. D. Carlson, Phys. Rev. **D34**, 1454 (1986).
113. S. P. Rosen and J. M. Gelb, Phys. Rev. **D34**, 969 (1986).
114. G. V. Dass and K. V. L. Sarma, Phys. Rev. **D30**, 80 (1984).
115. P. T. Leung, S. Boedo, and M. L. Rustgi, Phys. Rev. **D29**, 2655 (1984).
116. V. Barger *et al.*, Phys. Rev. **D22**, 1636 and 2718 (1980).
117. Lev Okun, private communication in Ref. 4 and Ref. 107.

Appendix

This Appendix includes the correspondence which brought forth the suggestion that long-baseline particle physics could be conducted between the Earth and the Moon, should a "mature" lunar base with a high-energy physics facility be established: (a) the letter of R. O. Hunter; (b) the letter of S. Wojcicki, and (c) the author's proposal.

(a)

Department of Energy
Washington, DC 20585

November 15, 1988

Dr. Francis Low
HEPAP Chairman
Laboratory of Nuclear Science
Massachusetts Institute of Technology
Cambridge, Mass. 02139

Dear Dr. Low:

The Department of Energy has announced the preferred site for the Superconducting Super Collider in the State of Texas. With the present schedule, construction of the SSC will be completed in 1996 and operation for physics experiments will begin shortly thereafter. The Ronald Reagan Center for High Energy Physics will provide an enormous new capability for the U.S. and the world high energy physics communities. We need to formulate plans to ensure maximum utilization and productivity of this facility. In addition, we need to ensure that present U.S. accelerators are fully used in the most productive manner between now and the time the SSC begins operation. Furthermore, we need to begin to consider the composition of the U.S. high energy program after the SSC begins operation. For these reasons, I would like HEPAP to provide advice on some elements of the high energy physics program that will be crucial for the next decade.

Specifically, I would like you to address the following questions taking into consideration not only U.S. facilities but existing and future accelerators in other countries.

1. What are the most important high energy physics problems, including advanced accelerators and detector R&D, to be addressed in the period between now and SSC operation? What steps should be taken to ensure that the operation of these facilities remains productive?

2. How should the SSC program be structured to ensure the greatest involvement with and most benefit to the U.S. scientific and technical community? How should the SSC facility operation be structured to maximize its use by the entire high energy physics community? For example, in its consideration of this question HEPAP might wish to consider the role of existing or planned communications networks in facilitating sharing of data, remote access to experiments and generally improving communication among scientists throughout the U.S.

3. What criteria should be used to determine the content of the U.S. program after the SSC begins operation?

Celebrating the U.S. Constitution Bicentennial — 1787-1987

Appendix

Your advice on these important questions will be very helpful in determining the future directions of the SSC and other activities in the high energy physics program. I would very much appreciate having your responses to these questions by February 15, 1989.

Sincerely,

Robert O. Hunter, Jr.
Director
Office of Energy Research

(b)

December 8, 1988

Dear Colleague:

As you can see from the enclosed attachments, HEPAP has recently formed a subpanel in response to Dr. Hunter's request to "provide advice on some elements of the high energy physics program that will be crucial for the next decade." I have been asked to serve as chairman of this subpanel. The major meeting of the subpanel will take place on January 18-22, 1988, at which time a report will be written for presentation to the full HEPAP at its meeting of February 6 and 7.

Input from the U.S. high energy community is crucial to having a properly informed debate during the subpanel's deliberations. Accordingly, I would like to invite you to send us a written summary of any thoughts you may have on the questions put forth in Dr. Hunter's letter of November 15, 1988. Because of the very compressed time scale during which the work of the subpanel must be completed, I would urge you to send us any comments you may have as soon as possible and certainly well before the subpanel's major meeting. It would be easiest if your letters were forwarded to me c/o Enloe Ritter, Executive Secretary of the subpanel, at the DOE HEP office (ER-221, Division of High Energy Physics, Washington, D.C. 20545; Bitnet address DOEHEP @ BNLVMA), who will then transmit them to all the panel members.

Of course, I and the other subpanel members would be happy to discuss these issues with you in person. However, the impacts of your thoughts will be much stronger if they are put down on paper.

To facilitate transfer of this information, we plan to have "town meetings", attended by some of the subpanel members, at several different locations. The location, tentative time, and subpanel member responsible for each meeting are given below:

Fermilab	Steve Errede	January 6
SLAC	Fred Gilman	January 5
Boston Area	Francis Low and Marjorie Shapiro	January 3 or 5
Cornell	Robert Siemann	Early January
Brookhaven	Jack Sandweiss	January 5

Additional details about these meetings will be publicized internally in the labs hosting the meetings or can be obtained from the relevant subpanel members.

Appendix

2

In addition, the chairs of the Users' organizations from the major accelerator laboratories will be asked to deliver (or organize) presentations to the full subpanel at its January meeting. The intent is that these presentations express the desires of the user community at that laboratory. Accordingly, you might want to get in touch with the chair of your user organization to express to them your views.

The Superconducting Super Collider will provide the U.S. high energy community challenging new opportunities for deeper exploration of inner space. We welcome your thoughts about how we should proceed during the coming years to take maximum advantage of the opportunities presented both by the SSC and by existing facilities.

Sincerely,

Stanley Wojcicki

Enclosures:
Hunter to Low Letter of
  November 15, 1988
Subpanel Membership List

(c)

Re: Space Initiatives
    & The SSC

JAN 6 1989

Dr. Stanley G. Wojcicki
c/o Enloe T. Ritter, Executive Secretary
U.S. Department of Energy, GTN
Division of High Energy Physics
Washington, DC 20545

Dear Dr. Wojcicki:

Thank you for an opportunity to comment on Dr. Hunter's questions, as well as your own letter, regarding the proposed Superconducting Super Collider (SSC) in the State of Texas.

Impromtu, as my remarks must appear due to the brief time allowed for response, I would like to remind everyone that while the U. S. is taking on a major initiative such as the SSC, it is also proposing bold new initiatives in space. Phase C/D has begun on the Space Station Freedom, which may serve as a testbed for a man-tended lunar base. Workshops are being organized for determining physics experiments which have a lunar-base justification, or alternatively might influence the selection of a lunar base over other alternatives as the next major space initiative. DOE and NASA are establishing major new initiatives in the same timeframe, and a man-tended planetary base lies somewhere in the future of U. S. high-energy physics facilities addressed by Hunter's letter.

Appendix

Clearly, a community of interest exists between these initiatives. These interests have proceeded in harmony, and continue to do so. But their common interests have been sadly neglected. I propose that the SSC working groups occassionally keep in mind that a man-tended lunar physics base will eventually exist, and SSC architectural design must not preclude the possibility of SSC-lunar base collaborative experiments. Obviously, long-baseline particle beam experiments can be performed across an Earth-Moon distance, but their scientific merit is unexplored. Let's consider an example. Because a lunar-based neutrino detector is futuristic (due to its mass or size, let's say), consider putting a neutrino source on the lunar surface to study neutrino oscillations across the Earth-Moon distance, and through the Earth and Moon. This might not be so far-fetched, simulating the Sun by using the Moon base as a fiducial companion. Solar neutrino studies are perplexed by the fact that the data are solar-model dependent, a problem which might be circumvented by an Earth-Moon laboratory using a known neutrino source. Standard theory predicts that the oscillations occur while traversing matter, which might be further tested with the advantage of Earth+Moon mass and distance as opposed to the Earth alone. Such a venture already suggests one result: The lunar base should be put on the back side in order to maximize the column of matter traversed by the neutrinos. The counter suggestion is to go with a lunar-based neutrino detector, and serve the dual purpose of supernova astrophysics while taking advantage of 20 TeV neutrinos from the SSC with the increased Earth-Moon opacity at that energy. Other gedanken experiments might be better suited for the front side. And other examples (e.g. interferometry, coincidence detectors in astrophysics, ..) remain unexplored.

The SSC community could conceivably cast a deciding vote on NASA's next major initiative in space (lunar base versus manned-Mars) by proposing an experimental justification for an Earth-Moon laboratory, defined as a long-baseline SSC-lunar-base link. Also, this community has the expertise to address issues of radiation safety standards that must be met by such a laboratory, which in turn influence many of the trade studies regarding space-based particle accelerators, space-based radioactive particle sources, and space-based particle detectors.

Treating this Earth-Moon laboratory as a future prospect, the Hunter-type issues appear to be what (are the experiments), when, why the SSC, and why the lunar base - as opposed to other facilities.

Respectfully yours,

Thomas L. Wilson

Dr. Thomas L. Wilson
NASA
Johnson Space Center
Houston, TX   77058
(713) 483-2147

cc: AT/J. Loftus

SEARCH FOR PROTON DECAY AND SUPERNOVA NEUTRINO BURSTS WITH A LUNAR BASE NEUTRON DETECTOR

David Cline
Physics and Astronomy Departments
University of California Los Angeles
405 Hilgard Avenue
Los Angeles, CA 90024-1547

ABSTRACT

We describe the current status of the search for proton decay on Earth, emphasizing the decay mode $P \to K^+ \bar{\nu}$ and discuss the possibility of detecting this mode with a simple detector on a lunar base station. The same detector could be used to search for neutrino bursts from distant supernova using the neutral current signature $\nu_{\mu,\tau} + N \to n + \nu_x$ by detecting the produced neutrons. The key advantage of the lunar experiment is the low neutrino flux and possible low radioactive background.

© 1990 American Institute of Physics

The search for proton decay on Earth is nearly stalled. The current lifetime limits on key decay modes are shown in Table 1. It is generally assumed that

(a) SU(5) is ruled out by the large lifetime of P → $\pi^\circ e^+$ and by the current value of $\sin^2\theta_W = 0.23 \pm 0.0048$ compared to the SU(5) prediction of 0.214.

(b) Supersymmetric SU(5) is still viable and one expects that the P → $K^+\bar{\nu}$ will prevail.

The best chance to extend P → $K^+\bar{\nu}$ to lifetimes beyond 10^{32} years on Earth appears to be the ICARUS detector being constructed for the Gran Sasso Laboratory. However, it is unlikely that any Earth bound experiment can reach limits of $\tau > 10^{33}$ years for relatively unconstrained decays such as P → $K^+\bar{\nu}$ due to large neutrino backgrounds. The neutrino background on the moon should be reduced by more than 200 over that on Earth. The limit for proton decay on Earth is due to neutrino backgrounds and is given in Table 1.

A lunar based experiment that utilizes moon dust for the bulk of the detector and a low mass sensor system could extend the search to the 10^{35} year level as shown in Tables 2 and 3.

The basic idea is to search for proton decay through the process

$${}^Z_A N \to {}^{Z-1}_{A-1} N^* + \begin{bmatrix} \pi^\circ e^+ \\ K^+\bar{\nu} \end{bmatrix}$$

$$\hookrightarrow \text{neutrons}$$

The proton decay products are being detected by a charged particle detector and the spallation neutrons by a BF_3 neutron detector. Due to the possible low radioactivity of lunar dust, the neutron detection could be background free. However, measurements of the flux are inconclusive at present. The relative neutrino background can be estimated from the ratio of the density of the Earth's upper atmosphere to the density of moon dust where the neutrinos would be produced on the lunar surface. This ratio is approximately 10^{-3}. A more precise estimate has been carried out in Reference 5 and is shown in Figures 1 and 2 as a function of energy. Additional, more precise calculations are needed.

The detector principle used could be similar to that required to detect supernova neutrino bursts following a scheme by D. Cline et.al. (paper reproduced in Appendix A). Figure 3 shows a schematic of the detector concept that utilizes an inexpensive neutron/tracking detector. In such an experiment, proton decay would be signaled by coincidence between the neutrons spallation due to the sudden breakup of the nucleus and the decay final state of K^+ followed by $K^+ \to \pi^+\pi^0$ or $K^+ \to \mu^+\nu$. A key problem is the ratio of mass of the sensor detector to the total mass. For present detectors, this ratio is usually $> 10^{-3}$, whereas the lunar detector would require $\sim 10^{-4}$. Such low mass structures will require an R & D program of study.

In Table 2, we have estimated the background rejection that can be achieved using a lunar base detector. The cuts applied are those that are applied usually in current proton decay experiments to reduce the atmospheric neutrino background. As we have seen, there is still appreciable background of this type on the moon and similar cuts would be required.

In Table 3, we estimate the sensitivity that could be achieved in a 10^6 ton lunar detector. The use of neutron spallation could improve the rejection by an additional factor of 5 or more. The most serious detector problem is the compacting of the lunar dust into a suitable structure for the detector configuration.

Monte Carlo calculations will be required to calculate the backgrounds to such a signature due to the breakup of the nucleus. It is possible that the neutron energy distribution would be characteristically different for the nucleon decay and the neutrino interaction due to the large energy the recoil nucleus gains from the neutrino interaction.

The lunar base proton decay detector could also be used for a Supernova neutrino burst detector. It would be possible to detect one or more of the neutrino interactions from the neutrino burst. For example, as shown in Appendix A, the detection of the neutral currect reaction

$$\nu_x + N \to N^* + \nu_x$$
$$\hookrightarrow \text{neutron}$$

is very sensitive to μ and τ neutrinos. However, it is difficult with a single detector to measure all the properties of the neutrino burst (direction, individual neutrino species, energy spectrum, etc.). We attempt to summarize the limitation of the neutrino telescopes that will be available in the 1990's in Table 4 to emphasize this point.

ACKNOWLEDGEMENTS

This paper was written while visiting the International Center for Theoretical Physics, Trieste, Italy. I wish to thank A. Salam for the hospitality of ICTP.

REFERENCES

1. J.C. Pati, Abdus Salam and B.V. Sreekantan, Modern Physics, **A1**, 147 (1986).

2. D. Cline, et. al., "A New Technique to Detect Neutrino Bursts From Distant Supernova", UCLA preprint and submitted to Astrophysical Letters and Communications, 1989.

3. See the papers in the Proceedings of the 9th Workshop on Grand Unification, Aix Les Bains, 1988 (World Scientific).

4. C.L. Bennett, "Limitations on Proton Decay Modes From a Passive Detector Scheme", published in the 2nd Workshop on Grand Unification, Birkhauser, 120, 1981.

5. R. Cherry and K. Lande, private communication.

TABLE 1

Current Experimental Limits on Proton Decay
Lifetime and Future Prospects

Mode	Limit	Ultimate Limit on Earth	Detectors
$p \to \pi^0 e^+$	2 - 3 $\times 10^{32}$y*	$\sim 10^{33}$y	IMB* KAMIOKA – Superkamioka
$p \to K^+ \bar{\nu}$	(1-6) $\times 10^{31}$y*	5×10^{32}y	FREJUS* KAMIOKA – ICARUS
$n \to 3\nu$**	$\sim 10^{28}$y	$\sim 10^{30}$y	Radiochemial – neutron burst detection

* See Reference 3
** This is an example of a nuclear decay mode that depends on a radiochemical or inclusive search. Such modes may be searched for in the future by using a signature that arises from the disintegration of the nucleus after nuclear decay liberating a burst of neutrons. See Reference 4 for more information.

TABLE 2

A Comparative Summary of Earth vs. Lunar Nuclear Decay Detectors
Background Rejection
(Decay Mode Independent)

Lifetime	Nuclear Decays in 10^6 Tons	ν Induced Events With Topological Cuts (Earth) $E_\nu = 1 \pm 0.2$ GeV	ν Induced Events With Topological Cuts (Lunar Base) $E_\nu = 1 \pm 0.2$ GeV
10^{33}y	600	500	2.5
10^{34}y	60	500	2.5
10^{35}y	6	500	2.5

TABLE 3

p DECAY (Future Prospects)

SITE	MODE	DETECTOR SIZE	τ_p LIMIT
EARTH	$p \to \pi^0 e^+$ (present)	10^4 Tons	$6 \times 10^{32} \sim 10^{33}$y $\sim 3 \times 10^{33}$y
(1995)	$p \to K^+ \bar{\nu}$ (present)	3×10^3 Tons	$\sim (1-6) \times 10^{31}$ $\sim 3 \times 10^{32}$y
LUNAR BASE*	$p \to \pi^0 e^+$	10^6 Tons	$6 \times 10^{35} \sim 10^{35}$
(2010)	$p \to K^+ \bar{\nu}$	10^5 Tons	$6 \times 10^{34} \sim 10^{34}$

* The ν flux is $\sim 10^{-3}$ less on the moon. Thus, if sufficient mass is available, it should be possible to increase limits by at least 10^2.

TABLE 4
Supernova Neutrino Detection in 1990's
SUMMARY

PARAMETERS MEASUREDReaction →	$\bar{\nu}_e p \to e^+ n$	$\nu_x e \to \nu_x e$	$\nu_x N \to \nu_x N$	$\nu_x N \to \nu_x N n_0$
CROSS SECTION	LARGE (KII, SK, IMB, LVD)	SMALL $\sim E_\nu^2$ (ICARUS)	LARGE For Coherent Process	LARGE At High E_{ν_x} SNO
NEUTRINO ENERGY	YES $\sim E_e$	PARTIAL $E_{\nu_e} \sim f(E_e)$	NO	NO But a threshold may set $E_{\nu_{min}}$
ν DIRECTION	NO	YES	NO	NO
TIME	YES	YES	YES	YES
DOWN TIME (guess)	$\gtrsim 10\%$	$\sim 30\%$	?	(could be small)
MAXIMUM DETECTOR SIZE	2×10^5 Tons (H_2O) LENA $\lesssim 10^4$ Tons Liq. Scint. (LVD)	$\sim 2 \times 10^5$ Tons (H_2O) $\sim 10^3$ Tons Cryogenic (ICARUS)	? Kilograms No detector proposed so far	$\sim 10^6 - 10^7$ Tons of $CaCo_2$ OR $\sim 10^3$ Top D_2O
BACKGROUNDS	SMALL If e^+ and n capture detected - OK for H_2O Galactic Signal	SMALL If directional (?) used to reject background	?	DEPENDS on Radioactivity of material Can Be Large

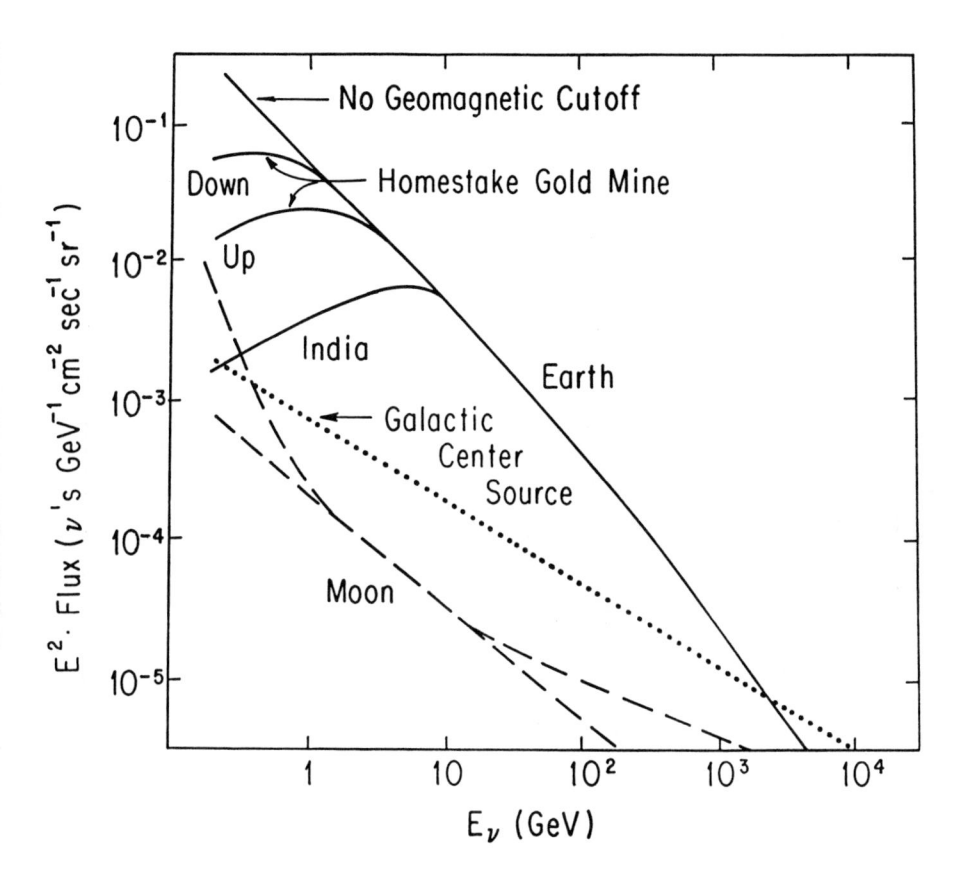

Figure 1 Flux of muon neutrinos and anti-neutrinos. The terrestrial flux is shown for the case of no geomagnetic field, for vertically downward and upward neutrinos at the Homestake and IMB sites, and for downward neutrinos in India in the direction of maximum geomagnetic shielding. The lunar flux is calculated (light line) for π and K meson decays; the heavy line takes into account the effect of low energy K and high energy charm decays.

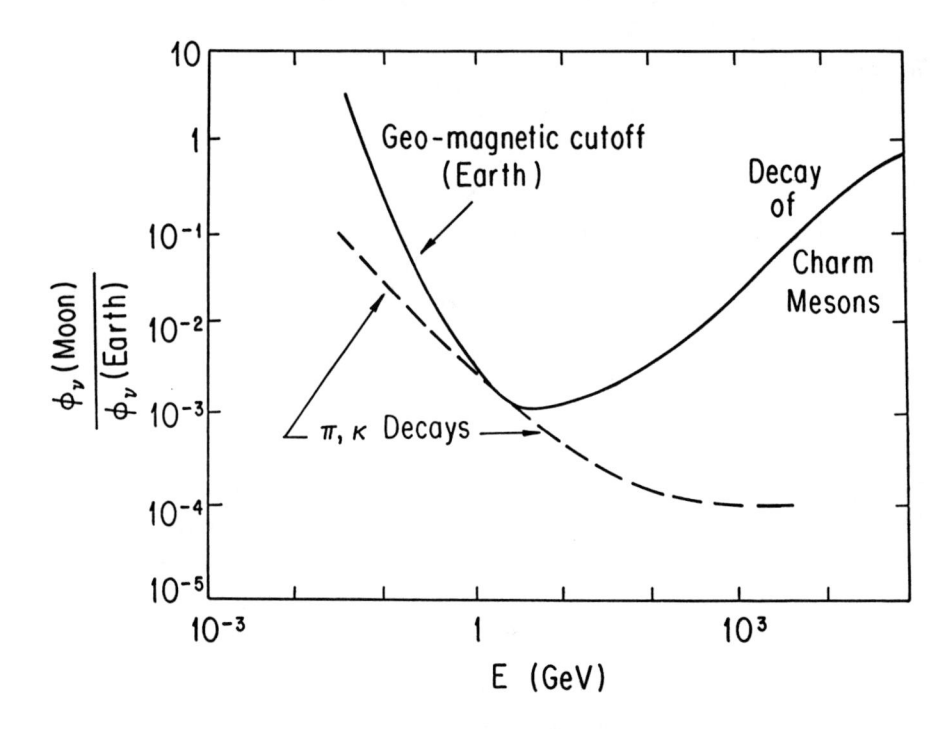

Figure 2 Ration of the neutrino flux on the moon to that on Earth as a function of the neutrino energy.

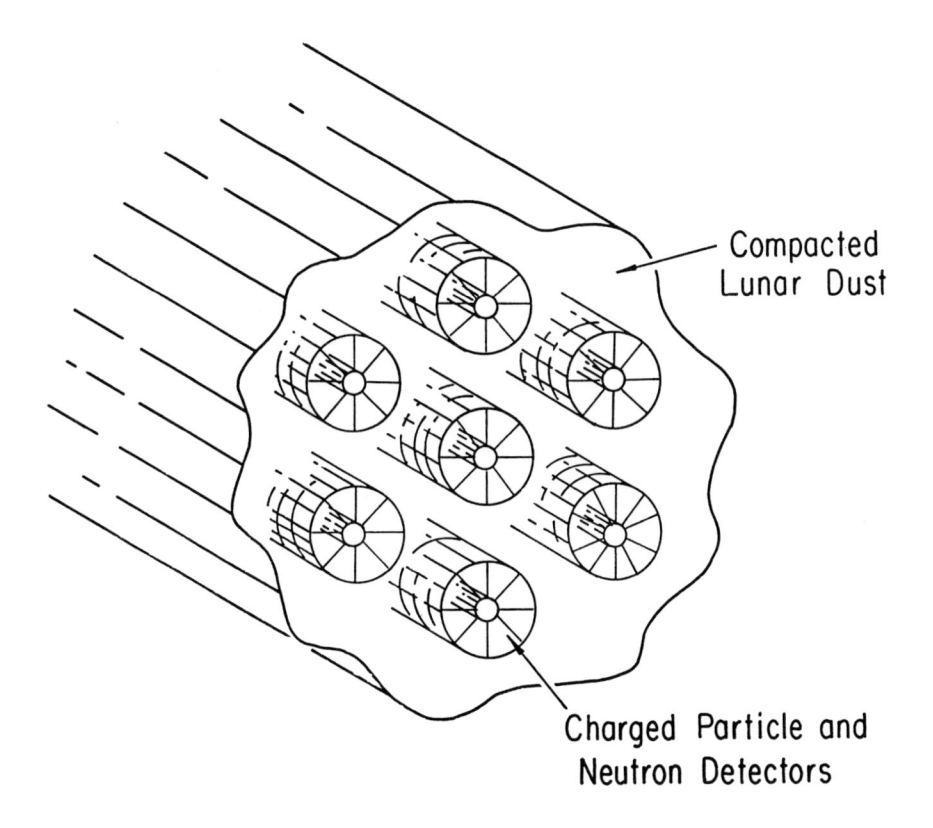

Figure 3 A proton decay and supernova neutrino detector on a lunar base.

APPENDIX A

A New Method for Detection of Distant Supernova Neutrino Bursts

D. Cline[1], E. Fenyves[2], T. Foshe[1], G. Fuller[3], B. Meyer[3], J. Wilson[3]

1 Departments of Physics and Astronomy
   University of California at Los Angeles

2 Department of Physics
   University of Texas at Dallas

3 Lawrence Livermore National Laboratory

ABSTRACT

We study the feasibility of astrophysical neutrino detectors based on the detection of neutrons produced in neutrino-nucleus inelastic scattering events. We discuss how collective nuclear effects greatly enhance the relevant interaction cross sections over those of single particle interactions. These effects can help to reduce the mass required for neutrino detectors. We present an example of a simple detector based on $CaCO_3$ neutrino targets and BF_3 neutron counters. We discuss neutron background limitations and also consider the possibility of forming a coincidence between neutrino detectors and future gravity wave detectors.

© 1990 American Institute of Physics

The fledgling field of neutrino astronomy can count two major successes to its credit: the detection of solar neutrinos and neutrinos from Supernova 1987a.[1,2,3,4] However, the detectors used in these efforts and proposed for future experiments are based on weak interaction processes involving charged current neutrino capture or charged plus neutral current scattering on single nucleons, electrons, or deuterons. These fundamental processes have cross sections typical of single particle weak interactions ($\sigma \lesssim 10^{-42}$ cm^2), and this represents an inherent limitation in expanding the size and efficiency of the current detector designs.

By contrast, it has recently been shown that collective effects in nuclei can increase the neutral current neutrino inelastic scattering cross sections on nuclei by at least one order of magnitude (for $T \geq 10$ MeV neutrinos) over single particle values.[5] We therefore propose to use neutrino inelastic scattering on nuclei together with subsequent neutron emission from the excited nucleus as the basis of an astrophysical neutrino detector. The process involved would be

$$\nu_x + A(Z,N) \to A(Z, N-1) + n + \nu'_x \qquad (1)$$

where ν_x is either a ν_e, ν_μ, or ν_τ since we use the neutral current channel, $A(Z, N)$ is a nucleus of mass number A with Z protons and N neutrons, n is the final state free neutron, and ν'_x is the scattered final state neutrino.

The advantage of such a detector over current designs rests on three points. The first is of course the enhancement of neutrino scattering cross sections due to nuclear collective effects. The second point involves using neutrons as the tracer of neutrino induced events. The efficiency of neutron detection (~20%) is inherently smaller than that of the Cerenkov detector schemes (~100%) currently used in water detectors, but this decrease in efficiency is offset by the enhancement in the scattering cross section discussed above. We will argue that it will ultimately be cheaper to scale up our proposed detector than an equivalent water detector because of the relative inexpensiveness of BF$_3$ detectors as compared to Cerenkov counters. We note, however, that the neutron detector efficiency is limited by the natural background. Finally, our third point is that geologically large, relatively pure deposits of suitable neutrino detector material (CaCO$_3$ or Na Cℓ) may exist in nature in shielded sites.

A key element in our scheme is the expected relatively high energy spectrum of the ν_μ and ν_τ neutrinos in the stellar collapse, being well above the threshold energy for reaction (1). In contrast, the ν_e neutrinos generated in supernovae have lower energy spectrum, below the threshold energy of reaction (1). This leads to a selective detection of ν_μ and ν_τ supernova neutrinos.

Detailed shell model calculations of neutrino-nucleus inelastic scattering are now becoming available.[5] These calculations show that the interaction cross sections in such events are enhanced by collective nuclear effects. In principle it may also be possible in the future to use electron scattering experiments to obtain the electromagnetic form factors for various nuclei. A minimum of theoretical input then allows a calculation of the appropriate weak neutral current form factors.[5] Thus, eventually, it should be possible to get accurate, semi-empirical results for the inelastic cross sections of various nuclei.

For the purpose of discussing astrophysical neutrino detectors, we have modeled inelastic neutrino-nucleus scattering with a scaled-up version of charged current interactions of neutrinos and nuclei and fitted our model to the shell model calculations of Haxton.[5] We then use this model to compute cross sections for neutral current interactions on various target nuclei. Figure 1a shows the cross section per nucleon computed for the excitation of ^{40}Ca to an energy of 24.15 MeV (corresponding to two oscillator levels) by inelastic neutrino nucleus scattering.[Note the rapid rise of the cross section above 15 MeV neutrino energy ($T_{\nu_\mu,\nu_\tau} \simeq 5$ MeV).] This cross section folded with a neutrino energy distribution emitted from a black body at T=10 MeV is shown in the curve labelled $\sigma_T \times f_\nu$ of Figure 1b. The corresponding cross sections for carbon and oxygen are close to that of calcium.

The nuclei excited by neutrino inelastic scattering are then allowed to decay via proton, neutron, or gamma emission. The spectrum of neutrons emitted from ^{40}Ca is shown in Figure 1c. We then compute an average cross section for a $CaCO_3$ molecule by summing the individual nuclear cross sections weighted according to their frequency of occurrence in the molecule. For a spectrum of neutrinos coming from a black body at T=10 MeV, we find a flux-averaged total inelastic cross section for a $CaCO_3$ molecule of $\sigma_T \sim 1 \times 10^{-42}$ cm^2 per nucleon and a cross section for neutron production of $\sigma_n \sim 2 \times 10^{-43}$ cm^2 per nucleon.

We consider for illustration a possible detector configuration using a geological $CaCO_3$ formation as the neutrino interaction medium (Figure 2). Cylindrical BF_3 detectors of radius r_1 are inserted into the $CaCO_3$ material and the neutrinos are detected by the reaction

$$n + B^{10} \to Li^7 + He^4 \tag{2}$$

A neutron thermalizing material $[n(CH_2)]$ is inserted into the $CaCO_3$ material around the BF_3 detector with radius r_2. We have calculated the efficiency of neutron detection by varying the parameters r_1, r_2 and the radius r_3 of the $CaCO_3$ medium that is the primary neutrino target. A detailed neutron propagation computer program for all the materials was developed at LLNL and used for the calculation. For example, the calculations give an overall neutron detection efficiency of $\sim 20\%$ for the array of Figure 2, with $r_1 = 20$cm, $r_2 = 22$cm, and $r_3 = 60$cm. We can now calculate the number of detected neutrons (N_{counts}) from a supernova at distance R (kpc). For an overall detector of volume V (m^3) of $CaCO_3$, assuming $N_\nu = 6 \times 10^{57}$ from the supernova and $\sigma_n \sim 2.0 \times 10^{-43}$cm^2 per nucleon for $CaCO_3$ we find the following expression for the number of counts detected, accounting for neutron detector efficiency:

$$N_{counts} \simeq \frac{20}{R^2} V \tag{3}$$

For supernovae in our galaxy ($R \simeq 20$ kpc) we find

$$N_{counts} \simeq 0.05 \times V \tag{4}$$

A modest detector of size 10m × 10m × 10m detects about 50 counts. These counts occur in a time interval of (10 – 20) seconds. For the local group of galaxies

$$R \simeq 1000 \text{ kpc} \tag{5}$$

and we find

$$N_{counts} \simeq 2 \times 10^{-5} V \tag{6}$$

A detector of 100m x 100m x 100m gives about 20 counts. However, this signal may or may not be above background (to be discussed below).

We have also considered NaCℓ and SiO$_2$ as possible detector media. NaCℓ is somewhat poorer regarding neutron transport than CaCO$_3$, but is better than CaCO$_3$ for neutron production by inelastic neutrino scattering. SiO$_2$ is much better than CaCO$_3$ as regards neutron transport, and about the same for neutron production. We note that, as regards the NaCℓ detector, (Ref. ?), some salt mines have very low neutron backgrounds.

We now turn to the question of backgrounds. The most important characteristics of the neutron detection is its time correlation to the neutrino burst. We have calculated the time of propagation of the neutrons in the detector and find that 90% of the neutrons are counted in the first millisecond after production. The neutrino signals are expected to have durations of about ten seconds with the intensity peaked towards early times. Thus, the detector time response is quite adequate. However, because the neutrino signal is spread out over several seconds it is necesary to have a very small neutron background rate.

The detection of a supernova neutrino event would rely on slow coincidence counting of neutrons from the neutrino interactions in more than one BF$_3$ detector. High energy cosmic ray muon induced photo-nuclear excitations, nuclear fission of heavy elements and other sources of background could be rejected because the neutrons produced in these events would be localized in one or a few BF$_3$ detectors and not in coincidence in many detectors, and would not show the time structure of a supernova neutrino event.

In order to estimate the rate of background events from the heavy element impurity in the CaCO$_3$ we use recent measurements of the neutron flux at the Gran Sasso Laboratory in Italy (the Gran Sasso Mountain is largely made of CaCO$_3$)[6]. From these measurements we estimate that the neutron count rate in a 10 m length of BF$_3$ detector is $(10^{-2} - 10^{-3})$ sec^{-1}. However, this estimate may not apply to the material in undisturbed sites and the background could be lower there.

A 10^3 m^3 detector would then give an average background rate of $(10^{-1} - 1)$ sec^{-1}. Over a ten second interval this gives between one and ten counts. The signal from a galactic supernova is about 50 counts, which is above the background. For a 10^6 m^3 detector, the background is $(10^2 - 10^3)$ sec^{-1} or 10^3

to 10^4 counts in a ten second burst. A galactic supernova would give 50,000 counts, an impressive number to study ν_μ and ν_τ supernova neutrinos. This background would, however, swamp the 20 count signal from the local group. Clearly, backgrounds are the major limitation to the power of the $CaCO_3$ detector.

If a purer target material is found, however, the distance out to which one could observe a supernova with such a neutrino detector could be greatly increased. It is clearly important to carry out experimental studies of the backgrounds in various geological formations. If materials of low backgrounds can be found, this technique could lead to the detection of supernova neutrino bursts from the local group with the possibility of detecting a cosmologically significant ν_μ or ν_τ mass.

These considerations show that the overall detection efficiency for supernova neutrino bursts using inelastic neutrino scattering detectors is roughly comparable to that of the present water detectors. We feel that the advantage of our design is in its possible inexpensiveness compared to the water detectors. Although BF_3 counters on the scale contemplated here have not been produced, the materials are readily avaliable. We thus expect that if neutrino detectors on scales considerably larger than the currently existing ones are to be built in the future, it will be less expensive to produce and insert BF_3 counters into the target material than to use many new expensive photomultipliers for water detectors. Furthermore, large deposits of target materials exist (e.g. $CaCO_3$ in the Grand Sasso and $NaC\ell$ in salt mines) so that preparation of the target material may be relatively inexpensive. We also note that if the detectors are set up in the Grand Sasso tunnel or deep in salt mines, they would be well-shielded from the bulk of cosmic rays.

There is going to be considerable improvement in gravitational wave detectors in the near future for both the large bar detectors and the laser interferometer detectors.[7] These detectors may in the next decades become sensitive to stellar collapse in the local group. In this case we propose to form a coincidence between gravitational wave detectors and the detectors of a correlated neutrino/neutron burst using the technique proposed here. If a gravity wave is detected, then using that event to determine the initial time will greatly enhance the chances of picking up the neutrino signal. Conversely, if a clear neutrino signal is observed, the search for a gravity wave in the gravitational wave detector signal would be more reliable.

We have illustrated a neutrino detector technique in this letter. Detailed calculations and study of various types of natural materials must be carried out before the construction of such a detector could be proposed. However, we believe our preliminary calculations demonstrate the feasibility of this approach. If geological formations of low backgrounds can be found, this technique could lead to the detection of supernova neutrino bursts from the local group with the possibility of detecting a cosmologically significant ν_μ or ν_τ mass.

ACKNOWLEDGEMENTS

We would like to acknowledge most valuable discussions with Drs. J. Bahcall, S. Woosley, J. Ferguson, R. Bauer, W. Haxton.

REFERENCES

1. For a review see: K. Lande, "Recent Results in Solar Neutrinos", 14th Texas Symposium on Relativisitc Astrophysics, Dallas, Texas, 1989, (to be published by the New York Academy of Sciences.)
2. K. Hirata, et al. Phys. Rev. Lett. 58, 1490 (1987).
3. R. Bionta, et al. Phys. Rev. Lett. 58, 1494 (1987).
4. For a review of the neutrino emission from Stellar Collapse and SN 1987A, R Mayle and J. R. Wilson, Ap. J. sub 1987, and A. Burrows and J. Lattimer, Ap. J. (Letters) 318, L63 (1987), and D. Joutras and D. Cline, Astro. Lett. and Communications, 1988 Vol. 26, pp 341-347 (1988).
5. T.W. Donnelly, Phys. Lett., 43B, 93 (1973), J. D. Walecka, in "Muon Physics", Vol. 2, ed. V. W. Hughes and C. S. Wu (Academic Press, New York, 1975), M. Fukugita, Y. Kohyama, K. Kubodera, IAS, preprint, 1988, W. Haxton, U. W. preprint, 1988.
6. E. Bellolti, et al. "New Measurements of the Rock Contamination and Neutron Activity in the Gran Sasso Tunnel", preprint, Univ. Milano, (1985).
7. For a review see: W.O. Hamilton and W. Johnson, "Present and Future Bar Detectors I", W. Fairbank and P. Michelson, "Present and Future Bar Detector II", R. Drever, "Laser Gravity Wave Detectors", Proc. Symposium on the Next Supernova: Astrophysics, Particle Physics, and Detectors, Santa Monica, California, 1989.

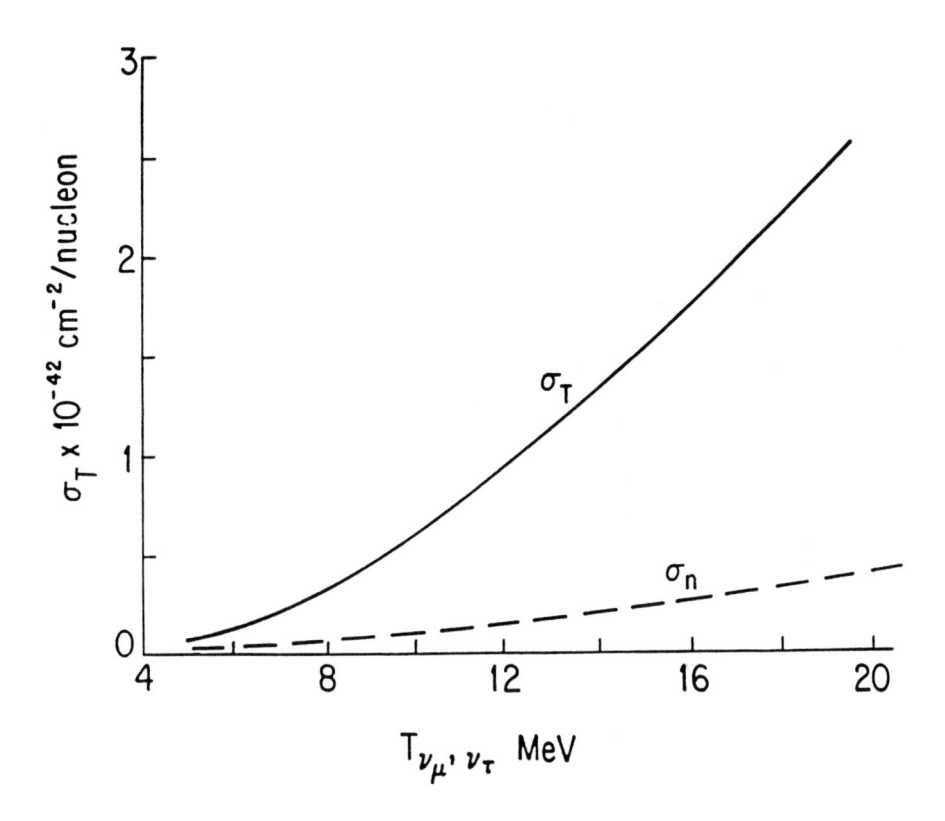

Figure 1a. $\nu_\mu, \nu_\tau +{}^{40}$Ca cross sections per nucleon for total neutral current scattering (σ_T) and single neutron production (σ_n), averaged over neutrino black body spectra at temperature T_{ν_μ, ν_τ}.

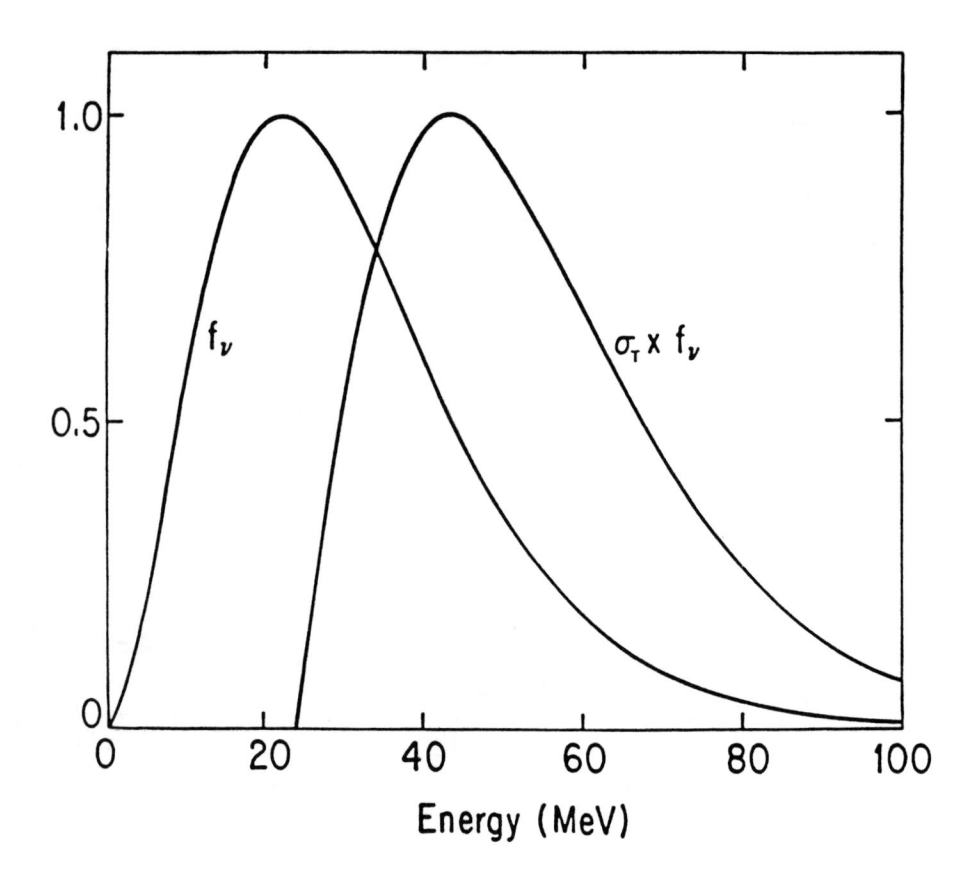

Figure 1b. Neutrino spectrum, f_ν, from a black body at T = 10 MeV and total inelastic neutrino scattering cross section off ^{40}Ca multiplied by f_ν versus neutrino energy adjusted to peak at one.

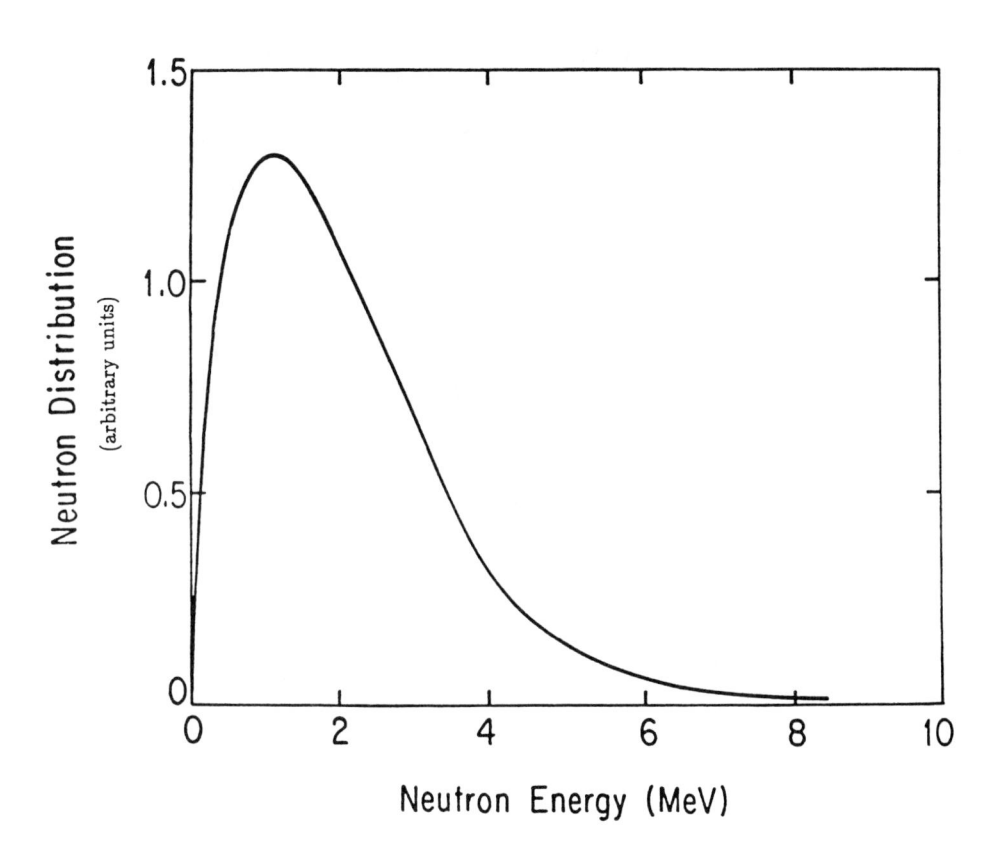

Figure 1c. Energy spectrum of neutrons emitted from excited ^{40}Ca.

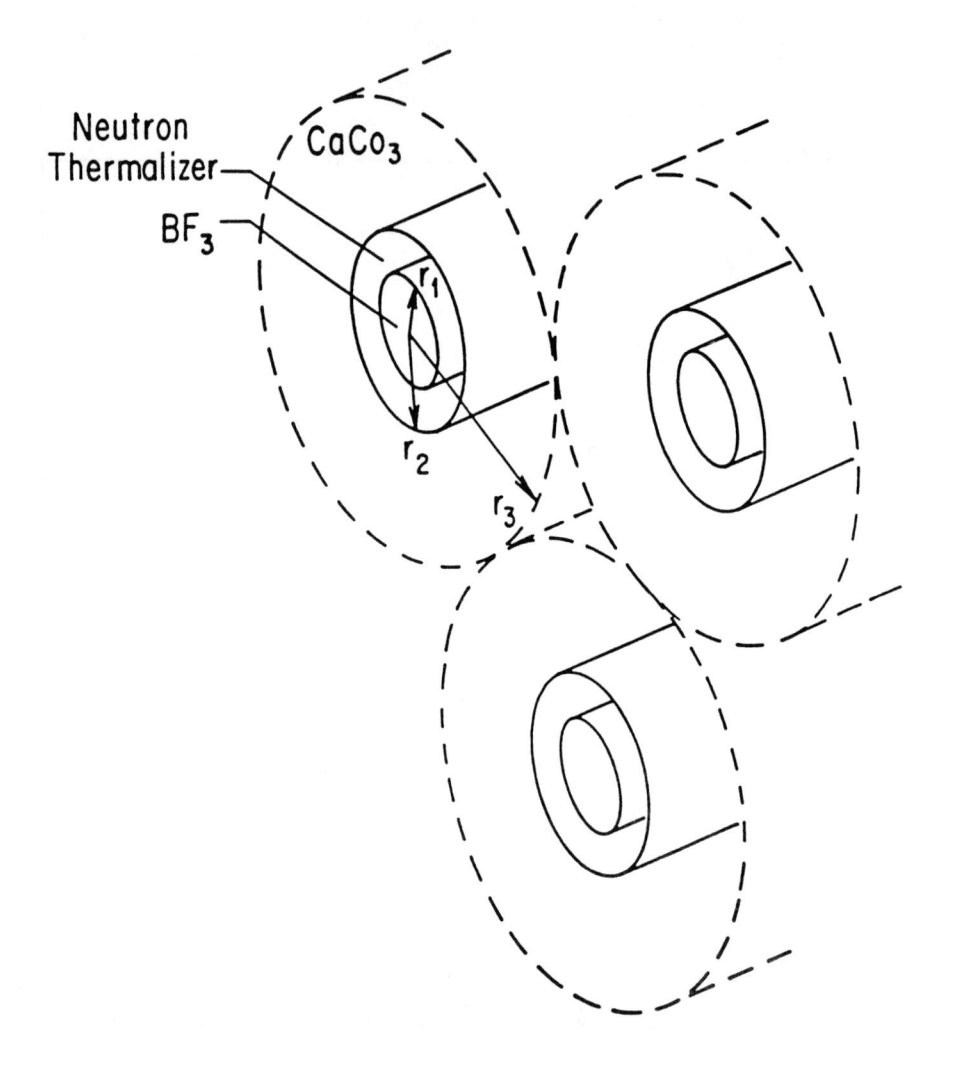

Figure 2. Concept of a distant supernova neutrino burst detector.

MUONS ON THE MOON

V.J. Stenger
University of Hawaii

ABSTRACT

Neutrino astronomy on the earth is currently signal-limited, rather than background-limited. Thus the absense of atmospheric muon background on the moon does not provide any obvious advantage in the search for point sources, although some advantage may exists for diffuse sources. A lunar detector for neutrino astronomy will still have to be as large as any on earth. The earlier suggestion that a window around 1 GeV exists, where backgrounds on earth are large, is shown to not provide for likely detectable sources.

INTRODUCTION

The Deep Underwater Muon and Neutrino Detector (DUMAND) has now received the endorsement of the High Energy Physics Advisory Panel (HEPAP) to move on to its second phase.[1] The DUMAND collaboration* has proposed to deploy an array of 216 photomultiplier tubes at a depth of 4.8 km off the coast of the island of Hawaii. This array, dubbed the Octagon, will have an effective area of 20,000 m^2, angular resolution of 1°, and some ability to discriminate energy by measuring mean dE/dx for throughgoing muons produced by the interaction of muon neutrinos in the water or earth below the array.

The Octagon will be more sensitive than any existing or planned underground detector and, unlike other detectors, will have essentially 100 percent sky coverage. Detectable neutrinos from sources such as Cygnus X-3 will occur, provided that the neutrino flux is enhanced by at last a factor of three over the observed γ-ray flux above 1 TeV. Such an enhancement is highly plausible, with some models predicting even more.

In thinking about any experiment on a lunar base, one must look for advantages over doing the experiment on earth. In the case of extra-solar neutrino astronomy, very high energy muon neutrinos represent by far the most promising avenue of search, first because

---

* Aachen, Bern, Caltech, Hawaii, Kiel, Kinki, Kobe, Okayama, Scripps, Tokyo, Vanderbilt, and Wisconsin

they are likely to be more plentiful, second because the cross section increases with energy, and third because the long range of high energy muons enable the material of the earth, or moon, to comprise the major portion of the detecting material. Any experiment able to detect muons will be far more sensitive than an experiment of comparable size designed to detect electrons or other particles produced by the interaction of electron neutrinos. In DUMAND the ocean provides the main detecting medium, allowing for very large detector dimensions. A detector on the moon will have to similarly rely on the moon as the main medium for detection.

On the earth, a large background of cosmic ray muons produced by primary cosmic rays hitting in the upper atmosphere exists. These will swamp any neutrino-induced muons over most of the overhead celestial hemisphere. Nevertheless, in the case of DUMAND sufficient solid angle remains to enable the experiment to explore virtually the entire celestial sphere at last half of the time. A background of upward-going muons from neutrinos produced by cosmic rays on the other side of the earth also exists, but these will be less than one per year in each one degree circle on the celestial sphere being searched for sources.

Previous studies of the possibilities for neutrino astronomy on the moon have suggested that a substantial advantage would result from the absence of these atmospheric neutrinos.[2,3] In particular, it was argued that, since no muon background in the range 1-1000 GeV exists on the moon, a lunar observatory could search for sources with steep spectra and fluxes too small at higher energies to be found by experiments such as DUMAND.

In this paper I basically dispute this conclusion, at least for point sources. Some advantage may exist in the search for diffuse sources, but the detector in any case will still have to be so large that neutrino astronomy on the moon can be made feasible only if the material of the moon itself can be used.

THE DETECTION OF MUON NEUTRINOS

In Fig. 1, a plot is shown of the muon spectra that would result for muons produced in the body of the moon below a muon detector at or near the surface, for neutrinos with power-law integral spectra E_ν^{-3} and E_ν^{-1}. The latter is closer to what is expected from the most likely sources such as Cygnus X-3, the former is comparable to the spectrum of atmospheric muons and neutrinos on earth. For the flatter neutrino spectrum, we see that most of the events occur with $E_\mu > 50$ GeV and no advantage is offered by being able to search for muons down to as low as 1 GeV. The steeper spectrum, however, would appear to provide such an advantage. However, another factor must be considered.

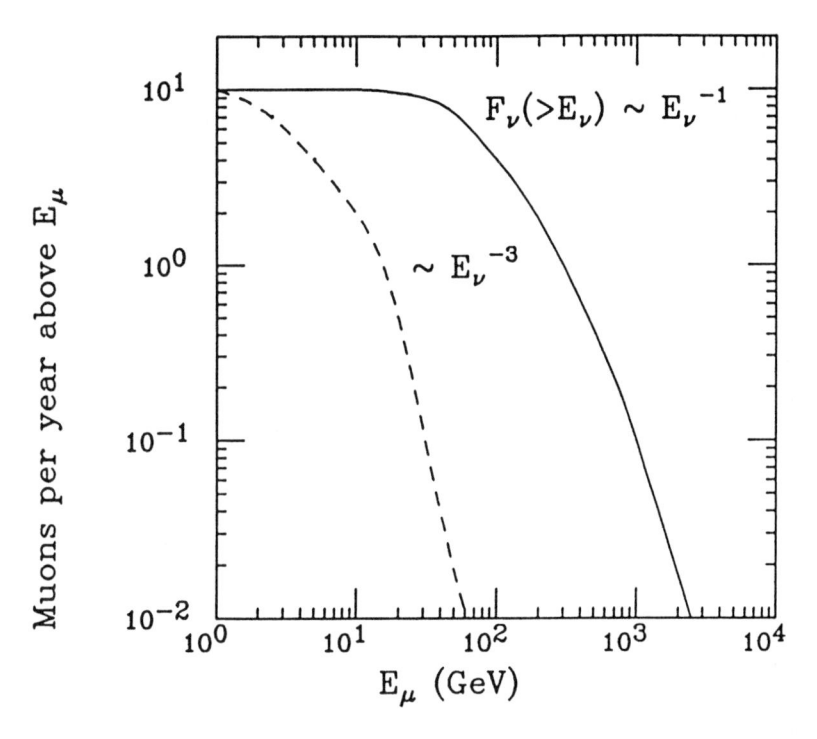

Fig. 1. The muon spectra that would be observed for muons from neutrinos produced with two different power law spectra. Both curves have been normalized to give the same number of events above 1 GeV.

In cases where the background in the region being searched for a source is negligible, a signal of ten events per year can be regarded as detectable. In Fig. 2, the <u>minimum detectable flux</u> of muon neutrinos from an astronomical a source required to produce ten events per year is presented for a muon detector with an arbitrary average effective area of 1000 m^2. The minimum detectable flux can be scaled up or down with that area. The flux is presented as the integral flux above 1 GeV and plotted as a function of the integral spectral index. That is, the integral neutrino flux is given by

$$F_\nu(>E_\nu) = F_\nu(> 1 \text{ GeV}) E^{-\gamma} \qquad (1)$$

where E is in GeV.

Note that the minimum detectable flux is some six orders of magnitude greater for $\gamma = 3$ than for $\gamma = 1$. That is, a muon detector of a given area has a sensitivity that is markedly better for flatter spectra than steep spectra. The reason for this large difference is the multiplying effect of the factors mentioned earlier: Both the νN interaction cross section and effective detector volume increase with energy.

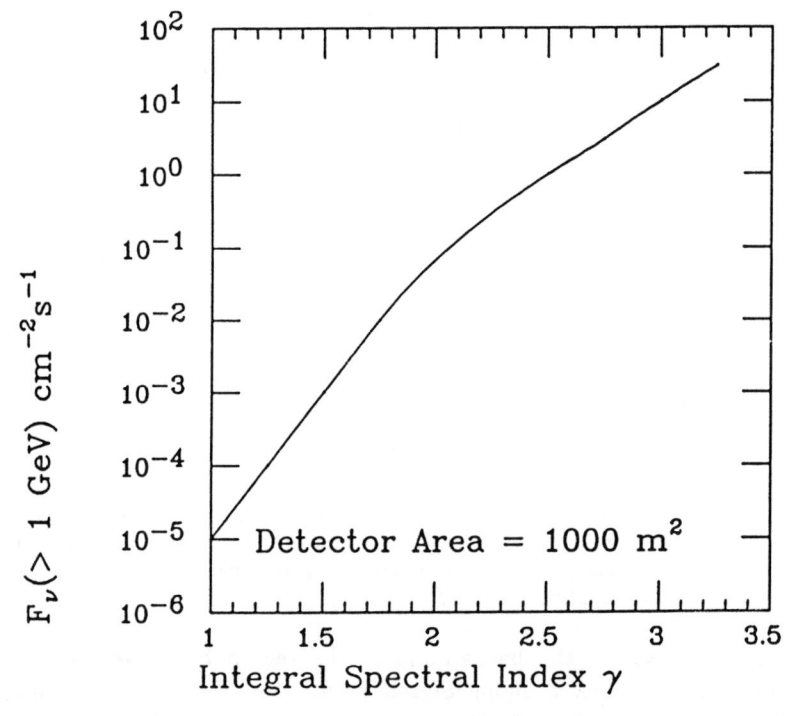

Fig. 2. The minimum detectable flux of neutrinos above 1 GeV that is required to give 10 events per year, as a function of the integral spectral index γ, for a detector area of 1000 m^2. The required flux for other detector areas may be scaled accordingly.

Fluxes higher than about 10^{-3} cm^{-2} s^{-1} above 2 GeV with integral spectral index $\gamma = 2$ are already ruled out over much of the sky by previous underground experiments.[4,5] So the best hope for neutrino detection remains sources with flat spectra. These are also the most likely. In the search for point sources, all current and planned experiments are signal-limited rather than background-limited. However, this is not the case for possible diffuse sources, such as the galactic center. In this case, the window that must be opened on the celestial sphere will let in considerable background from atmospheric neutrinos on the earth, and a lunar neutrino observatory would have certain advantages.

With the whole Pacific Ocean available, the DUMAND array will be readily expandable to even greater dimensions, should the science warrant it. At some point, perhaps above about 10^5 m^2, the atmospheric background to point sources will become important and a lunar observatory could provide some advantages.

BACKGROUNDS ON THE MOON

The background of muons on the moon is small, but not zero. Muons will be produced by the so-called "prompt" neutrinos that are produced in the decay of short lived particles with heavy flavor, such as charm. The pions and kaons that are primarily produced when cosmic ray protons bombard the moon interact in the rock before having a chance to decay to muons and neutrinos. However, the heavy flavor particles produced have such short lifetimes that they decay first. I estimate that the flux of upward neutrinos from heavy flavors will be about 4 m^{-2} y^{-1} and that this will lead to an upward muon flux on the surface of the moon of 100 km^{-2} y^{-1}. These muons will all be above several TeV.

So, the background for point sources in a one degree circle will be only about 1 per 100 km^2 per year, negligible for all practical purposes. In a 20°x70° region that encompasses the galactic nucleus where neutrino production is expected from cosmic rays interacting with the dense matter, the background would be 2 km^{-2} y^{-1}, much smaller than the corresponding flux on earth.

CONCLUSIONS

The moon offers some advantages over the earth for neutrino astronomy, but they are not so obvious. Basically neutrino astronomy is now, and is likely to remain for some time, signal-limited rather than background-limited. The fact that the muon background is essentially zero down to 1 GeV range is of little consequence. If the neutrino spectra from astronomical sources is as flat as expected, then most of the events will have $E_\mu > 50$ GeV where the atmospherically-produced backgrounds on earth are already small. If sources exist with steep spectra, or spectra that cut off above of few tens of GeV, the sensitivity of any muon detector will be far lower than for flat spectrum sources. Of course such sources could exist, and it may someday prove important to try to detect them on the moon. However the detector to do this will have to have an effective area average over all directions of at least 100,000 m^2 and maybe more.

A lunar neutrino detector would be superior to earth-bound detectors of the same effective area in the search for diffuse sources such as the galactic center. Again such sources could prove important. Perhaps an isotropic high energy neutrino background exists (the 1/4000 eV relic neutrinos from the Big Bang are hopeless to detect, at least by any means thought of so far). This would be of great cosmological interest. However, it is amply clear that any such detector on the moon would have to be of such great size that the detector medium could not be transported to the moon. Thus the material of the moon itself must be used, either directly or to manufacture the detector medium in great quantities. Unless glass can be made from green cheese, optical detection techniques are unlikely. Acoustic detection may be the only viable alternative.[6]

REFERENCES

1. DUMAND II Proposal, Hawaii DUMAND Center report HDC-2-88 (1988).
2. M.M. Shapiro and R. Silberberg in "Lunar Bases and Space Activities in the 21st Century, W.W. Mendel ed., Lunar and Planetary Institute, Houston (1985), p. 329.
3. M. Cherry and K. Lande, ibid., p. 335.
4. R. Svoboda et al., ApJ **315**, 420 (1987).
5. Y. Oyama et al., Phys. Rev. **D39**, 1481 (1989).
6. J.G. Learned, this conference.

Lunar Neutrino Physics

John G. Learned
Department of Physics and Astronomy, University of Hawaii, Manoa
2505 Correa Road, Honolulu, HI 96822

Abstract

The possibilities of the use of the moon as a base for conducting neutrino physics are examined, emphasizing neutrino astronomy. The principle advantage of the moon for this research is freedom from the atmospheric layer of the earth: cosmic rays hitting the atmosphere generate a rather copious source of neutrinos, which are a terrestrially inescapable diffuse background to neutrino astronomy. The cosmic ray generated neutrinos on earth are also a limiting background for other sensitive particle physics experiments, typically those performed underground.

The most severe limitation of conducting this type of physics research on the moon seems to be the pragmatic one of mass transport to the moon: many of the immediately obvious research initiatives which could benefit from lower backgrounds than possible on earth (eg. proton decay searches, low energy neutrino experiments) involve massive (megaton size) detectors. These experiments will thus not be practical until substantial manufacturing capability exists on the moon.

We do however identify two very different prospects for neutrino astronomy, in very different energy regions, which are worthy of more immediate study: 1) a 1 km^2 detector using the moon as the cosmic ray shield and the target for TeV neutrinos, and which observes the product muons emerging from the surface; and 2) more speculatively, an EeV detector employing acoustic detection to probe an otherwise inaccessible energy domain using the entire moon's core as the target. It is suggested that the first detector might perform experiments detecting a neutrino beam generated by a terrestrial accelerator, such as the SSC, permitting an interesting exploration of muon neutrino oscillations.

Several followup initiatives are suggested on low mass planar detectors and on lunar acoustic properties. It seems possible that the future of very high energy neutrino astronomy is on the moon.

17 July 1989
Hawaii Preprint number HDC-9-89

© 1990 American Institute of Physics

Introduction

First, an apology: this paper is not meant to be definitive in any way, but is only a tentative exploration of possibilities, and even those experiments considered are inadequately treated herein. Moreover, there is an inherent problem in such speculation as this in that it is very difficult to guess what physics will be relevant and what technology will be practical some years from now. That said we march bravely ahead.

Surely we can count upon the convergence now manifest between physics, astrophysics, cosmology, and astronomy, to continue for some time. In particular it would seem reasonable that any endeavor in neutrino astronomy which leads to a more sensitive detector than possible on earth will be at least as interesting as now, both in terms of particle physics and neutrino astrophysics.

Double beta decay, proton decay, dark matter searches, and generally, all the physics experiments that are performed underground on earth may benefit from similar sites on the moon. For example a low energy (few MeV) neutrino experiment may benefit from a lunar location in order to escape the background neutrino flux from terrestrial reactors, as pointed out by A. Mann in his paper in these Proceedings[1]. One needs to examine the experiments individually, however, because the low energy radiation may not be better in particular instances (eg. if the limitation is due to local radioactivity of the surrounding rock, as is the case for double beta decay). Another concern for some of these, such as dark matter searches, is whether they will still be of high interest when a moon base becomes available.

For neutrino astronomy, the attraction of an experiment on the moon is that the background for high energy neutrino research will be $10^{-(3-4)}$ of it's level on earth. The background neutrino flux on earth is generated by the cosmic rays impinging upon the earth's atmosphere. The secondary pions and kaons produced in the atmosphere frequently decay before coming to rest. The cosmic rays striking the moon, and resulting secondary mesons, however, are generally absorbed prior to producing neutrinos. Very short lived particles, for example mesons composed of heavy quarks, decay so speedily that they result in neutrinos. The latter causes the limiting neutrino background on the moon. Since this flux is a result of heavier quarks, study of this 'direct production' (a misnomer) is of physics importance, particularly at extremely high energies, because it probes for new structure, for example a new generation of quarks, or for a new layer of matter (preonic structure). Thus just the study of the background flux of neutrinos on the moon has significant particle physics interest.

As Stenger has pointed out in his contribution to this Workshop[2], it appears that escaping the terrestrial cosmic ray neutrino flux provides no strong motivation for going to the moon to do neutrino astronomy at energies of a few GeV, nor does it do so for TeV energy range detectors of the size under construction in the ocean (DUMAND II for example, with 20,000 m^2 muon detecting area[3]). However, when moving to next generation detectors, with muon collecting area in the range of a few km^2 or more, backgrounds within a resolution circle of typically 1° (determined by the neutrino-lepton scattering angle at a TeV) will become important, and a detector on the moon may then offer significant scientific advantage over deep ocean experiments.

First we discuss a neutrino detector in the TeV range, then we discuss the potential for detection of an SSC (or LHC or UNK or Eloisatron) type of neutrino beam in such a detector, and finally we examine the potential for a detector of much higher energy employing the acoustic radiation from 10^{18} eV neutrinos in striking the moon's core.

TeV Neutrino Astronomy

This discussion begins where Stenger's[2] ends, so the reader may want to review that paper first. However, this author is more optimistic about the potential of the moon for future neutrino astronomy. For the following discussion we take as a goal a minimum detector size of 10^6 m^2 for muon detection on the moon. The reader is further referred to the DUMAND II proposal[3], and to the various proceedings of DUMAND Workshops[4] to look at the prospects for neutrino astronomy for detectors in this size class. It appears that, in fact, detectors in the multi-km^2 class are going to be needed before we can really begin what astronomers would consider regular astronomy: the ability to look at sources at will, examining details of spectra and temporal behavior. The detectors on earth, proposed or currently under construction (eg. DUMAND II), will probably find a few sources, and may make great discoveries, but surely will not achieve the status of something like contemporary X-ray or gamma ray astronomy until at least another generation of instruments, and probably further yet in the future.

To give an example of possible rates: everyone's favorite X-ray binary system, Cygnus X-3, has been the subject of much study as a potential generator of neutrino fluxes. It is generally believed now that it may produce about 1 muon/1000 m^2/year[5] from neutrino interactions in the earth. The same flux would apply to muons emerging from the moon, and in our hypothetical 1 km^2 lunar array we could expect then as many as 1000 neutrino events per year from such a source. Moreover, this object, and similar objects, are well known as episodic emmitters, with outputs blooming by factors of a hundred or more for short times (typically one to a few times per year for bursts of a few minutes to a few days[6]). Particularly through the study of the temporal structure of such emissions one can begin to learn details of the astrophysics of such objects. One example of this is the potential to "neutrino-ray" the density profile of the companion star.

Assuming that the reader is convinced of the scientific worth of such an endeavor, let us examine some of the practical considerations. Taking as a given the necessity to utilize a minimum of material for such a detector, we are restricted to considering the use of the moon as target and shielding medium. Should one find substantial quantities of subsurface water, or develop a way to make glass in quantity on the moon, we could consider many attractive possibilities. We will assume that not to be practical for the present, but that we must import the detectors, which must detect and track muons emerging from the moons surface.

As an example, the technology of transition radiation detectors (TRDs) appears to have promise for this application. TRDs have the property of requiring only a small amount of mass per unit area, they are inherently directional, and produce signals proportional to a high power of the relativistic gamma factor of the traversing charged particle. This would permit measuring the energy of the emerging muon. If we could construct counters with 1 kg of mass per 10 m^2 of detector, then the total detector mass to cover 1 km^2 of area would be 100 metric tons, a plausible mass to consider sending to the moon. This seems to be an extremely low figure for mass per unit area, however. Note also that such a detector would almost surely have to be covered with shielding material against the downgoing cosmic rays (the amount needs study, and depends upon detector technique, but would be in the range of a few meters of moon dust). Clearly the engineering will require great cleverness.

The location for such a detector might be in a flat bottomed crater, with high walls (for maximum shielding). For good sky coverage (looking through the moon) the crater should not be far from the equator, so that the detector field of view would rotate over the whole sky once per 28 days. (Indeed, for the further future, 3 such detectors would cover the entire sky all the time). For purposes of detecting a terrestrially generated neutrino beam (see below), the location should be on the outer lunar surface.

The author suggests that it may be worth convening a technology study group to examine the projected limits of technology for such an endeavor. This might be followed by the investment of some funds to pursue likely techniques.

10 TeV Accelerator produced Neutrinos to the Moon

The idea of detecting a neutrino beam at great distance from an accelerator has been around for quite a while, but has not really been taken seriously for two reasons. First, the flux limitation has made detection rates very low, and second, the cost of bending the proton beam downwards into a target and subsequent decay tunnel has deterred even simple experiments using existing underground detectors. Both these objections may not apply to the hypothetical lunar neutrino detector.

First let us consider the rate question. A quick way to scale a rate is to employ the flux calculated for a proposed beam at the planned 150 GeV Fermilab Main Injector. We found that we might expect as many as 100 events of 20 GeV or more muons per week in DUMAND at 6000 km distance from Fermilab[7]. At 400,000 km on the moon we loose rate by a factor of 2.25×10^{-4}, but gain by area ratio of 50 for the 1 km^2 detector, in net being down to 1.125% of the rate in DUMAND. This is overwhelmed by the increase in rate due to larger machine energy, from which we get two exponents due to kinematics (neutrinos into a smaller solid angle), and one each from neutrino crossection and muon range. This would get us a factor of 2×10^7 for a 10 TeV machine, for a net gain of about 20,000 over the proposed DUMAND experiment. This is an overestimate because the Fermilab beam and duty cycle are probably several orders of magnitude higher than practical for the 10 TeV machine, and because a decay tunnel scaled from Fermilab is not practical (26 km). Even giving away a factor of 1000 for these would leave us with a rate of order of 1000 events/week, an eminently practical rate (with completely negligible background too).

The other problem for previously proposed remote neutrino experiments is also fortuitously answered because the neutrino beam naturally emerges tangent to the earth's surface, and it only remains to point it East or West. However, the alignment would occur only once per day, when the moon is rising (or setting). The dwell time of the beam on the lunar detector would only be a few seconds without beam steering; this is the biggest problem foreseen for such an experiment. This can be overcome by sweeping the beam to track the moon. It is difficult to imagine doing this over an angle of more than several degrees, corresponding to a dwell time of a few minutes. Beam steering would be needed anyway, because of the tilt of the lunar orbit relative to the equatorial plane. I estimate a rate of a few events per week in such conditions.

Is the physics of any interest? The important parameter for neutrino oscillations is the ratio of distance to energy, about 1300 km/GeV for the presently considered experiment, which corresponds to a δm^2 in the 10^{-4} eV^2 range. The latter is unchallenged by any other suggested experiment of which the author is aware for ν_μ oscillations. It is about a factor of 10 lower than the proposal of detecting a Fermilab neutrino beam at DUMAND, and is about a factor of 10^4 below existing accelerator based experimental limits. Given this unique opportunity it would seem to merit more careful consideration, though the beam steering requirements may prove fatal.

EeV Neutrino Astronomy

Another possibility for lunar neutrino research is to use the acoustical technique to search for neutrino interactions deep inside the moon. The idea is to detect a neutrino interaction via the feeble acoustical pulse produced when the neutrino generates a cascade of particles. The particles heat a long (10 m) thin (few cm) region of matter, essentially instantaneously on an acoustic time scale, leading to a miniscule radial expansion, and consequent radiation of a bipolar pressure wave. The outgoing pulse has a pancake beam pattern, being strongest in the direction perpendicular to the initial particle direction. The initial characteristic frequency is determined by the velocity of sound and the transverse shower dimensions, leading to typically a 20–50 kHz peak. In typical liquids (the moons core?), where the attenuation varies with frequency squared, the acoustic pulse amplitude possesses the peculiar property of decreasing as the inverse square of the distance from the source, including attenuation from the medium (no exponential decrease)[8]. (The signal-to-noise ratio, in a thermal noise dominated medium decreases as the distance to the −5/2 power, again with no exponential fall off). At large distances the directionality decreases too, which could make reconstructing the original neutrino direction impossible if no wavelengths less than about 20 m remain in the detected signal. The flip side of this potential problem is that as the signal becomes less directional it is detectable, if at all, in a larger solid angle (fewer detectors needed to intercept the signal).

We have explored this detection method in the past, both theoretically[8] and experimentally[9], with the idea to employ it in the ocean. Our conclusion was that the threshold for detection of neutrino interactions was so high, around 10^{16} eV in the best circumstances, over a volume of ocean water in the few km^3 range, that no flux was likely. Of course the existence of such a flux would be a spectacular discovery, but the experiment such a long shot as not to be justifiable. The use of the moon may open up a new avenue however, in the following way.

Suppose that we are able to place geophones into the moon at a number of locations, so as to be able to image acoustic pulses from deep within the moon. Such pulses coming outwards from the core will have little background, because most of the lunar seismic activity will excite high order modes of the moon, and will not consist of solitary outwards-going pulses with known high frequency content. The same thing goes for meteorite impacts, from which one also probably gets substantial shielding by the thick dust layer on the moon's surface. With a sufficient number of such detectors (and we are not able to say now what such a number is), we can triangulate on the outgoing pulse and determine the location of the neutrino interaction, and the direction and energy of the cascade.

The target volume potentially available is quite amazing: about 10^{18} Tons (assuming about 1% of the mass of the moon can be monitored), roughly equal to the mass of the oceans! The threshold for detection needs study, but may be (author's guess) about 10^{18}eV. The cross section for neutrino interactions at 10^{18}eV may be about 10^{-33}cm^2(10) (an extremely interesting quantity which itself could be measured, in principle, by such an experiment, via the interaction depth distribution of events). The total effective cross section of the detection region would be then 6×10^{14}cm^2.

As an example to give some scale, let us calculate an event rate using the reported flux from the Fly's Eye experiment for particles coming from Cygnus X-3 with energy greater than 0.5×10^{18} eV, of 2×10^{-17}/cm^2/sec(11). These particles may well be neutrons, and if so, neutrinos are present in similar numbers. This flux falls approximately on a linear extrapolation of observations at lower energies, so we can take the flux above 10^{18}eV as 10^{-17}/cm^2/sec, so that the interaction rate from Cyg X-3 would be 200,000 events per year! Taking another example, the flux of neutrinos from our galaxy's central region may be about 10^{-19}/cm^2/sr/sec above 10^{18}eV(12), which would yield an event rate of 2000 events per year. Many other interesting phenomena might be observed, such as neutrinos from cosmic strings(13). Our estimates are very crude, but still probably alright at the order of magnitude level. The physics and astrophysics potential is thus enormous. The big uncertainty, upon which all this speculation hangs, is the detection threshold energy which as of this writing is little more than a guess, due to lack of lunar acoustic data.

An amusing possibility comes from considering the potential to acoustically detect a terrestrial neutrino beam interacting inside the moon. While individual interactions would not be detectable, a great many occurring simultaneously might be heard. For example, assume that a hypothetical beam from a 100 TeV machine might yield 10^{10} neutrinos of energy in the 10 TeV range. About 4% of these would be absorbed in traversing the moon, depositing 600 J of energy in a cylindrical region about 6 km in radius. Given the exact knowledge of time of deposition, and of beam impact one should be able to detect such a beam readily if the acoustic detection threshold is as guessed above. One could obviously make use of such a signal to map the interior of the moon.

Unfortunately, we can not go any farther without knowing information not presently available (to the author anyway): What are the acoustic properties of the deep moon in terms of noise, and attenuation versus frequency (the two most important questions)? We also need some material properties (specific heat, volume expansivity, speed of sound, energy loss rate) to calculate the pulse amplitude. The previous work on acoustic pulse generation by elementary particle cascades has (to this author's knowledge) only been concerned with pressure waves, but some radiation will occur in the solid by shear wave. This should be investigated, particularly since it could lead to interesting information (such as reconstructing the range, and perhaps angle, of the cascade from one observation point). Also, one must consider the Landau-Pomeranchuk-Migdal effect, which operates at extremely high energies, lengthening cascades. How practical is it to install geophones around the moon, and can they be placed sufficiently deep to have good acoustic coupling to the core region? Is the core region sufficiently homogeneous to have predictable acoustic propagation? How many such geophones could we reasonably expect to install? Is there sufficient interest in other science communities to support this effort as a cross-disciplinary endeavor (after all one would learn an enormous amount of geophysical information from such studies too)?

Summary

Two potential neutrino experiments for the moon have been identified, and both deserve further consideration in the author's opinion. (A number of others were considered and discarded too).

In the first case, that of a 1 km^2 TeV neutrino detector, practicality will hinge upon the technology of low mass large area detectors to be placed upon the lunar surface for recording upmoving muons. The author recommends that a study of potential detector technology be carried out. The physics is unquestionably unique and worth pursuing, though one must assess the competition from future deep ocean detectors. This experiment might also have the interesting ability to detect neutrinos generated in multi-TeV terrestrial particle accelerators.

In the second case, an acoustical experiment may be possible to probe ultra-high energies, using the moon's core as neutrino target. In this instance followup will require the interdisciplinary considerations of geophysicists and particle physicists to determine the likely lunar acoustic characteristics (most importantly noise and attenuation), and thus to determine if the detection threshold is really interesting (anywhere up to about 10^{20} eV would probably be worth pursuing). A study of the implantation of geophones is also needed. The experiment suggested would be truly spectacular, not approachable by any other means so far suggested.

It could be that the moon is indeed the location for the long range future of neutrino astronomy, and that the moon itself will be our neutrino telescope.

References

1) A. K. Mann, ' A Lunar-Based Detector to Search for Relic Supernovae Antineutrinos

2) V. J. Stenger, 'Muons on the Moon', this Workshop.

3) DUMAND Collaboration (P. Bosetti, et al.), 'DUMAND II – Proposal to Construct a Deep-Ocean Laboratory for the study of High Energy Neutrino Astrophysics and Particle Physics', HDC-2-88, University of Hawaii, High Energy Physics, August 1988.

4) there have been a total of 14 DUMAND Workshops for which there are Proceedings, from 1975 through 1988. See detailed References in 3), above. They are abrreviated DUMAND-nn below, where nn refers to the year of the workshop.

5) T. K. Gaisser and A. F. Grillo, Phys. Rev. $\underline{D36}$, 2753 (1987).

6) T. C. Weekes, Physics Reports 160, 1 (1988).

7) J. G. Learned and V. Z. Peterson, HDC-5-89, submitted to the Proceedings of the Workshop on Physcics at the Main Injector, May 18, 1989, Fermilab, to be published (1989).

8) J. G. Learned, Phys. Rev. D19, 3293 (1989).

9) S. D. Hunter, et al., J. Acoust. Soc. Am. 69(6), 1557 (1981).

10) V. S. Berezinsky and A. Z. Gazizov, DUMAND-79, 218 (1980).

11) G.L. Cassiday, et al., Phys. Rev. Lett. 62, 383 (1989).

12) V. S. Berezinsky and L. M. Ozernoy, DUMAND-80, II, 202 (1980).

13) C. T. Hill, D. N. Schramm, and T. P. Walker, Phys. Rev. D36, 1007 (1987).

A LUNAR-BASED DETECTOR TO SEARCH FOR RELIC SUPERNOVAE ANTINEUTRINOS[1]

A. K. Mann and W. Zhang

Department of Physics[2]
University of Pennsylvania
Philadelphia, Pennsylvania 19104

Abstract

We discuss the motivation for and advantages of a lunar-based detector capable of observing the relic antineutrino flux from all past supernovae at or close to the lowest theoretical estimates of the magnitude of that flux.

---

[1] Invited talk given at the NASA Workshop on Physics and Astrophysics from a Lunar Base, Stanford University, May, 1989.

[2] Supported in part by the U.S. Department of Energy under contract DE-AC02-76-ERO-3071.

Introduction

The recent observation [1, 2] of electron type antineutrinos ($\bar{\nu}_e$) emitted by SN1987A yielded direct experimental data in support of the fundamental assumptions and predictions of the current model of type-II supernovae [3]. The total number ($\sim 10^{57}$) and total energy ($\sim 3\text{--}5 \times 10^{53}$ ergs) of the emitted antineutrinos determine the temperature of the residual neutron star to have been approximately 4 MeV during the time of emission of neutrino-antineutrino pairs. This empirical and theoretical knowledge suggests the possibility of searching for the relic antineutrino ($\bar{\nu}_e$) flux from the totality of all gravitational stellar collapses that have ever occurred. A principal aim of such a search is to study cosmic evolution, e.g., to specify the epoch of (spiral and irregular) galaxy formation through determination of the value of the effective redshift factor (z) necessary to describe the present energy spectrum of the relic antineutrino flux from all past supernovae.

The resultant broadening and degrading of the energy spectrum of the relic $\bar{\nu}_e$ comes about from the different distances (and times) at which supernovae occurred during cosmological evolution. This has been studied for several cosmological models by Bisnovatyi-Kogan and Seidov [4] who concluded that the mean $\bar{\nu}_e$ energy is lowered from the value appropriate to that for a recent supernova by a factor between 0.5 and 0.6, depending weakly on the mass parameter of the universe. The flux of relic supernovae $\bar{\nu}_e$ is accordingly expected to peak between 6 and 10 MeV. The predicted magnitude of the flux is much less certain and ranges from an "admittedly conservative" estimate [5] of $\phi_0 \simeq 1\,\text{cm}^{-2}\text{sec}^{-1}$ to several orders of magnitude larger values [6, 7]. Much of the variation in these estimates of ϕ_0 lies in the assumptions relating to the current rate of type-II supernovae, some of which are based on luminosity densities and some on mass densities and supernova rates per galaxy [8].

In this note we briefly argue that a detector of appropriate design located on the moon would be capable of observing the relic supernovae antineutrino flux at or close to the lowest theoretical estimates of the magnitude of that flux, and indicate the essential characteristics of the detector.

Present Status of the Search

The known extra-terrestrial neutrino fluxes and the terrestrial nuclear reactor flux as they appear to an observer on earth are shown in Fig. 1, which shows also their approximate energy distributions. It is evident that the energy region in which a possible fruitful search for relic supernovae $\bar{\nu}_e$ might be made is occupied by other ν_e and $\bar{\nu}_e$ fluxes of larger intensities. The relic flux is expected to exhibit an energy spectrum that is a superposition of many initially approximate Fermi-Dirac distributions, and therefore to have a long tail stretching to higher energies. As a consequence, a search was made [9] for $\bar{\nu}_e$ in the energy region 19 to 35 MeV in the data from the 2.1 kiloton, imaging water Čerenkov detector, Kamiokande-II. Concentration on $\bar{\nu}_e$ was dictated by the fact that the cross section for the reaction $\bar{\nu}_e p_{\text{free}} \to e^+ n$ is roughly 20 times larger than any other neutrino interaction cross section in water at the low antineutrino energies expected; p_{free} refers to the two essentially free protons in the water molecule. The result of that search is given in Fig. 2 which presents the upper limit on ϕ_0 as a function of the present effective temperature, T_{eff}, of the sea of relic supernovae $\bar{\nu}_e$. The redshift factor is contained implicitly in T_{eff}, as is indicated by the equation describing the limiting curve in Fig. 2.

$$\text{Total Observed Events} = \sigma_0 L \phi_0(\bar{\nu}_e) \int_{E_1}^{E_2} \left[\epsilon(E_{\bar{\nu}_e}) dE_{\bar{\nu}_e} \int_0^{\infty} F(E, T_{\text{eff}}) E^2 G(E - E_{\bar{\nu}_e}) dE \right], \quad (1)$$

where $\sigma(\bar{\nu}_e p_{\text{free}} \to e^+ n) = \sigma_0 E^2$, L is the exposure in number of free protons times seconds, $\epsilon(E_{\bar{\nu}_e})$ is the detection efficiency, $F(E, T_{\text{eff}})$ is the normalized energy distribution of the $\bar{\nu}_e$, $G(E - E_{\bar{\nu}_e})$ is the energy resolution function of the detector, and $\phi_0(\bar{\nu}_e)$ is the total relic $\bar{\nu}_e$ flux in units of cm^{-2}sec^{-1}. Note that $E_{\bar{\nu}_e} = E_{e^+} + 1.3$ MeV.

Advantages of a Lunar-Based Detector

A lunar-based detector of relic supernovae neutrinos might be, for example, an appropriately shielded 200 ton scintillation detector, containing an element with a high neutron capture cross section. As noted, the cross section for $\bar{\nu}_e p_{\text{free}} \to e^+ n$ in the $\bar{\nu}_e$ energy region 3–20 MeV is significantly higher than any other neutrino cross section in that energy region. The event signature would be similar to that utilized in the first detection of antineutrinos ($\bar{\nu}_e$) [10], i.e., a low energy positron ($3 \leq KE_{e^+} \leq 20\,\text{MeV}$) in time coincidence ($\leq 10\,\mu s$) with a gamma-ray ($\gtrsim 5\,\text{MeV}$) from the capture of the final state neutron in the detector. This coincidence requirement is necessary to specify a genuine event and also to discriminate against background events.

The advantages of such a lunar-based detector are evident from a study of Fig. 1.

(i) The $\bar{\nu}_e$ from earth-based nuclear reactors will be completely negligible on the moon, but a power reactor on the moon might be a limiting factor in the search for relic supernovae $\bar{\nu}_e$, depending on the power produced and the reactor location relative to the $\bar{\nu}_e$ detector.

(ii) The flux of neutrinos and antineutrinos produced in the earth's atmosphere by the primary cosmic ray proton flux ($\lesssim 3 \times 10^3\,\text{m}^{-2}\text{sec}^{-1}\text{sr}^{-1}$ for $E_{\text{tot}} \geq 2\,\text{GeV}$) would be largely absent.

Consider a detector of roughly 200 metric tons enclosed in a cubic volume 10m on a side. Let the detector be buried under 5m of lunar rock. The mean free path for *inelastic* hadronic interactions in lunar rock should be $\approx 120\,\text{gm/cm}^2$, and for $\rho_{\text{rock}} \simeq 2.5\,\text{gm/cm}^3$, that thickness corresponds to $\lambda_{\text{inter}}^{\text{moon}} \simeq 0.5\,\text{m}$. The value of $\lambda_{\text{inter}}^{\text{earth}}$ is $\approx 5000\,\text{m}$ of the earth's atmosphere. Hence the ratio of neutrino fluxes ϕ_ν^{moon} and ϕ_ν^{earth} is just

$$\phi_\nu^{\text{moon}}/\phi_\nu^{\text{earth}} \simeq \lambda_{\text{inter}}^{\text{moon}}/\lambda_{\text{inter}}^{\text{earth}} \simeq 10^{-4}, \qquad (2)$$

more or less independent of hadronic energy.

The 5m rock shielding (10 nuclear inelastic interaction mean free paths) will reduce the incident proton flux by a factor of 4.5×10^{-5}, and will range out all remaining protons with $K.E. \leq 2.5\,\text{GeV}$. Protons that penetrate the shield will be vetoed in the anticounter (say, 1 m thick) surrounding the detector with negligible dead time. Radioisotopes produced by primary protons that traverse the detector (spallation products) will be strongly rejected by the coincidence requirement for useful signal events.

From Fig. 1, in the neutrino energy region 1–20 MeV, we find an average interaction rate of 5×10^{-5} interactions per kiloton-year per MeV of atmospheric neutrino flux incident on an earth-based detector. Accordingly, the interaction rate in the lunar detector from neutrinos from the decays of π- and K-mesons produced by proton inelastic interactions in the 5m shield will be $\simeq 10^{-4} \times 5 \times 10^{-5}$(kiloton-yr-MeV)$^{-1} \times 0.2$ kiloton $\times 20$ MeV or $\simeq 2 \times 10^{-8}$ interactions/year which is negligible relative to the expected rate (see below) from the relic flux ϕ_0.

Note that almost all π^- which come to rest will be captured before decaying $(\pi^-\text{-decay}/\pi^-\text{-capture})_{\text{at rest}} \leq 10^{-4}$, while π^+ that stop and decay produce ν_μ predominantly, $(\nu_e/\nu_\mu \leq 10^{-4})$, and μ^+ decays yield $\bar{\nu}_\mu$ and ν_e. The ν_μ, $\bar{\nu}_\mu$, and ν_e are discriminated against by the relic antineutrino-induced event requirements specified above.

Antineutrinos from the prompt decays of heavy mesons, e.g. charm, are not expected to contribute appreciably because of the relatively small heavy meson production cross sections. The flux of antineutrinos from the decays of radioactive spallation products of the primary protons in the shield is unlikely to exceed 10^{-7}cm^{-2} sec^{-1}.

(iii) With regard to the background contribution of lunar radioactivity, it is known that the relative abundances of radioisotopes on the moon are generally similar (within factors of 2 or 3) to those on earth [11]. Consequently it seems unlikely that an antineutrino intensity (cm^{-2}sec^{-1}), with $E(\bar{\nu}_e) \gtrsim 3$ MeV, of magnitude to compete with hypothesized values of ϕ_0 can be generated by lunar radioactivity. For example, for a relative abundance of ^{238}U of 1 gm ^{238}U per metric ton of lunar surface material, the decay rate of surface material would be 0.02 sec^{-1}gm^{-1}. The value of the corresponding total emitted intensity (cm^{-2}sec^{-1}), after integrating over the moon's crust, is numerically equal to the above rate per gm-sec, again within a factor of order 2 or 3.

Gamma-ray background from lunar radioactivity needs more extensive study including a Monte Carlo calculation of the discrimination factors provided by the energy thresholds and the time coincidence requirement for the relic $\bar{\nu}_e$ signal. However, rough calculations do not suggest that the lunar gamma-ray background will be a serious obstacle.

Expected Count Rate

The expected signal count rate for three assumed values of the relic supernovae flux may be calculated from eq. (1) and is given in Table I. For simplicity we set the integrals equal to unity, use an average value of the cross section for $\sigma(\bar{\nu}_e p_{\text{free}} \to e^+ n)$ of 1.5×10^{-40} cm^2/ free proton–MeV in the interval 3 to 20 MeV, and 2×10^{31} free protons (in a 200 ton scintillation detector).

It might be expected that the detector would be operated for a several year period. Accordingly, Table I suggests that values of $\phi_0 \gtrsim 50$ cm^{-2}sec^{-1} would yield a significant signal in such a period, and, depending on the actual background level, would determine T_{eff} unambiguously.

Summary

It appears that a detector of relatively modest size (~ 200 metric tons) located on the moon would be capable of observing or at least setting relevant limits on the relic supernovae neutrino flux at values close to the lowest theoretical estimates of that flux. Location of the detector on the moon has the advantage of significantly reducing the principal backgrounds in an earth-based detector, namely backgrounds from atmospheric antineutrinos and nuclear power reactor antineutrinos. A positive observation of relic supernovae neutrinos would yield a value of T_{eff}, the effective, present temperature of the relic supernovae neutrino sea. From T_{eff} a value of an effective red shift factor, integrated over the entire period of cosmic evolution, might be obtained. The value of z would cast light on the history of supernovae occurrences, i.e., their rate as a function of time, for comparison with recent observations and with models of nucleosynthesis and galaxy formation.

It is worth noting that a lunar detector composed of scintillating material viewed by photosensitive devices, and with an appropriate cellular structure, would be suitable for the precision measurement by ionization and range of the primary components of cosmic radiation. It is possible that with careful design the neutrino detector described here, with part of its vertical shielding removed, would serve also as a detector of primary cosmic rays.

One of us (AKM) is grateful for the hospitality of the National Laboratory for High Energy Physics of Japan (KEK) where some of this work was done.

Table I. Approximate signal in a 200 ton neutrino scintillation detector for different assumed values of the relic supernovae neutrino ($\bar{\nu}_e$) flux ϕ_0.

ϕ_0 $(\text{cm}^{-2}\text{sec}^{-1})$	Signal (counts/yr)
1	0.1
10	1.0
100	10.0

References

[1] K. Hirata et al., Phys. Rev. Lett. **58**, 1490 (1987); Phys. Rev. **D 38**, 448 (1988).

[2] R. M. Bionta et al., Phys. Rev. Lett. **58**, 1494 (1987).

[3] S. A. Colgate and R. H. White, Astrophys. J. **143**, 626 (1966); S. E Woosley, J. Wilson, and R. Mayle, Astrophys. J. **302**, 19 (1986); A. Burrows and J. M. Lattimer, Astrophys. J. **307**, 178 (1986); R. Mayle, Ph. D. Thesis, University of California, (Berkeley) 1985; and S. E. Woosley and Thomas A. Weaver, in Ann. Rev. of Astron. and Astrophysics **24**, 205 (1986).

[4] G. S. Bisnovatyi-Kogan and Z. F. Seidov, Ann. N. Y. Acad. Sci. **422**, 319 (1984).

[5] S. E. Woosley, James R. Wilson, and Ron Mayle, Astrophys. J. **302**, 19 (1986).

[6] G. S. Bisnovatyi-Kogan and Z. F. Seidov, Astron. Zh. **59**, 213 (1982) [Sov. Astron. **26**, 132 (1982)].

[7] J. N. Bahcall et al., Astrophys. J. **267**, L77 (1983); L. M. Krauss et al., Nature (London) **310**, 191 (1984); G. V. Domogatskii, Astron. Zh. **61**, 51 (1984) [Sov. Astron. **28**, 30 (1984)]; R. Mayle, James R. Wilson, and David N. Schramm, Astrophys. J. **318**, 288 (1987).

[8] Sidney van den Bergh, talk given at the Tenth Santa Cruz Summer Workshop "Supernovae," University of California, Santa Cruz, July, 1989.

[9] W. Zhang et al., Phys. Rev. Lett. **61**, 385 (1988).

[10] F. Reines and C. L. Cowan, Phys. Rev. **92**, 830 (1953); C. L. Cowan et al., Science **124**, 103 (1956).

[11] See, for example, A. E. Metzger et al., Science **179**, 800 (1973); S. R. Taylor, "Lunar Science, A Post-Apollo Perspective," pp. 242–243, Pergamon Press

(1975); H. C. Urey et al., Phil. Trans. Roy. Soc. London **A285**, 23 (1977). We are grateful to Thomas Wilson for references to lunar radioactivity.

Figure Captions

Fig. 1. (a) Neutrino fluxes and (b) interaction rates in earth-based detectors of neutrino fluxes from known and potential extraterrestrial sources.

Fig. 2. The 90%-C.L. upper limit on the total relic-supernova-antineutrino flux $\phi_0(\bar{\nu}_e)$ as a function of the effective equilibrium temperature T_{eff} of the relic-antineutrino sea. Also indicated are regions suggested by theory, and a numerical value of the number density at the "extreme theoretical upper bound."

Fig. 1a

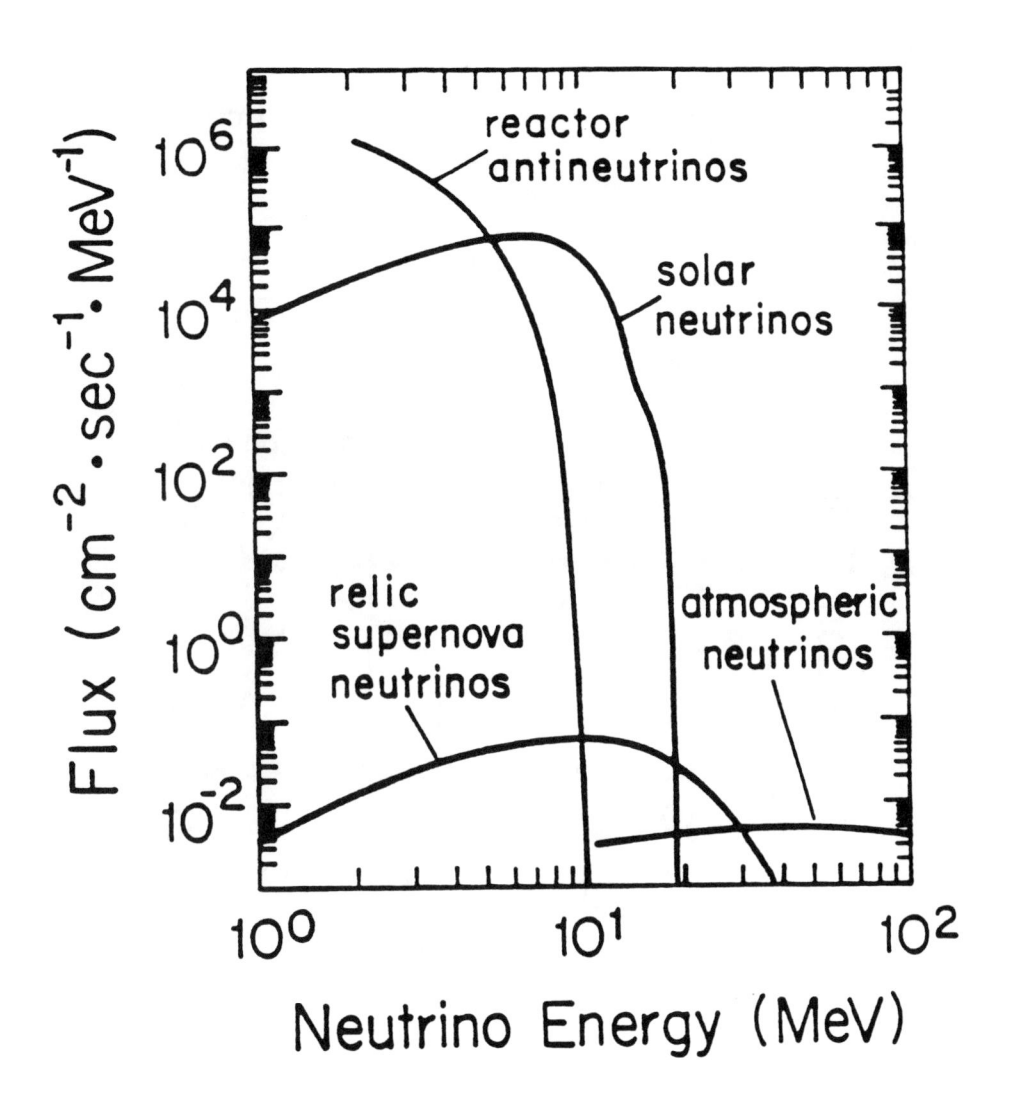

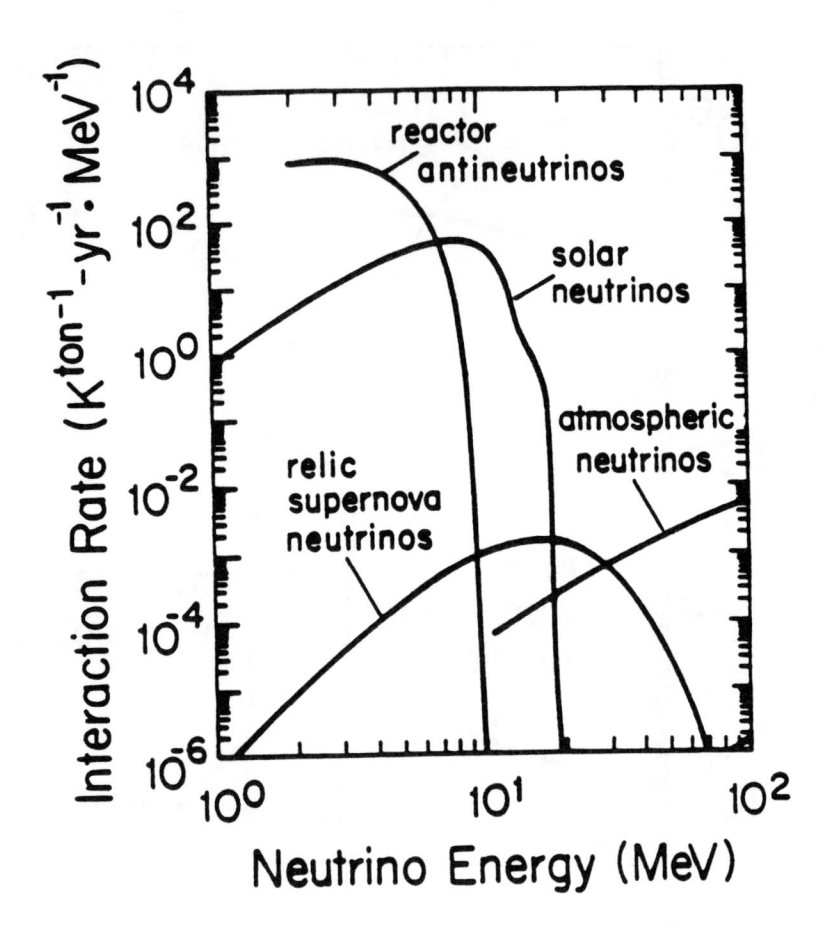

Fig. 1b

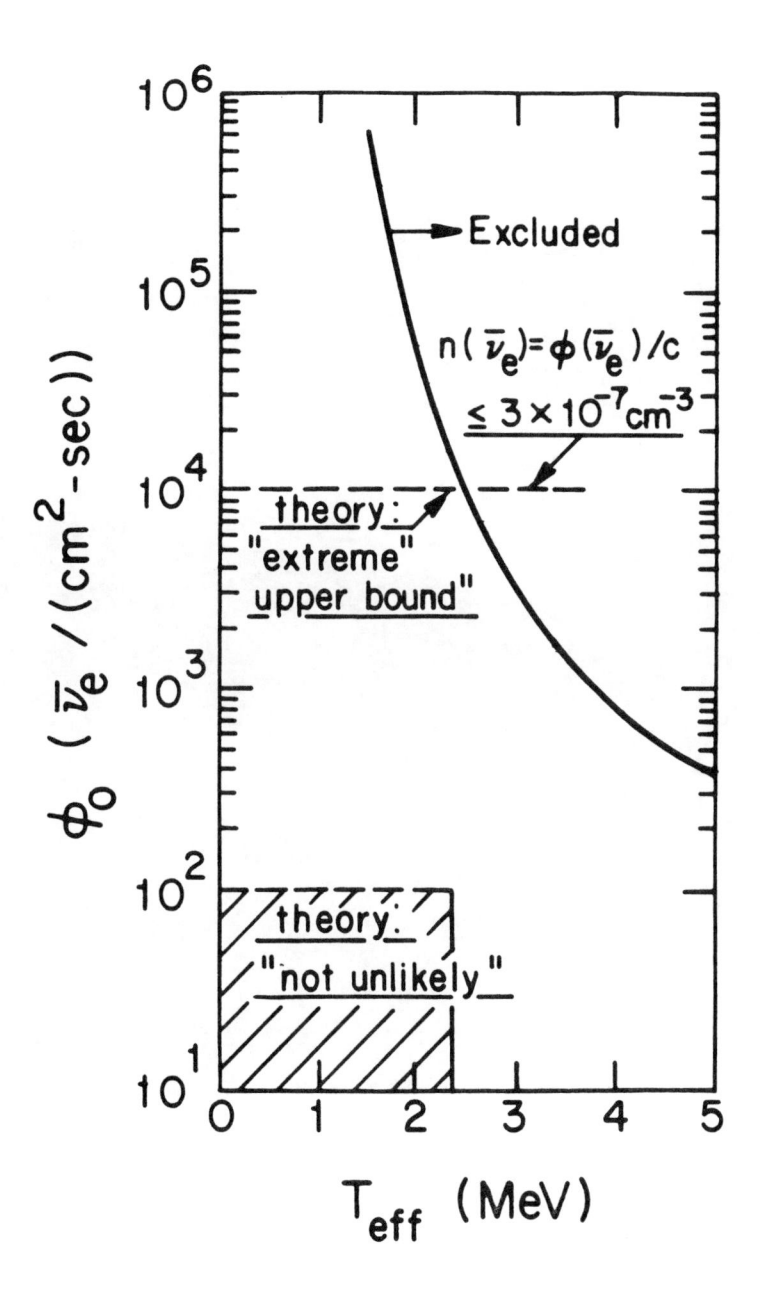

Fig. 2

Distinctive "wishbone" feature of King crater [5.0° north, 120.5° east: 76 km diameter], taken from 98 km altitude. Principal point is 4.9° north, 119.7° east, and crater Abul Wafa [1.0° north, 116.6° east: 55 km diameter] lies beyond. (AS16–120–19268).

Experiments In Gravitation & General Relativity

View of Mare Humorum (Sea of Moisture) from 122 km altitude, and crater Mersenius [21.5^0 south, 49.2^0 west: 84 km diameter]. Principal point is 22.0^0 south, 45.0^0 west. (AS16–120–19338).

On the Use of Clocks in Satellites Orbiting the Moon for Tests Of General Relativity

Arnold Rosenblum

International Institute of Theoretical Physics, Utah State University,
Logan, Utah 84322

Abstract

With the enormous improvement in the accuracies of clocks and the potential use of these clocks for tests of general relativity theory, relativity physics is probably at peak interest with the physics community. It is shown that the use of very accurate clocks around the Moon makes possible the testing for dragging of inertial frames, the oscillations of the sun, and gravitational radiation predicted in relativity.

1. Introduction: New Macroscopic Tests of General Relativity

To date, neither the Lense-Thirring effect nor gravitational radiation, both predicted by General Relativity, have been observed directly. In this presentation, I should like to propose how, by using the "synchronization gap" in closed paths of clocks in accelerating coordinate systems or gravitational fields, it is possible to test for these effects.

2. Clock Synchronization in General Relativity

I would first like to start with a discussion of synchronization in special relativity. Assume that as in Figure 1 we have two clocks at rest at points a and b, respectively. If we wish to synchronize the clock at a with the clock at b using light signals, we first send from a light signals to b' and then a light signal from b' to a'. We next send a message to clock b, setting

$$t_{b'} = t_{a''} = \tfrac{1}{2}(t_a + t_b) \quad . \tag{1}$$

Still another way of synchronizing clocks at rest, the method of slow-clock transport, is shown in Figure. 2. We can simply take a third clock, compare its time with clock *a* and slowly transport it to clock *b* and then synchronize clock *b*. One might immediately ask if this slow–

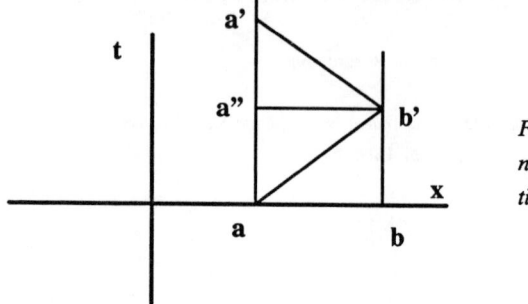

Figure 1. Clock synchronization in special relativity.

clock transport synchronization is equivalent to light-signal synchronization for the case of two static clocks. I will show that this is the case. We have

$$d\tau^2 = dt^2 - (dx^2 + dy^2 + dz^2)/c^2 \quad , \tag{2}$$

and then

$$d\tau = \sqrt{(1 - v^2/c^2)} dt \quad . \tag{3}$$

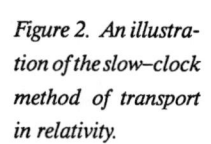

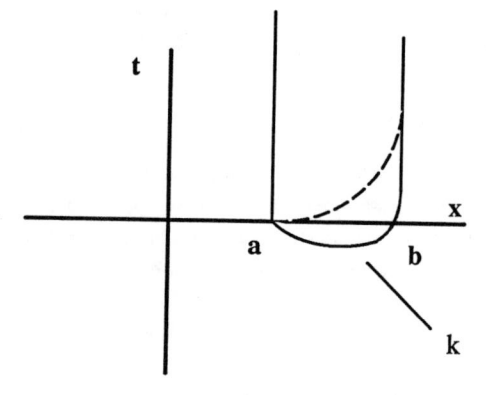

Figure 2. An illustration of the slow–clock method of transport in relativity.

We have

$$d\tau - dt = [\sqrt{1 - v^2/c^2} - 1]\, dt \quad, \tag{4}$$

and $dt = k/v$ from Figure 2. Therefore, it follows that

$$d\tau - dt = [\sqrt{1 - v^2/c^2} - 1]\, k/v \quad, \tag{5}$$

and in the low velocity limit we have

$$d\tau = dt \quad, \tag{6}$$

or

$$t = \tau = t_{b'} = t_{a''} \quad. \tag{7}$$

3. Clock Synchronization in General Relativity

I would now like to discuss the problem of clock synchronization for the general case. We have

$$ds^2 = g_{\alpha\beta}\, dx_\alpha\, dx_\beta \quad, \tag{8}$$

or

$$ds^2 = g_{oo}\, (dx^o)^2 + 2 g_{oj}\, dx^o\, dx^j - g_{ij}\, dx^i\, dx^j \quad. \tag{9}$$

If we simply complete the square in expression (9), we obtain

$$ds^2 = g_{oo}\, (dt + g_{oo}^{-1} g_{oi}\, dx^i)^2 + (g_{ij} - g_{oo}^{-1} g_{oi} g_{oj})\, dx^i\, dx^j \quad. \tag{10}$$

The second term is the well-known Landau–Lifshitz spatial line element associated with local light-signal synchronization. The first term acts as the effective time or

$$\text{`}dt\text{'} = dt + g_{oo}^{-1}\, g_{oi}\, dx^i \quad. \tag{11}$$

We now integrate expression (11) along a closed path and obtain

$$`\Delta t` = \oint g_{00}^{-1} g_{0i} dx^i . \tag{12}$$

Let us apply the above formula to the case of clocks synchronized by light signals on a circle in a rotating frame. We have

$$ds^2 = -\gamma^{-2} c^2 (dt - r^2 c^{-2} \omega \gamma^2 \sin^2\theta d\phi)^2 + dr^2 + r^2 d\theta^2 + r^2 \gamma^2 \sin^2\theta d\phi^2 , \tag{13}$$

with $\gamma^2 = 1 - r^2 \omega^2 c^{-2} \sin^2\theta$.

If we use expression (13) in formula (12) we obtain

$$`\Delta t` = 2\pi r^2 \omega c^{-2} , \tag{14}$$

Using the radius of a geosynchronous orbit, r = 42,000 km and the ω of the Earth $\omega = 7 \times 10^{-5}$ sec^{-1}, we obtain

$$`\Delta t` = 9\mu \text{ sec} , \tag{15}$$

which is an easily measurable quantity.

Let us now examine the case of clock transport synchronization in an even more general context. We have

$$ds^2 = -A^2 dt^2 + B^2 dr^2 + r^2 d\theta^2 + r^2 \sin^2\theta \, d\phi^2 , \tag{16}$$

with $A^2 = B^{-2} = 1 - 2m/r$,

We re-write the above as

$$d\tau^2 = \left\{ A^2 - B^2 \left[(dr/dt)^2 - r^2 (d\theta/dt)^2 - r^2\sin^2\theta \, (d\phi/dt)^2 \right] \right\} \tag{17}$$

Taking the weak-field, slow-motion limit of the above, we obtain

$$d\tau - dt = [\phi - v^2/2] , \tag{18}$$

with ϕ the Newtonian potential. Integrating the above along a closed path, we obtain

$$\Delta (t - \tau) = c^{-2} \int_{t_b}^{t_a} (\tfrac{1}{2}v^2 - \phi) \, dt \quad , \tag{19}$$

with $v \ll 1$ and $\phi c^{-2} = -GM/rc^2 \ll 1$. If we apply the above result for one revolution about a circular orbit, we obtain [1]

$$\Delta (t - \tau) = 3\pi r^2 \omega c^{-2} \quad . \tag{20}$$

This synchronization gap clearly differs from that obtained by using light-signal synchronization. If we use clock transport synchronization around an ellipse we obtain

$$\Delta (t - \tau) = 3\pi/c^2 \, (GMa)^{1/2} \quad , \tag{21}$$

where a is the semi-major axis. For the case of light-signal synchronization, we would clearly like to relax the restriction that the clocks must be infinitesimally close to one another, which was assumed in the derivation of formula (12). We notice that for clocks on the vertices of an N-sided regular polygon rotating with angular velocity ω we obtain [2]

$$\Delta = r^2 \omega c^{-2} \, N \sin(2\pi/N) \quad , \tag{22}$$

which of course reduces in the large N limit to the case of the circle.

4. Dragging of Inertial Frames: Detection With Orbiting Clocks

To discuss the case of using light-signal synchronization to detect dragging of inertial frames, we start from the Cohen-Brill metric

$$-d\tau^2 = -A^2 \, dt^2 + B^2 \, dr^2 + r^2 (d\phi - \Omega dt)^2 \quad , \tag{23}$$

where $A^2 = B^{-2} = 1 - 2m/r$, $\Omega = 2J/R^3 - \omega_o$, and $J = K M_E R_E^2 \omega$, with M_E and R_E the mass and radius of the Earth respectively, with $K \simeq 1$, and with ω_o the rotation rate of an inertial frame as measured by an observer at infinity. J is the body's angular momentum. The synchronization gap expressed in terms of proper time for orbiting satellites in circular equatorial orbits is

$$\Delta S = 2\pi r^2 \omega [1 + 3M/2r - (2KM)r^{-1} (R_E/r)^2] \quad , \tag{24}$$

where the first term in the parenthesis is the special relativistic one, the second term arises from a term similar to the gravitational red shift, and the third term from the dragging of inertial reference frames.

5. The Case for Lunar Orbit

For experimental purposes, it is clearly better to eliminate the lower order terms in expression (24). To do this we imagine that we send up clocks in satellites both in geosynchronous and anti-geosynchronous orbits. We synchronize them using light signals, first the clocks in the geosynchronous orbits and the using light signals going the other way. The simple addition of the synchronization gap of both the geosynchronous and anti-synchronous orbits will isolate the effects of dragging of inertial frames. With absolute clock accuracies of 1 part in 10^{18} possible in the next five years and stability over a two year period [3], this leads to the feasibility of experimental determination of dragging of inertial frames. Such a clock is already being built in the Max Planck Institute for Quantum Optics in Munich under the supervision of Professor Herbert Walther. In geosynchronous approximately 1.92×10^{-17} seconds. Three orders of magnitude can be gained by orbiting the Sun or Jupiter.

Putting the clocks in satellites around the Moon leads roughly to the same effect as putting them around the Earth. The fact that there is no atmosphere around the Moon means that the satellites can be brought closer to the surface (although the mass concentrations or mascons on the near-Earth side begin to introduce orbital instabilities and decay below 60 km altitudes). However, the angular velocity of the Moon is only 1/27 that of the Earth.

To summarize, I have presented a way using the latest in clock technology, to measure dragging of inertial frames. It should be very strongly pointed out that the use of light signals is not necessary when using the path dependence of clock synchronization because the method of clock transport synchronization can be used which requires, in principle, no light signals.

6. Gravitational Radiation Bursts & Orbiting Clocks

In addition, the improvement in clock technology leads to the belief that very accurate clocks might have the potential of serving in the detection of non-periodic gravitational waves ("bursts"). One way, in principle, to try to detect gravitational radiation is to place several clock-carrying satellites in nearly identical circular orbits around the Earth and to track the times kept by these clocks with a clock on the Earth. Similarly, one can conceive of such an array or configuration orbiting about the Moon tracking the times kept by those clocks with an identical clock at the fixed-base surface facility on the Moon. The pattern of these times will be disturbed in a characteristic way by the passage of gravitational impulse waves, and moreover, the disturbance grows with time after the event. As a simple hypothetical case, consider four such satellites circling the Earth (or Moon) at the corners of a square whose diagonals just happen to be aligned with the direction of polarization of the wave as it passes. They go more slowly relative to the higher-flying ones, and we can let this time difference accumulate over many orbits.

Let us assume the Schwarzschild metric

$$ds^2 = A^2 dt^2 - A^{-2} dr^2 - r^2 (d\theta^2 + \sin^2\theta \, d\phi^2) \quad , \tag{25}$$

where $A^2 = (1-2m/r)$ and, as usual, we choose θ to be $\tfrac{1}{2}\pi$ in the "plane" of the orbits. For a satellite circling the Earth with angular velocity ω it is convenient to introduce a new azimuthal coordinate

$$\phi' = \phi - \omega t \quad , \tag{26}$$

which remains constant for that satellite. Accordingly, we now have

$$ds^2 = [A^2 - \omega^2 r^2] dt^2 - 2r^2\omega \, dt \, d\phi' - A^{-2} dr^2 - r^2 d\phi'^2 \quad . \tag{27}$$

This is a stationary metric, and the satellite under consideration is one of its "fixed points." Its worldline will therefore be geodesic if the gradient of g_{oo} vanishes. This leads to the associated "Kepler's Law"

$$\omega^2 = 2m/r^3 \quad . \tag{28}$$

From (27) the proper time on the satellite is then seen to be given by

$$d\tau = [1 - 3m/r]^{1/2} dt \approx [1 - 3m/2r] \quad , \tag{29}$$

which of course is a function of r.

Suppose the two satellite pairs experience respective strain ratios $\pm \delta r/r$, as is characteristic of an impulse wave. Their new proper time $d\tau_+$ and $d\tau_-$ will be given by

$$d\tau_\pm = [1 - 3m/2r \pm (3m/2r)(\delta r/r)] dt \quad , \tag{30}$$

and consequently we have the change in clock readings $\delta(\Delta\tau)$ over finite time Δt

$$\delta(\Delta\tau) = (3m/2r)(\delta r/r) \quad . \tag{31}$$

In gravitational units, the mass m of the Earth is approximately 0.5 cm. For the special case of geostationary satellites, r is approximately 4×10^9 cm and for bursts from such violent events as the formation of giant black holes at the center of galactic nuclei, strain ratios as high as 10^{-17} have been suggested. This leads to $\delta(\Delta\tau) = 0.2 \times 10^{-26} \Delta t$. Similar numbers can be obtained for the Moon. This is, of course, far outside of present day clock accuracy. A way around this problem is to consider the binary pulsar. If we do the same calculation for the binary pulsar as was done for the Earth and satellite system using approximate values for the parameters we obtain

$$\delta(\Delta\tau) \approx 10^{-22} \Delta t \quad . \tag{32}$$

This value is again outside the accuracies of current clock technology. However, we have the possibility of using the millisecond pulsar 1937+21 as the clock. If we assume that it has an accuracy of 1 part in 10^{-22} then we can use it directly as the clock to detect the signature of a gravitational wave. Since it is not known if this is really the accuracy of the millisecond pulsar, another idea is to integrate the pulses of the millisecond pulsar and the binary pulsar together over long periods of time to check for inherent differences in the priodicities. In any case, the discovery of the millisecond pulsar leads to the possibility of its use as a clock.

7. Solar Oscillation Tests With Orbiting Clocks

It is also possible to use the synchronization gap to test for oscillations of the sun. Following the work of H. Hill [4], the Sun's gravitational field can be written as

$$\phi = -GM_0 r^{-1} \left\{ 1 + \sum_n \sum_{l=2} \sum_m J'_{n,l,m} (R/r)^l P_l^{|m|} \cos[\omega_{n,l,m} t + m(\lambda - \lambda_{n,l,m})] \right\}, \tag{33}$$

where $J'_{n,l,m}$ and $\lambda_{n,l,m}$ are constants, $\omega_{n,l,m}/2\pi$ is the eigenfrequency of the oscillation, and λ is the heliocentric longitude. The short period changes in longitude δL_{sp} can be obtained and can be written as

$$\delta L_{sp}/\delta L = 2(GM_0/a)^{\frac{1}{2}} R^{-1} \sum_n \sum_{l=2} \sum_m m(l+1)(\omega_{n,l,m}+m\lambda)^{-1}(R/r)^{l+1} J'_{n,l,m} P_l^{|m|}(\sin D) \cdot$$
$$\cos[\omega_{n,l,m}t + m(1+\tfrac{1}{2}\delta L - \lambda_{l,n,m})] \quad , \quad (34)$$

where δL is the difference in longitude of the free masses not including the short period contributions, a is the semi-major axis of the orbit of the two masses around the Sun, and D is the declination of the masses at time t. The contributions of D have been neglected in this expression. It has also been assumed that $\delta L \ll 1$ and higher-order terms in the eccentricity have been omitted.

The amplitude of the strains $\delta L_{sp}/\delta L$, defined as $h_{n,l,m}$, for a given perturbing field is therefore given by

$$h_{n,l,m} = 2(GM_0/a)^{\frac{1}{2}} m(l+1)(\omega_{n,l,m}+m\lambda) J'_{n,l,m} P_l^{|m|}(\sin D_0) \quad . \quad (35)$$

These strains mimic the effects of a gravitational wave and must be seriously considered in all investigations of gravitational radiation.

8. Conclusions

It is in the interest of relativity physics that as many different approaches as possible for detecting the "dragging of inertial frames" and gravitational radiation be attempted. Hopefully, even coincidences between the different methods could be used to test the reliability of the results. Clearly, a case can be made for orbiting high precision clocks about the Moon since, unlike the Earth, it is free of atmospheric drag at low altitudes. Because lunar orbiting satellites will be employed for both scientific study as well as for relaying telecommunication data back to Earth from a lunar base, accurate clocks of the type proposed here appear to be a sensible candidate for such an initiative in space. Precise timing would benefit both the network data system and the scientific investigations as well.

Acknowledgement

The above was done with Professor Wolfgang Rindler and Professor Jürgen Ehlers as consultants. A more rigorous treatment is in progress. The author would also like to thank Dr. Thomas Wilson of NASA who edited and prepared the final draft of this manuscript for publication.

References

1. J. Cohen, H. Moses, A. Rosenblum, Phys. Rev. Lett. **51**, 1501 (1983).
2. J. Cohen, H. Moses, A. Rosenblum, Class. Quan. Grav. **1**, L57–59 (1984).
3. H. Dehmelt, Annales de Physique (Paris) **10**, 777–795; D. Wineland (personal communication. See also H. Dehmelt and G. Janik, Bull. Amer. Phys. Soc. **27**, 481 (1982); and H. Dehmelt, Physica Scripta **T22**, 102 (1988) for additional references.
4. H. Hill, to appear in Proceedings NATO Advanced Workshop on "Mathematical Aspects of Gravity and Supergravity."

The Apollo Retroreflector Arrays Revisited: A Lunar Beaconed Array

James E. Faller[*]

Joint Institute for Laboratory Astrophysics
University of Colorado and National Institute of Standards and Technology
Boulder, CO 80309-0440

Following the placement on the lunar surface of the Apollo 11 retroreflector array in July of 1969, and the subsequent placement of the Apollo 14 and 15 arrays, lunar laser ranging has been successfully carried out from the McDonald Observatory in Fort Davis, Texas; the Haleakola (LURE) Observatory on Maui, Hawaii; and from a number of foreign lunar laser ranging stations. The range data from the ground stations are used in an ongoing analysis effort -- an effort that continues to extract new and exciting science as a result of the increasing data span and the improving accuracy of the range measurements.

Though this is already a highly successful program, two things would further improve on the present situation. These are: (1) more and higher quality ranging data and (2) more uniformity in the data coverage during the lunar cycle. The present gaps in the data result during periods when the retroreflector site is poorly illuminated (lunar shadows are not always as crisp as one would like) or in the dark, which then requires the use of offset or absolute pointing -- both of which are considerably more difficult than visual guiding.

[*]Staff Member, Quantum Physics Division, National Institute of Standards and Technology

The suggestion being made here is to take advantage of the proposed lunar base by locating, at any earthward-facing lunar base, a new Apollo-type retroreflector array <u>together</u> with an earth-pointing laser beacon. The latter would greatly simplify guiding on this site by the ground stations. The retroreflector array would be similar to those placed at the Apollo sites. It could, however, be somewhat larger (e.g., 400 cube corners).

The most important feature of this proposal is the location of a guide beacon near the retroreflector array. Its placement there would eliminate all of the difficult guiding problems. As a result, a dramatic increase in the data rate from ranging to this array as well as a filling-in of the data gaps would result. Furthermore, the existence of a beacon-guided retroreflector array at the lunar base would greatly simplify the periodic monitoring of a ground station's performance, since guiding questions would no longer be a factor in a station's ability to acquire data. Thus, it would provide a marvelous means for verifying and improving ground station performance; and, as a result, an increase in the data rate from the existing Apollo (and Lunahod) arrays would result. Furthermore, once vehicles for long distance "transportation," operating out of the lunar base, are available on the moon, remotely operated laser beacon devices and new (larger) retroreflector arrays could also be taken to the existing sites. This would further enhance the return rate from these sites.

A lunar beacon would need to be aimed so as to illuminate the whole earth -- in that way all of the ground stations could use it simultaneously. This will require an angular beam diameter of 1/30 of a radian. This degree of "aiming" could be accomplished with a simple drive system. (Adjustments in the pointing direction every few days would be adequate.) The actual beacon device would need to have only a one or two inch aperture. Short laser diode pulses would be transmitted at some <u>known</u> rate so as to permit time filtering as well as spacial and wavelength filtering to be used by the ground stations to discriminate against background light. A small array of detectors would be located in the focal plane of each receiving telescope. (The array would provide directional information with which to control the pointing.) A 1 W flux of pulsed photons transmitted to the earth would -- for an earth-based receiving aperture of 100 cm (40 inches) -- result in a guide signal of about 2000 photons per second! Also, the manner in which the diode light distributes itself between the elements of the pointing array sensors, would provide a quantitative measure of the atmospheric seeing during the periods of lunar ranging.

In discussions subsequent to the Stanford meeting, P. L. Bender has pointed out the value of placing three optical or microwave transponders on the lunar surface with separations of 300 to 1000 km. Ranging from a single earth-based "telescope" to these transponders would make possible differential ranging (libration studies) with an accuracy of between 30 and 100 μm. (Because the earth's atmosphere would be sufficiently common over the angles subtended by the proposed lunar transponder array, atmospheric propagation delays cancel out so far as libration studies and lunar tide measurements are concerned.) Though this suggestion of lunar transponder ranging will require considerably more development effort than the lunar beacons, the resulting libration and tide data would provide an absolutely unique "window" into the details of the moon's internal structure and liquid core. Accordingly, though one would initially plan to deploy a beaconed array at the lunar base to be followed by locating beacons and improved arrays at the Apollo 11, 14 and 15 plus Lunahod sites, a strong case can be made to develop a transponder-type system for future deployment.

In summary, locating an Apollo-like array together with a simple laser beacon at the site of any future lunar base would greatly improve the ranging success rate with the existing laser ranging ground stations. Locating additional beacons and new arrays at the other laser ranging sites would further enhance the data rate using multiple array ranging. In addition, the development of lunar transponders (optical or possibly microwave) would permit a spacially separated transponder array to be located on the moon which would allow differential ranging measurements to be made at the 30 to 100 μm level of accuracy. This would result in a more than two orders of magnitude improvement in our ability to study the free librations and tides of the moon and through studying them provide a better understanding of the moon's internal structure and core.

Acknowledgment

The author acknowledges helpful discussions with Peter Bender as well as the support and encouragement received from the Lunar Laser Ranging Working Group.

Session On Gravitational Radiation Physics

Joseph Weber, University of Maryland, has long proposed that the Moon's quadrupole oscillation modes could be used to measure gravitational radiation. This full-disk view shows the Mare Crisium (Sea of Crises) on the upper horizon (10:30 o'clock). North is along the diagonal toward the upper right-hand corner, and northwest is top center. (Apollo 16 metric frame 3023).

Gravitational Radiation Antennas: History,
Observations, and Lunar Surface Opportunities

J. Weber
University of California, Irvine, California 92717
and
University of Maryland, College Park, Maryland 20742

Abstract

Elastic solid and interferometer antennas were developed at the University of Maryland beginning in 1958. Backgrounds of pulses were observed by the Universities of Rome and Maryland kilohertz antennas during the period 1969-1982, and during the Supernova 1987A period February 1987.

The quadrupole modes of the moon have unusually large cross sections. The lunar surface is an excellent site for a long period accelerometer to observe the quadrupole modes, and for low frequency bar and interferometer antennas to observe gravitational radiation.

© 1990 American Institute of Physics

Gravitational Radiation Antennas

Einstein unified physics and geometry, and this unification is among the greatest intellectual achievements.

In the absence of gravitation a triangle with sides which are light rays is described by Euclidean geometry, Figure 1

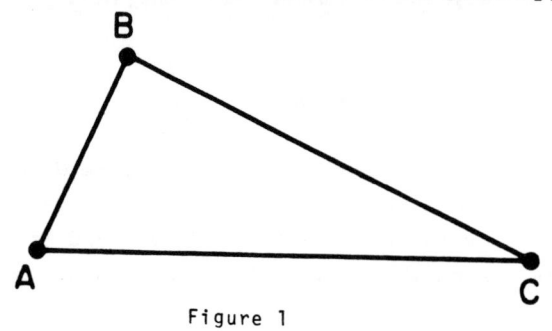

Figure 1

The sides are straight lines. The sum of the angles, A + B + C = 180 degrees.

If a massive body such as the sun is in the center of the triangle, the sides are curved, the sum of the angles A + B + C is greater than 180 degrees

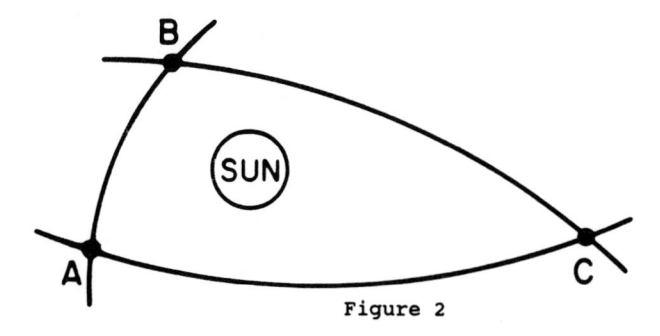

Figure 2

Figure 2 describes a curved, non Euclidean space.

Einstein's unification of physics and geometry describes gravitation as the spacetime curvature.

The General theory of Relativity predicts that changes in spacetime geometry may propagate as gravitational waves, Figure 3 shows such a wave field.

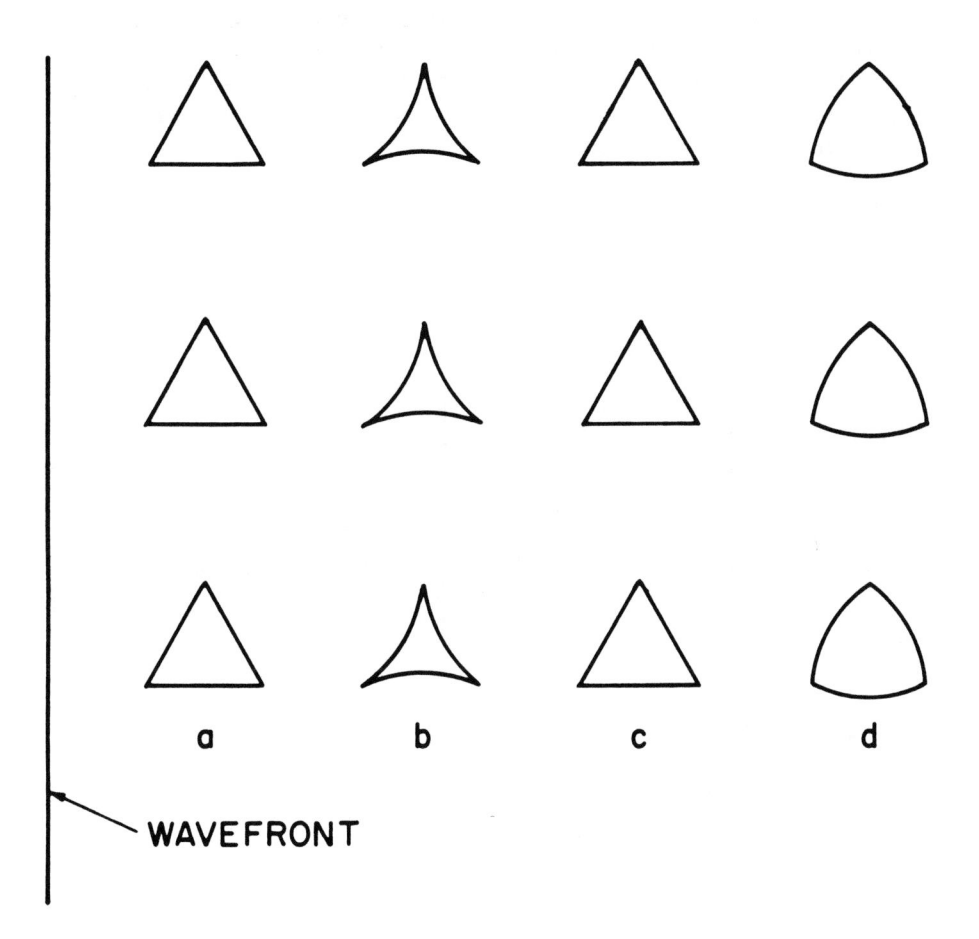

```
   △      △      △      ◠
   a      b      c      d
```

← WAVEFRONT

Figure 3

At the moment of observation, the region a is flat, region b is concave, region c is flat, region d is convex.

For the gravitational radiation expected from Supernova 1987A, for a laboratory size triangle, the sum of the angles differs from 180 degrees by about 10^{-24} degrees.

A quantity as small as 10^{-24} degrees is too small to measure directly with present technology. The new methods were proposed and developed during the period 1957-1959. One makes use of an elastic solid such as an aluminum cylinder. Figure 4 shows such a cylinder in a wave field

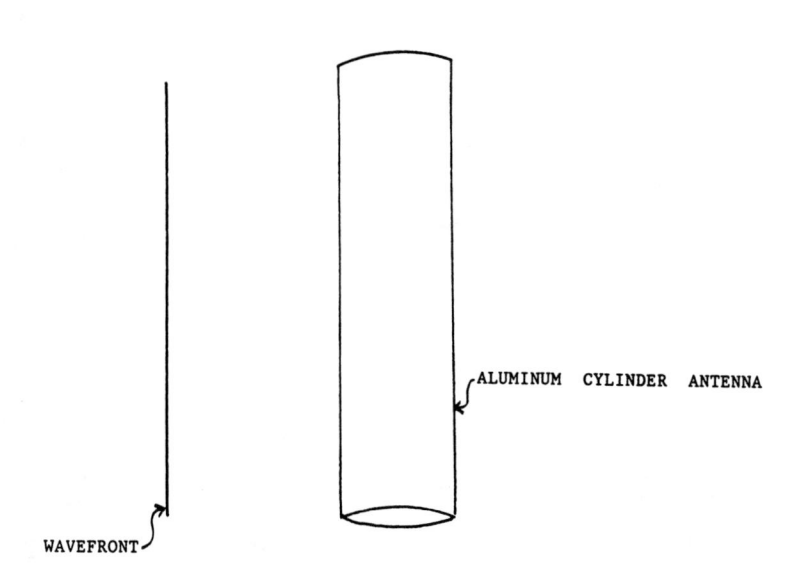

Figure 4

The changes of spacetime curvature cause the length of the cylinder to change. These changes are about 2×10^{-17} (length) for the gravitational radiation from supernova 1987A.

Another method proposed at that time involved use of a Michelson interferometer, as shown in Figure 5.

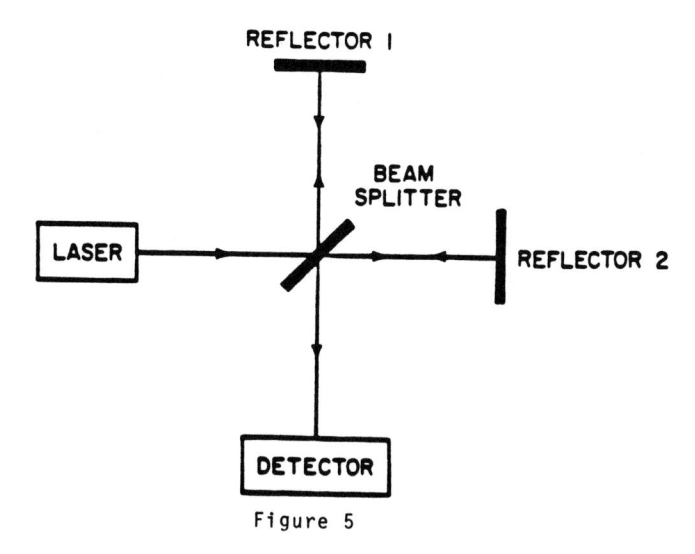

Figure 5

A gravitational wave moves in a direction normal to Figure 5. At one instant mirror 1 has moved up, and mirror 2 has moved to the left. At the next half wavelength period mirror 1 has moved down and mirror 2 has moved to the right. For supernova 1987A the fractional changes in length for the two paths shown in Figure 5 is about 2×10^{-19}. These fractional changes in length are very much larger than the fractional changes in angle sum for the light ray triangles.

Figure 6 shows the first Maryland gravitational radiation antenna, now on permanent display at the Smithsonian Institution in Washington, D.C.

Figure 7 is a schematic diagram showing what has been accomplished. The gravitational radiation antennas are a kind of bridge which enables the 20th century instrumentation to observe the world of Einstein.

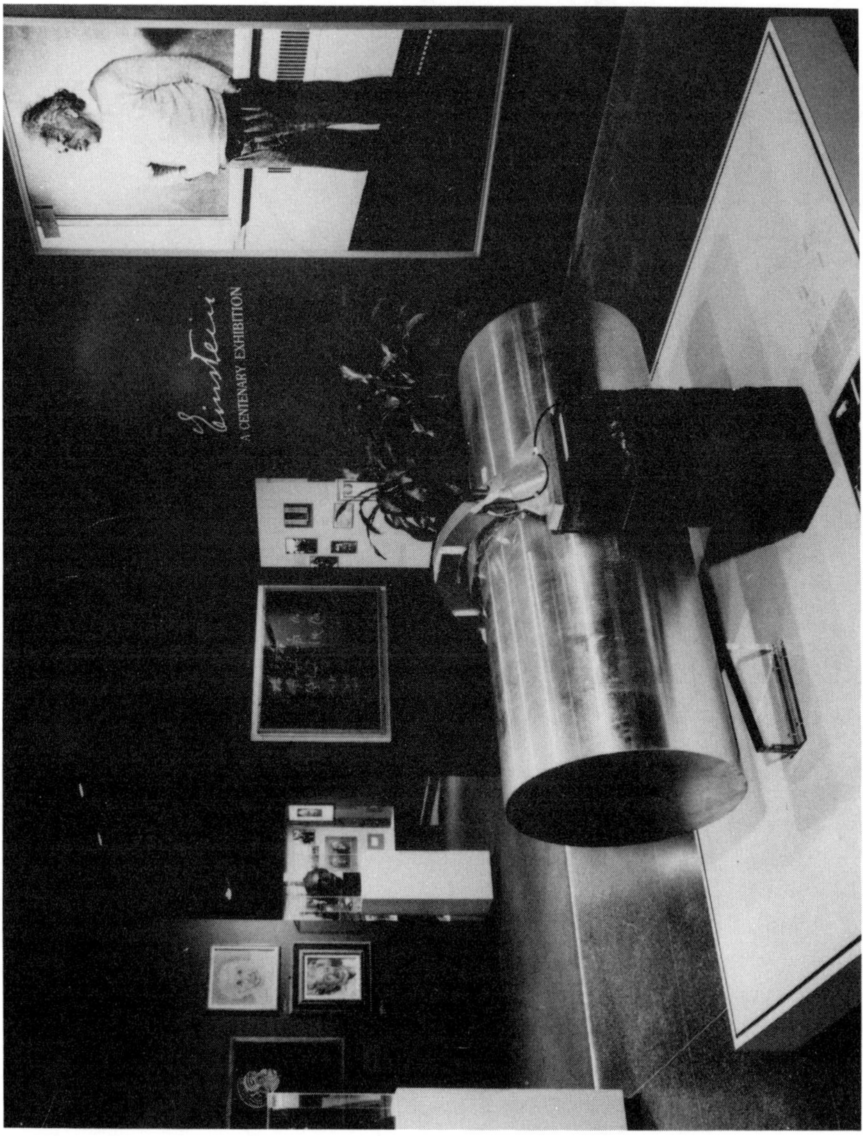
Einstein
A CENTENARY EXHIBITION

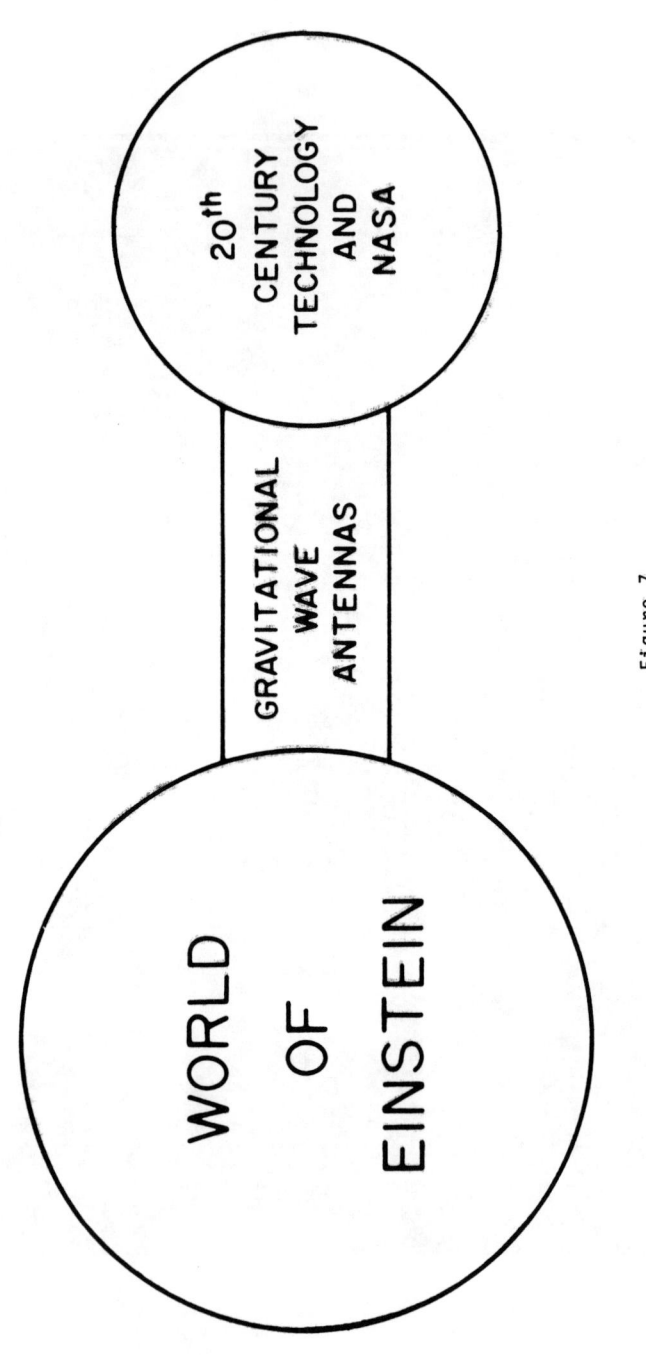

Figure 7

OBSERVATIONS

Coincident pulses were observed on widely separated gravitational wave antennas beginning in 1968. After an extended period of controversy, all of the early observations were confirmed.[1]

Antennas at the Universities of Rome and Maryland were operating during the rapid evolutionary phase of Supernova 1987A.[2] Large correlations were observed with the two widely spaced gravitational radiation antennas and the neutrino detectors -- under Mont Blanc in the Alps, at Kamioka in Japan, and at Baksan, in the U.S.S.R.. An abstract of one already published paper is shown as Figure 8

IL NUOVO CIMENTO　　　VOL. 12 C, N. 1　　　Gennaio-Febbraio 1989

Analysis of the Data Recorded by the Mont Blanc Neutrino Detector and by the Maryland and Rome Gravitational-Wave Detectors during SN1987A.

M. AGLIETTA, G. BADINO, G. BOLOGNA, C. CASTAGNOLI, A. CASTELLINA
W. FULGIONE, P. GALEOTTI, O. SAAVEDRA, G. TRINCHERO and S. VERNETTO
Istituto di Cosmogeofisica del CNR - Torino
Istituto di Fisica Generale dell'Università - Torino

E. AMALDI, C. COSMELLI, S. FRASCA, G. V. PALLOTTINO
G. PIZZELLA, P. RAPAGNANI and F. RICCI
Dipartimento di Fisica dell'Università ·La Sapienza· - Roma
Istituto Nazionale di Fisica Nucleare - Roma

M. BASSAN, E. COCCIA and I. MODENA
Dipartimento di Fisica dell'Università ·Tor Vergata· - Roma
Istituto Nazionale di Fisica Nucleare - Roma

P. BONIFAZI and M. G. CASTELLANO
Istituto di Fisica dello Spazio Interplanetario del CNR - Frascati (Roma)
Istituto Nazionale di Fisica Nucleare - Roma

V. L. DADYKIN, A. S. MALGUIN, V. G. RYASSNY, O. G. RYAZHSKAYA
V. F. YAKUSHEV and G. T. ZATSEPIN
Institute of Nuclear Research, Academy of Sciences of USSR - Moscow, USSR

D. GRETZ, J. WEBER and G. WILMOT
Department of Physics and Astronomy, University of Maryland, USA

(ricevuto il 6 Settembre 1988)

> **Summary.** — The data recorded by the gravitational wave and the neutrino detectors mentioned in the title have been analysed over a period of several days that includes the Mont Blanc 5ν burst occurrence time. A correlation is found during a period of about two hours roughly centred on the 5ν burst, independently between Maryland and Mont Blanc and Rome and Mont Blanc. The probability that these two correlations be due to chance is of the order of between 10^{-6} and 10^{-5}. It is found that this effect is mainly due to a dozen of large Maryland and Rome events distributed during the above two-hour period.
>
> PACS 97.60.Bw – Supernovae.
> PACS 04.80 – Experimental tests of general relativity and observations of gravitational radiation.

Figure 8

A summary of the observations follows:

Rome Maryland Gravitational Radiation Antenna Correlations
with Neutrino Detectors at Mont Blanc, Kamioka, and Baksan,
Associated with Supernova 1987A

Gravitational Radiation Antennas at Rome and Maryland, and several neutrino detectors, were operating during the rapid evolutionary phase of Supernova 1987A. 5 neutrinos were observed at about 0245 Universal time 23 February 1987, on the Mont Blanc neutrino detector. Pulses were observed on both Rome and Maryland gravitational antennas 1.2 seconds earlier than these Mont Blanc events.

The analyses therefore begin with the assumption that the neutrinos and gravitational detectors might have correlated outputs for a 1.2 second earlier gravitational antenna time. The two gravitational antenna outputs are at first combined and treated as one.

MARYLAND-ROME GRAVITATIONAL ANTENNA CORRELATIONS

In two hours there were about 100 neutrino events. The total event sum of the gravitational antenna outputs, is obtained for times of the 100 neutrino detector events minus 1.2 seconds. If there are correlations, the gravitational antenna output sum will be much larger than for other times. A large number of time delays δ, between neutrino detector events and gravitational outputs, in increments of one second, are introduced and the gravitational antenna sums again carried out. The number of time delay values giving gravitational antenna sums larger than the zero δ value is studied, as the center time of the two hour interval is changed. Let this value be n. If the central point

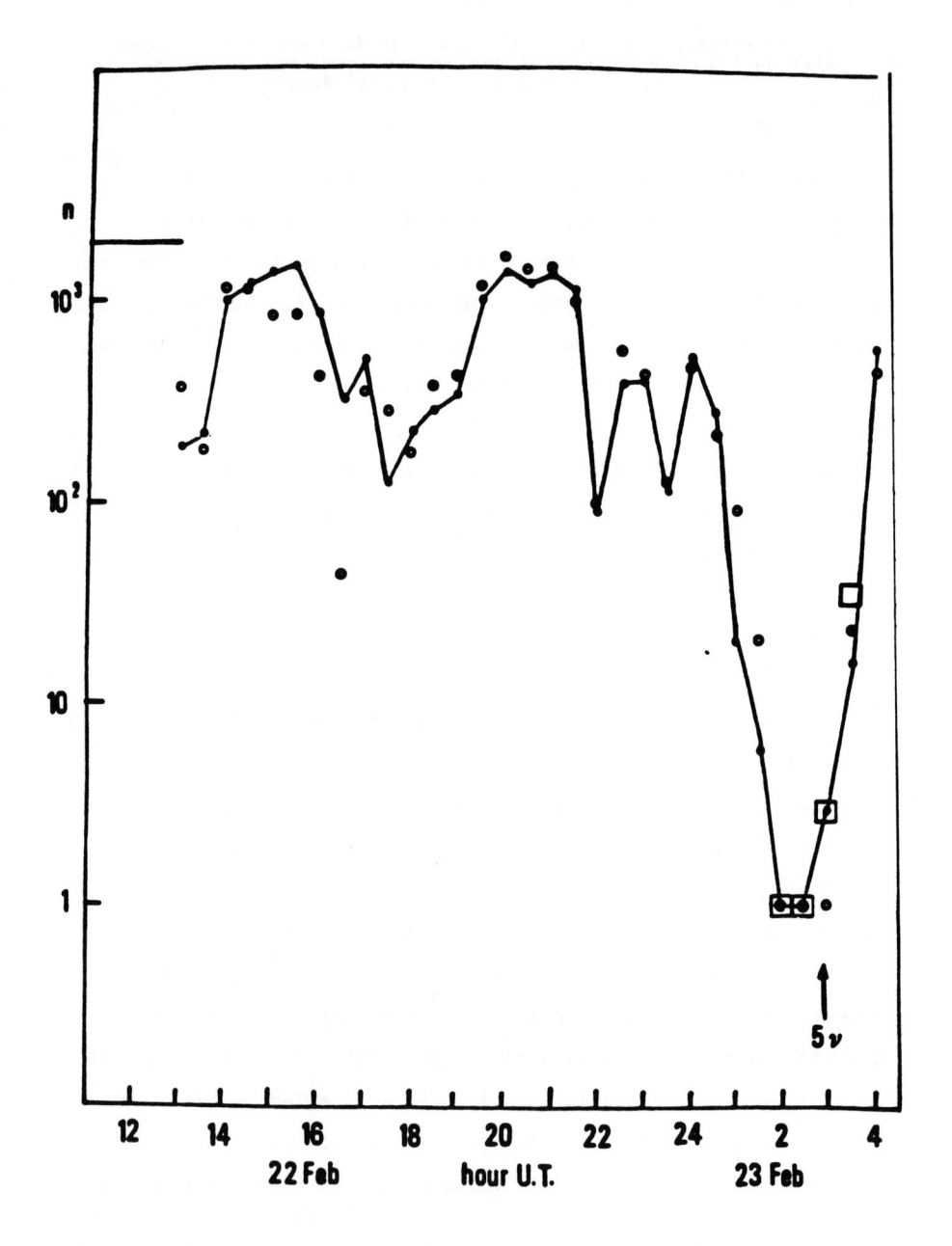

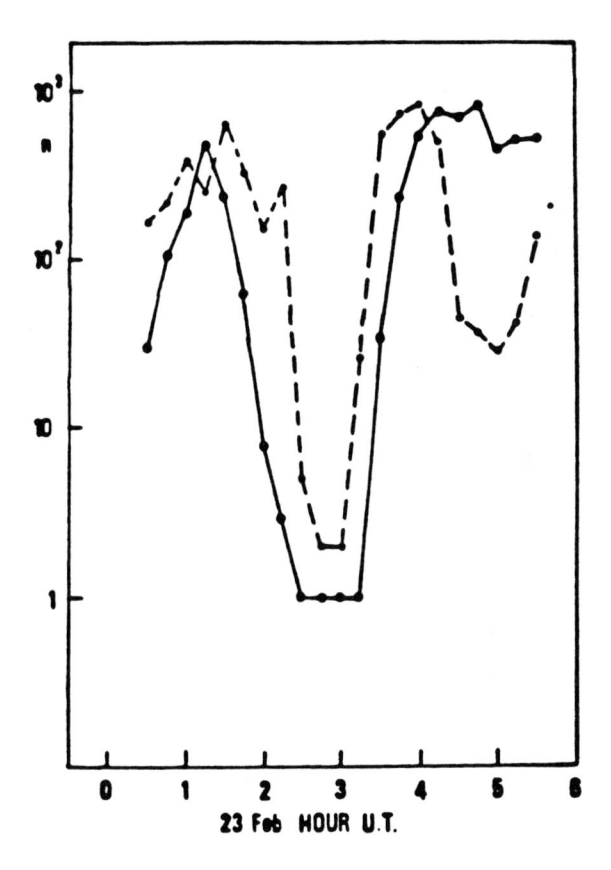

Figure 10

Comparison of the correlation previously found between the sum of the Rome and Maryland data with the Mont Blanc signals (continuos line) and the present correlation with the Kamioka signals (dashed like). $N=10^3$.

$T = 1$ h
$\Delta t = \pm 1$ sec.

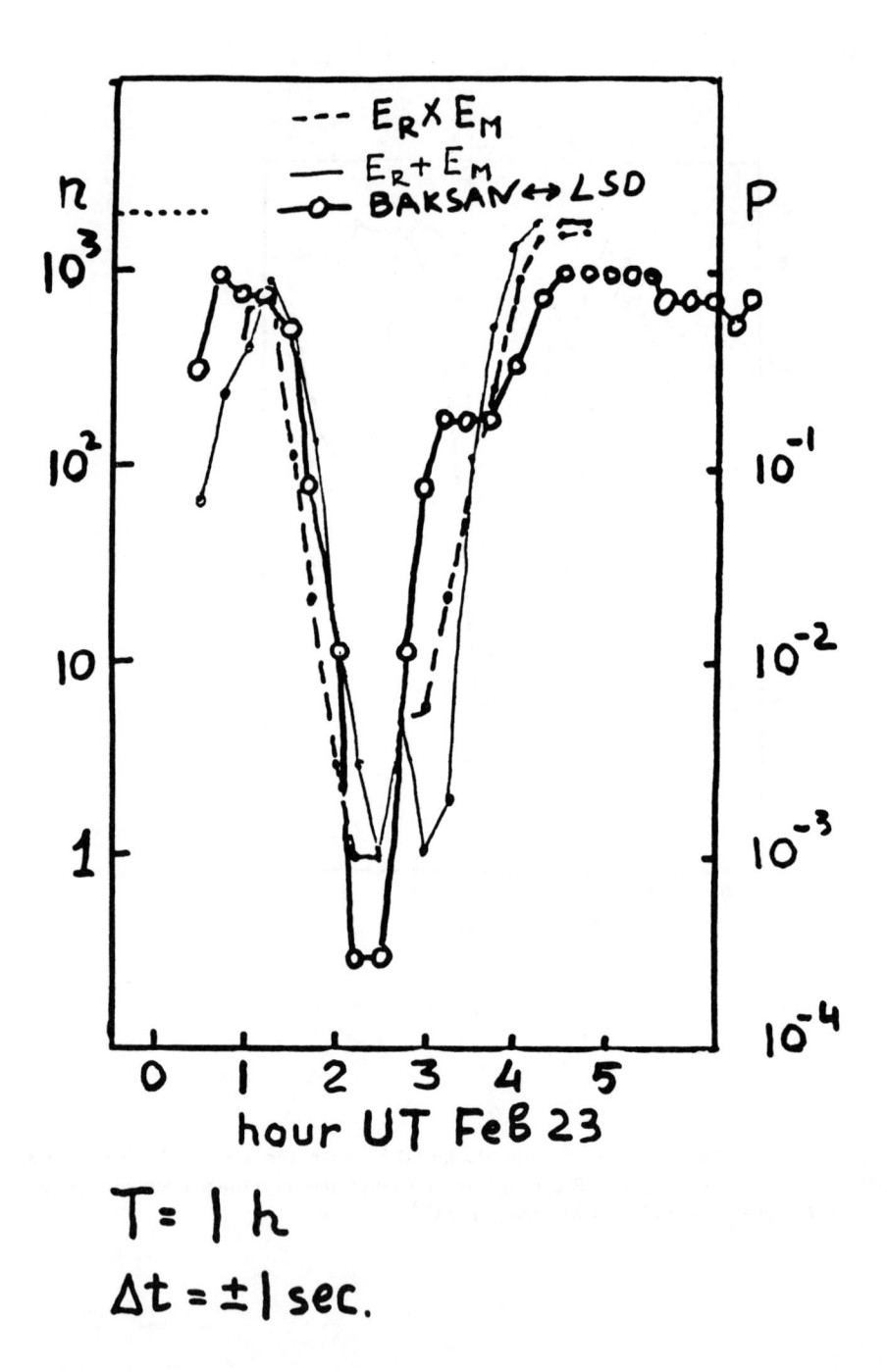

Figure 11

Figure 12

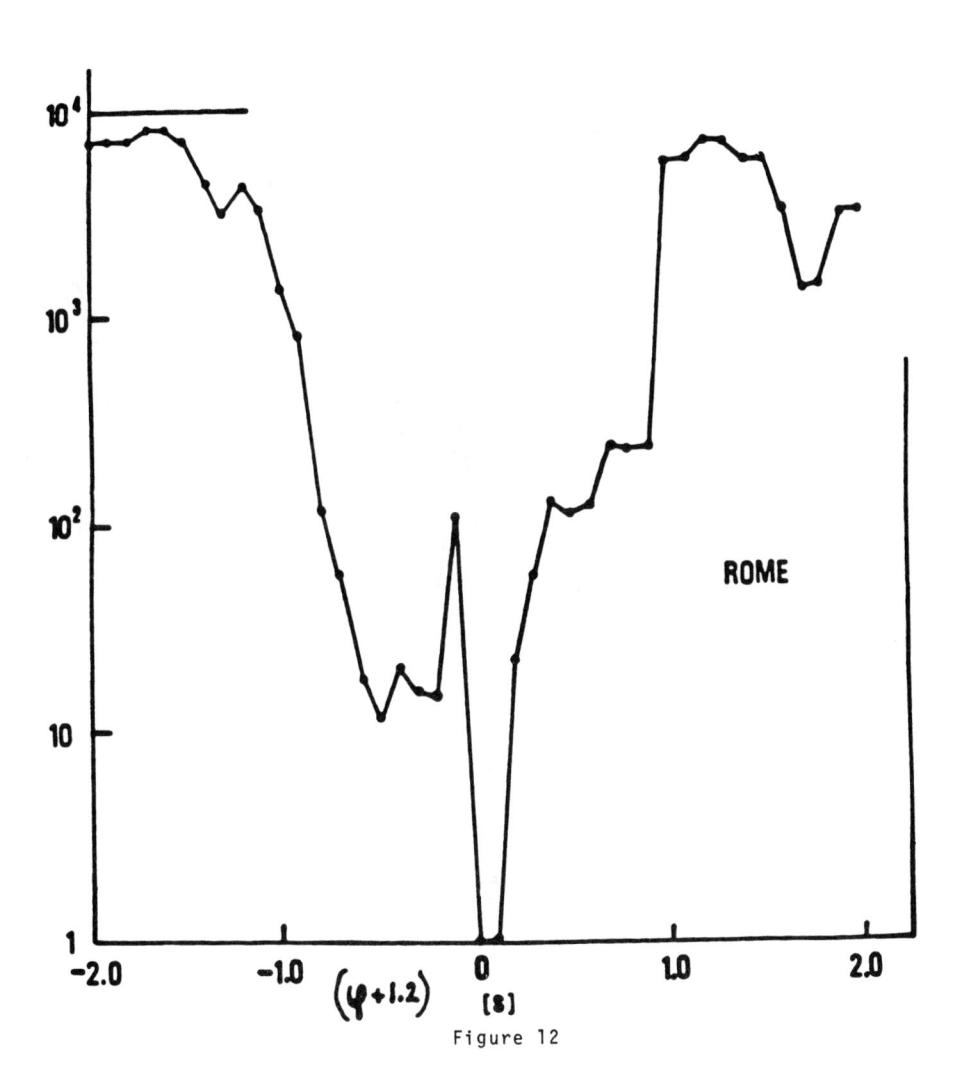

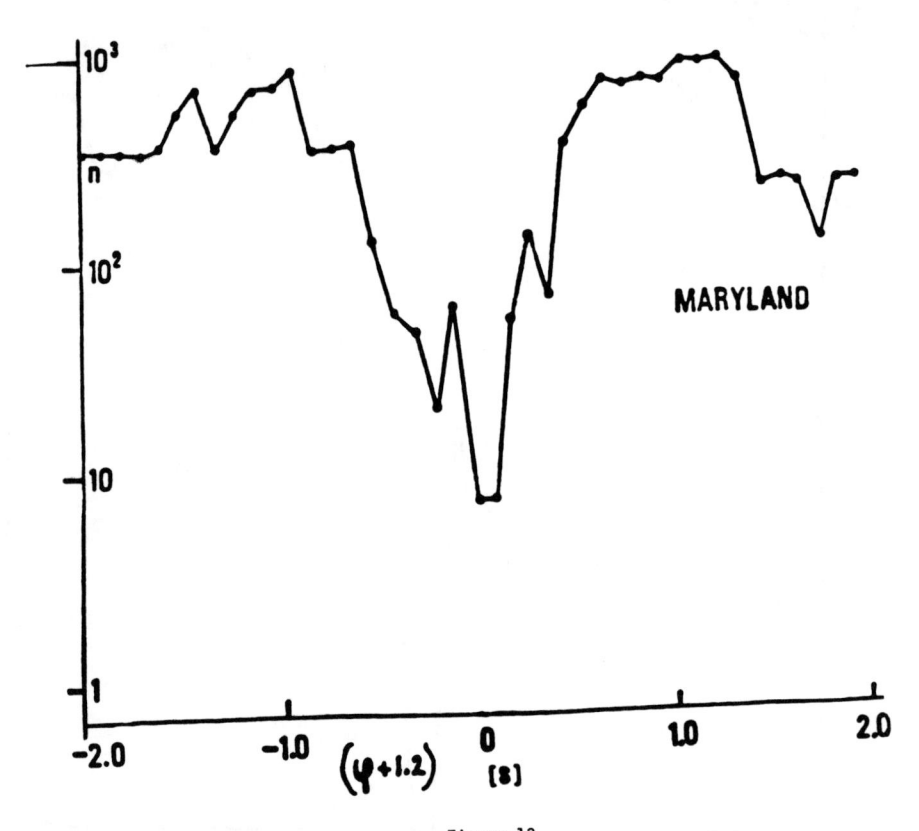

Figure 13

($\delta = 0$) of the interval is a time of large gravitational antenna output, n will be small. If the central point is a time of close to average gravitational antenna outputs, n will be large.

Rèsults are plotted in Figure 9 , as the central point is moved in increments of 1/2 hour. The minimum occurs close to the Mont Blanc 5 neutrino detector event burst.

The same analyses have been carried out for the Kamioka and Baksan neutrino detectors. There were uncertainties in the Kamioka data because of timing problems. Since large correlations were reported between the Kamioka and the IMB (Irvine Michigan Brookhaven) neutrino detectors, the IMB times were employed to correct the Kamioka data. Figure 10 is the result for Rome-Maryland and Kamioka, and Figure 11 is the result for Rome-Maryland and the Baksan neutrino detector.

Returning to the Mont Blanc data, 2 hours 45 minutes is near the center of the minimum of Figure 9 The Rome and Maryland gravitational radiation antenna data are then treated separately. The center of the observation period is kept fixed at 2 hours 45 minutes U.T. Delays are inserted into the Rome and Maryland antenna times in increments of 0.1 seconds. This gives the correlation curve Figure 12, for the Rome gravitational radiation antenna and the Mont Blanc neutrino detector. Figure 13 is the correlation curve for the Maryland gravitational radiation antenna and the Mont Blanc neutrino detector.

Figures 9 ,10, and 11 indicate that the Rome and Maryland gravitational radiation antennas correlate with the Mont Blanc, Kamioka, and Baksan neutrino detectors for about an hour, centered at 0245 U.T. February 23, 1987. Figures 12 and 13 state that the detailed correlation curves for the two gravitational radiation antennas, separated by 8000 kilometers, are essentially the same for the Mont Blanc data.

CROSS SECTIONS

In 1960 the gravitational antenna was studied, employing Einstein's equations and the classical theory of elasticity. For pulses the absorption cross section, σ_{1960}, was calculated[1] to be

$$\sigma_{1960} = \frac{8\pi^3 GML^2}{c^2 \lambda} \qquad (1)$$

G is Newton's constant of gravitation, M is the reduced mass, L is the effective length of the antenna considered as a single large mass quadrupole. c is the speed of light, and λ is the wavelength.

In 1984 and 1986 the problem was reconsidered.[2] It seemed more appropriate to treat the antenna as a large number of atoms coupled by chemical forces. An analysis employed the S matrix to calculate the cross section as a problem in modern elementary particle physics.

The observed cross section is an appropriate summation over all Feynman graphs. Some processes give small cross sections and others give large cross sections.

It is reasonable to assume that a graviton may be exchanged at any mass element. These mass elements may be considered as an

---

[1] J. Weber, Phys. Rev. 117, 306 (1960), General Relativity and Gravitational Waves, Wiley, Interscience New York, London 1961.

[2] J. Weber, Foundations of Physics 14, 1185 (1984); J. Weber and T.M. Karade Volume 3, Sir Arthur Eddington Centenary Symposium, pages 1-78, World Scientific, 1986.

ensemble of quadrupoles. The antenna is very short, compared with a gravitational wavelength. Phase shifts over the antenna are therefore very small. Under these conditions the total cross section is computed to be proportional to the square of the number of quadrupoles.

Detailed analyses show that some kinds of pairing give large cross sections and other kinds of pairing give small cross sections. For example, pairs of atoms do not have normal mode frequencies close to the lowest compressional mode. Analysis for a Raman process involving pairs of atoms gives the small 1960 cross section.

Plane slabs normal to the cylinder axis, approximately one atom in thickness, do have resonance at the normal mode frequency. Quadrupoles consisting of pairs of slabs give large cross sections. Another model which gives large cross sections consists of single atoms within a slab, coupled to the gravitational radiation. It is also important to consider effects of heat bath interactions on the total cross section.

Taking these issues into account gives

$$\sigma_{1986} = \frac{8\pi^3 GML^3 Q_\bullet}{c^2 \lambda L_a} \qquad (2)$$

As before, G is Newton's constant of gravitation, M is the reduced mass, L is the effective quadrupole length, λ is the gravitational wavelength, and c is the speed of light. $Q_\bullet$ is the quality factor of a single slab, much less than unity for present antennas. L_a is the length occupied by a single atom. The value of σ_{1986} is several orders greater than σ_{1960}.

The observed gravitational radiation antenna pulse heights are in good agreement with the 1986 published cross section given by (2). Each observed pulse requires total radiated energy less than 3×10^{-3} solar masses.

The 1984 and 1986 cross section analyses have been criticized. Careful study of the criticism indicates that it is not relevant. There are now other experimental data which support the S matrix treatment.

Similar analyses have been carried out for the near field, continuous experiments, done with laboratory sources of dynamic gravitational fields. These agree with observational data.

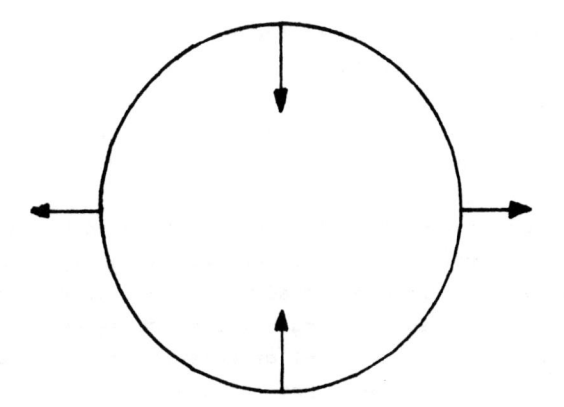

Figure 14

LUNAR GRAVITATIONAL RADIATION OBSERVATIONS

The earth and moon are elastic solids with well understood normal modes of oscillation. Einstein's General Relativity theory predicts that only modes with quadrupole symmetry will be excited by gravitational radiation. As already noted, new analyses of cross sections were published in 1984 and 1986.[3] These have given the cross section σ, for pulses, with

$$\sigma = \frac{8\pi^3 GMR^3 Qs}{c^2 \lambda La}$$

MR^2 is the quadrupole moment of the moon, R is the effective radius, Q_s is the quality factor of a spherical shell approximately one atom thick, La is the length occupied by a single atom, λ is the gravitational wavelength, c is the speed of light. Figure 14 shows quadrupole excitation of a sphere.

The lowest frequency quadrupole mode of the moon has a period about 20 minutes and a cross section approximately 70 square meters. The lowest frequency quadrupole mode of the earth has a period about 54 minutes and a much larger cross section. However the earth has a very large level of seismic activity, and large low frequency noise inversely proportional to frequency.

A vertical axis accelerometer was emplaced on the moon by the astronauts of Apollo 17. Correlations were observed between the lunar surface acceleration and a gravitational radiation antenna at the Argonne National Laboratory.

The Alsep Apollo seismology observations reported lunar surface displacements at relatively short periods. No displacements had Fourier components at periods longer than three seconds, and no normal mode (free oscillations) were observed.

These data imply that the lunar surface has a very small seismic background at very low frequencies. The reduced noise leads to very attractive possibilities for gravitational observations as follows:

A) Three axes accelerometers for observation of the lunar normal modes of oscillation. If these are being excited by gravitational radiation, the modes of quadrupole symmetry should have a higher energy than the modes of spherical symmetry. Earth observations imply that large seismic activity excites earth modes of all symmetries. If the lunar backgrounds are sufficiently low, these accelerometers might be cryogenically cooled instruments.

B) Large interferometer gravitational antennas. Two kinds are suggested. Firstly the low noise background and available high vacuum offer unusual opportunities for the kind of interferometers now being developed. These would employ systems acoustically isolated from the lunar surface and function as free mass antennas for gravitational radiation.

The second type of interferometer would attempt to measure very small lunar surface displacements, using the moon as an elastic solid gravitational radiation antenna.

Finally the low, low frequency noise backgrounds make the lunar surface an excellent site for bar gravitational radiation antennas at frequencies below 100 hertz, where isolation from earth seismic noise is very difficult. These would be at cryogenic temperatures. Sources such as the Crab nebula at 60 Hertz would be searched for.

1. Ferrari, Pizzella, Lee, Weber, Physical Review D, 25, 10, 2471, May 15, 1982. Proceedings of the Sir Arthur Eddington Centenary Symposium Volume 3, pages 1-78 World Scientific 1986.

2. Les Recontres de Physique de la Valleé d'Aoste, Supernova 1987A, One year Later, Results and Perspective in Particle Physics, Edited by M. Greco page 107, G. Pizzella, Analysis of the Data Recorded by the Mont Blanc Neutrino Detector and by the Maryland and Rome Gravitational Wave Detectors During SN 1987A. Il Nuovo Cumento, to be published.

3. J. Weber, Foundations of Physics December 1984, Proceedings of the Sir Arthur Eddington Centenary Symposium Volume 3, pages 60-65, World Scientific 1986.

4. Ph.D. Thesis R. L. Tobias, University of Maryland 1978, TR78-086, pp. 78-192.

The Moon as a Gravitational Wave Detector, Using Seismometers

Warren W. Johnson
Department of Physics and Astronomy
Louisiana State University
Baton Rouge, LA, 70803

Abstract

For most searches for gravitational waves, a lunar base does not have a compelling advantage over earth-bound or earth-orbit bases. The exception might be searches for *known, predictable* continuous wave sources with frequencies in the milliHertz region, which would use long period seismometers to detect the mechanical response of the whole moon both on and off its quadrupole resonance. An estimate of the two simplest sources of instrument noise indicates that an interesting sensitivity is tough but possible. Seismic noise is very uncertain, but might be prohibitive, in which case the network could turn to seismic studies.

The search for gravitational radiation has at least two aims : to find out if gravitational waves exist, and if so, to use them to do gravitational wave astronomy.

Searches can be classified by the expected temporal behavior of the wave-form. The most popular search has been for the gravitational collapse of a massive star, which is expected to produce a very short burst-type waveform, so that the emitted power is spread over a considerable range in frequency f about a peak in the kHz region. The 'opposite' type of search might be to look for continuous wave emission from sources of known frequency, such as rotating neutron stars (f > 50 Hz) or binary star systems with a very close orbit (f < 10 mHz).

The major potential advantage that earth orbit or a lunar base might provide is a instrument platform with tremendous reduced

© 1990 American Institute of Physics

nongravitational accelerations, i.e., vibrations. The isolation of the proof masses from the platform vibrations is a necessary and difficult task; 250 to 300 dB of isolation (sic) is a typical requirement for an earth-bound platform in the kiloHertz region.

The major reason that space or lunar bases are not absolutely required is that, in theory, a multistage mass+spring system ought to be able to provide the required isolation at frequencies higher than some cutoff f_c, whose value is perhaps a few hundred Hz. This is high enough to make kiloHz searches possible on earth. Then of course, the much greater access and much lower cost of earth-bound experiments makes them preferable. At lower frequencies, perhaps 10-100 Hz, isolation becomes extremely hard, and so the low vibration environment of a lunar or space base is probably our only hope for high sensitivity.

The unique advantage of a lunar base would be to use the whole moon itself as a gravity wave detector. The basic idea goes back to Weber and collaborators. One can use seismometers to monitor the normal modes of the earth, or the moon, to detect gravity waves at the mode frequency. The most thorough application of this idea is due to Boughn and coworkers, who are using the IDA global network to put upper bounds on the stochastic gravitational wave background at the earth's mode frequency (0.3 mHz). *

Stimulated by the topic of this workshop, I propose that the moon can also be a useful detector at frequencies well away from resonance. For example, if the frequency is higher than the moon's resonant frequency, and the seismometer proof masses are arranged to respond *in*-phase, then the approximate amplitude for the differential acceleration a between a proof mass and the surface would be

$$a \approx \omega^2 hR \approx \left(\frac{2\pi}{500s}\right)^2 (10^{-21})(2\times 10^6\,m) \approx 3\times 10^{-19}\,\frac{m}{s^2}$$

and differential displacement x would be

$$x \approx hR \approx (10^{-21})(2\times 10^6\,m) \approx 2\times 10^{-15}\,m$$

and where the gravity wave strain amplitude h and frequency are those predicted from optical observation of the close binary

AmCVn. The direction of the differential displacements between the lunar surface and the proof masses is illustrated by the following figure, assuming the wave is directed into the page :

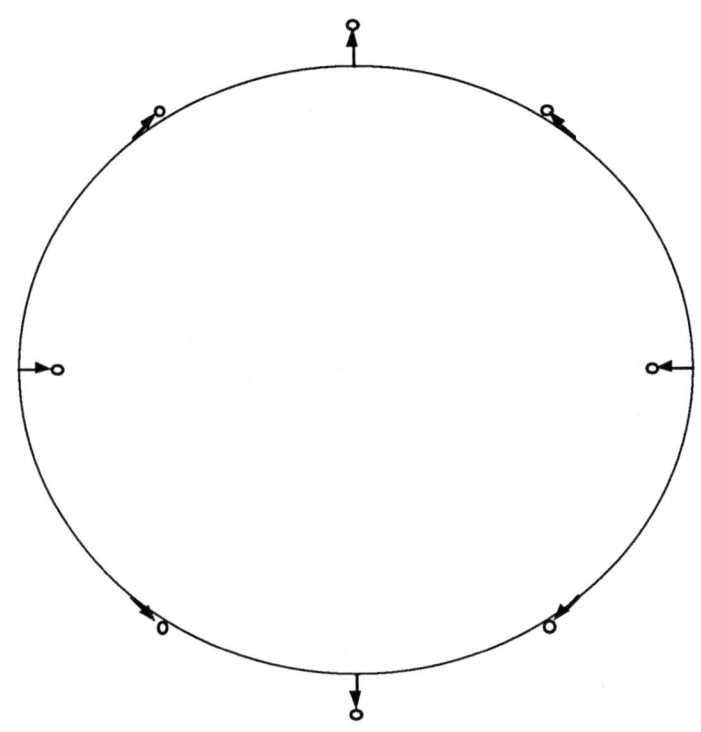

Nearby close binary stars are *the only certain sources of gravitational radiation with predictable strength, frequency, and location*, the only sources free from large astronomical uncertainties. They provide a definite test for the existence of gravitational radiation and the use of Einstein's General Theory of Relativity to calculate the intensity.

Can the instrumental noise be made small enough? The displacement noise of good inductive superconducting transducers with SQUID amplifiers can be less than 10^{-16} m/√Hz at somewhat higher frequencies; if 1/f noise is not too big, then these are easy displacements to measure in a cryogenic environment. The acceleration noise would probably be a bigger problem. A necessary condition is that the acceleration due to thermal excitation, within the observing bandwidth, be smaller than the signal. For mass m, temperature T, relaxation time τ^*, and observing time τ, it is

$$\delta a = \sqrt{\frac{4kT}{m\tau^*\tau}} = \left(10^{-19}\frac{m}{s^2}\right)\sqrt{\left(\frac{T}{2°K}\right)\left(\frac{100kg}{m}\right)\left(\frac{10^8 s}{\tau^*}\right)\left(\frac{10^6 s}{\tau}\right)}$$

which is a factor of 3 smaller that the signal, for the indicated assumptions about the parameters. The difficult parameter is the relaxation time $\tau^* = 10^8$ s, which is three years! This may seem unrealistic, but is a Q of 10^6, which is much smaller than Q's that have been achieved at cryogenic temperatures. All other sources of instrumental noise are unevaluated.

Seismic noise could well be a prohibitive noise source. On the earth, in this low frequency region (1-10 mHz), Boughn and coworkers[*] have estimated that the background acceleration noise density is 10^{-9} m/s$^2\sqrt{Hz}$, which is 7 orders of magnitude larger than the thermal acceleration density 10^{-16} m/s$^2\sqrt{Hz}$ assumed above. Generally, it is known the the moon is much more quiet than the earth, but unfortunately, there is no lunar data exists for frequencies below 0.25 Hz. Stebbins, in this workshop, has extrapolated moonquake frequency distributions to conclude that at higher frequencies (0.25-2 Hz) the acceleration density is 10^{-11} m/s$^2\sqrt{Hz}$. On the earth the density is known to fall very steeply between the high and low frequencies, but not the five orders of magnitude desired. A better estimate of the low frequency acceleration density would be very desirable.

A network of seismometers would be necessary. The gravity wave accelerations are coincident over the entire moon, and seismic noise would not. Therefore a network would provide a tool to distinguish the noise from signal.

I would guess that 'selenologists' would want a seismic network for higher frequencies; if so, then the incremental cost of gravity wave detection might not be very high. If seismic noise is too high for gravity wave detection, at least the instruments can be turned to collecting data to understand the geology behind it.

* S.P. Boughn, S. VanHook, C.M. O'Neill, and R. Rangarajan, "How Good is the Earth as a Gravitational Wave Detector?", pg. 358, Experimental Gravitational Physics, ed. Peter F. Michelson, World Scientific, Singapore, 1988.

A Lunar Gravitational Wave Antenna Using a Laser Interferometer

R. T. Stebbins and P. L. Bender[*]

Joint Institute for Laboratory Astrophysics, University of Colorado and National Institute of Standards and Technology, Boulder, Colorado, U.S.A.

A moon-based laser interferometer for detecting gravitational radiation could detect signals in the band 10^{-1} to 10^4 Hz. A preliminary evaluation of the noise budget for an optimistic antenna design is reported here and compared to that for other planned gravitational wave interferometers. Over most of the frequency range, the sensitivity is controlled by the thermal noise in the test mass suspensions. From roughly 3 to a few hundred Hertz, it is about the same as the sensitivity expected in terrestrial antennas of the same construction, which will have been operating for at least a decade. Below 0.3 Hz, a proposed space-based interferometer, designed for operation down to
10^{-5} Hz, would have better sensitivity. As pointed out by others, the most important role of a lunar antenna would be the improved angular resolution made possible by the long baseline to terrestrial antennas.

INTRODUCTION

The development of gravitational wave antennas based on laser interferometers has advanced to the point where prototypes are in operation, detailed engineering of full-scale ground-based instruments has begun, and an initial conceptual design of a space-based instrument has been made. In this

[*] Staff Member, Quantum Physics Division, National Institute of Standards and Technology.

© 1990 American Institute of Physics

paper, we consider the merits of a lunar interferometer. Except for the final section, we will discuss only the question of whether higher sensitivity can be expected for a lunar instrument in some frequency range. A lunar base makes feasible a lunar antenna derived from mature terrestrial designs, and the moon's seismic noise is lower than the earth's. We assume here that advanced ground-based detectors will have been in operation for a number of years before a lunar instrument comes online, and that the space-based instrument also may be providing results. Thus, we will consider the sensitivity of the lunar antenna in comparison with these other instruments.

The motivation for considering a laser interferometer at a lunar base can be seen against the backdrop of the major sensitivity considerations for the other instruments. The spaceborne version achieves its maximum sensitivity through its great length, 10^7 km, and the highly inertial "suspension" of the test masses within drag-free cavities. However, the long baseline begins to reduce the sensitivity above 0.01 Hz because multiple wavelengths of the gravitational radiation fit inside the interferometer. Ground-based interferometers will, in all likelihood, be in operation long before any other, and advanced terrestrial designs will probably approach their ultimate sensitivity before a lunar base is built. For a particular assumption about the achievable thermal noise level, that sensitivity is determined by the gravity gradient noise below 3 Hz and the thermal noise in the test mass suspensions just above 3 Hz. A lunar site potentially offers baselines comparable to the earth and a quieter seismic and gravity gradient environment.

The scientific role for a lunar laser gravitational wave antenna can be summarized as follows: The existence of gravitational radiation has already been demonstrated indirectly through the decay rate of the binary pulsar system. Direct detection is likely to be achieved by ground-based interferometric or bar antennas. Before a lunar instrument comes on line, these terrestrial antennas will probably have mapped the gravitational wave spectrum above 1 Hz, and the spaceborne interferometer may have begun mapping the spectrum from 10^{-5} to 1 Hz. A lunar instrument might have higher sensitivity at frequencies of roughly 0.3 to 3 Hz. In addition, as pointed out by others, it would certainly enhance the directional sensitivity of the network of antennas.

SCIENTIFIC GOALS

We can anticipate that a lunar interferometer will benefit from the extensive experience with terrestrial interferometers. Consequently, its design and performance and scientific goals will be much the same. The gravitational wave signals which might be detected by advanced terrestrial antennas has been extensively reviewed by Thorne.[1] Burst type radiation might be detected from coalescing of neutron star binaries, supernovae and stellar collapse to form black holes. Periodic radiation could be expected from pulsars (known and unknown), gravitational spindown of neutron stars and the Chandrasekhar-Friedman-Schutz instability. Finally, stochastic radiation from a primordial background and other less certain sources is possible. A lunar antenna would seek these same signal sources.

The end product of all gravitational wave observations is a picture of the gravitational universe which would supplement the electromagnetic one which we have now. An interferometer sited near a lunar base could provide high sensitivity down to below 1 Hz.

GRAVITATIONAL WAVE OBSERVATORIES

Gravitational wave interferometers detect the very faint strain induced by a passing wave through the interferometric comparison of the lengths of two arms oriented at right angles (Figure 1.). The ends of the arms are defined by test masses, inertially suspended, carrying the interferometer optics. A gravitational wave, impinging from above, causes the arms to expand and contract out of phase, thereby producing a shifting fringe pattern. The crux of the matter is the elimination or reduction of all other causes of apparent differential length change. Since the gravitational wave strain amplitude may be only as large as, say, a few times 10^{-21} at 1 kHz for the coalescence of a neutron star binary 10 Mpc away, the experimentalist needs to be concerned with spurious accelerations like seismic or thermal noise and measurement noise, such as shot noise.

The noise budget for terrestrial and lunar interferometers is dominated by the laser shot noise, ground motion (i.e. seismic noise), the thermal noise in the suspension of the test masses, the gravity gradient noise (the spurious acceleration of the test masses induced by gravitational coupling to moving masses nearby) and, in the lunar case, the acceleration imparted to the test masses by cosmic rays. It is important to note that these noise sources are associated with the endpoints of the interferometer, whereas the response is dependent on the length of the interferometer. Consequently, the signal-to-noise ratio increases linearly with the length of the interferometer.

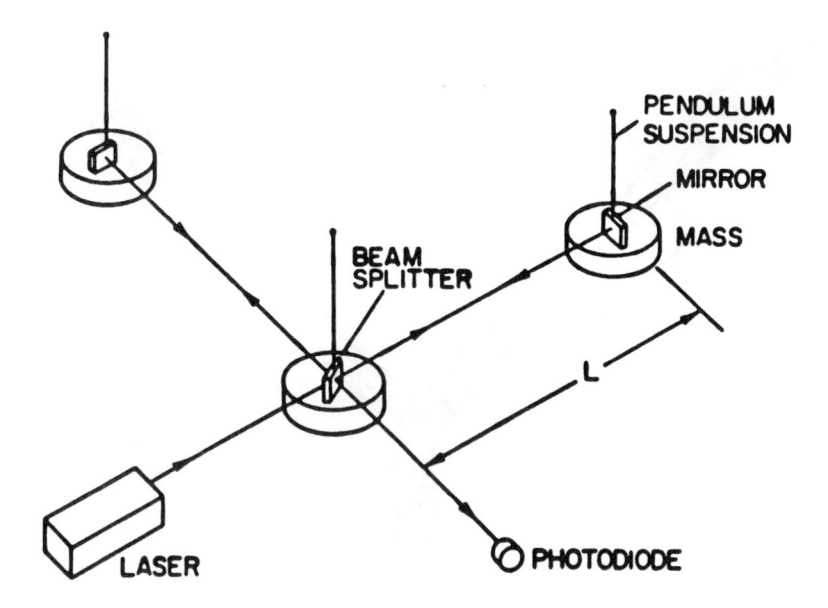

Figure 1. Conceptual arrangement of an interferometer for detecting gravitational waves.

The spaceborne interferometer[2,3] is different from the terrestrial and lunar versions in that the test masses bearing the optics are in drag-free chambers within three different spacecraft. The three spacecraft are put into solar orbits similar to the earth's so that they maintain a relatively constant separation of 10^7 km apart (see Figure 2). The instrument works in the same general manner as its terrestrial counterparts, that is, the passage of a gravitational wave is detected by measuring the changing separation of test masses with an interferometer. A space instrument has three essential advantages. It is free from the external gravity gradients and mechanical noise which limit ground-based interferometers to operating at higher frequencies.

Interferometric antennas, in which the signal scales as the length and the noise is dominated by fixed end effects, can be made much larger in space than on Earth to achieve a high signal-to-noise ratio at low frequencies. And, the test masses can be supported in a more nearly inertial manner, since they are in free-fall.

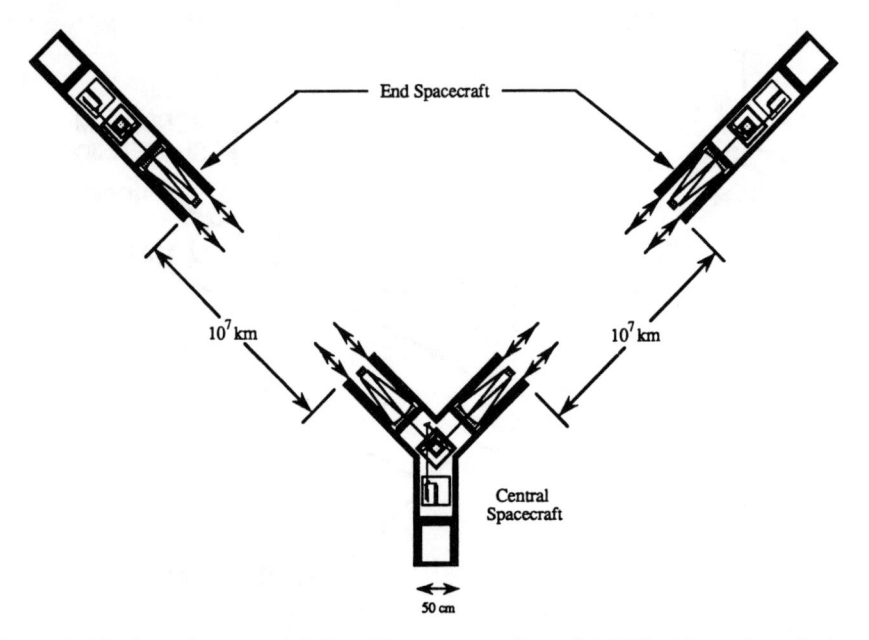

Figure 2. The Laser Gravitational-Wave Observatory in Space (LAGOS). A beam from the laser in the central spacecraft is split , reflected off of optics mounted in the test mass and sent through transmitting telescopes to the end spacecraft. In each end spacecraft, a laser transponder phase locked to the incoming beam returns a beam, reflected off of the local test mass to the central spacecraft for interference.

While the space environment may be free of the dominant terrestrial noise sources, it has its own unique disturbances. There are also forces which compromise the free-fall condition. And, it has special demands on engineering and cost. Any strain measuring instrument which hopes to achieve a sensitivity of $5 \times 10^{-21}/\sqrt{Hz}$ from 0.001 to 0.01 Hz and useful sensitivity down to 10^{-5} Hz will require careful design.

ANTENNA PARAMETERS

For the purposes of analyzing a noise budget, we have chosen a set of design parameters which describe a lunar interferometer. Our choices in some cases are similar to those of the Caltech/MIT Project for a Laser Interferometer Gravitational Wave Observatory (LIGO), which has looked at advanced designs for terrestrial instruments. In the case of the suspension Q's, we have chosen a value which is much more optimistic than has been demonstrated in prototype instruments; this is an attempt to anticipate technical advances which might occur before the design of a lunar instrument is finalized.

We assume an interferometer with 5 km long arms and thermal insulation over the end stations. In the absence of a lunar atmosphere, no buried vacuum pipe is needed, eliminating the costliest part of a terrestrial interferometer. However, the end stations need to be designed to improve the thermal stability. The optical layout would follow that of the LIGO design, two Fabry-Perots fed from a common beamsplitter. With light recycling, a laser power of 1 watt should be sufficient. The end mirrors would have a reflectivity suitable for containing up to 10^4 bounces in each arm, or a 20 msec storage time.

Although the critical parts of the instruments could be cooled, our analyses of the lunar, spaceborne and terrestrial antennas are all based on a temperature of 300° K. The lunar test masses have been chosen to be 1000 kg to aid in reducing the thermal noise in the support pendulums.

As with the advanced LIGO design, the test masses are made "inertial" by their suspension on a combination of active and passive isolation devices to isolate them from seismic noise. The moon's lower gravity and advances in vibration isolation techniques may produce a final passive suspension which has a resonant frequency as low as 0.1 Hz and a Q as high as 3×10^8. Also in keeping with the LIGO design, there will be a secondary interferometer servoing the support points of the final suspensions together so that both arms experience the same seismic input to a high degree. To the extent that the main suspensions' response can be made identical, the seismic input will be rejected as common-mode in both arms. This should provide an additional isolation factor of perhaps .001. Finally, we anticipate that a supplementary seismic isolation system, comprising two isolation stages each stabilized along six axes with sensor resonant frequencies of 0.01 Hz, would further reduce the seismic input at the support points of the final suspensions. A system of this kind, but with 0.1 Hz sensor resonant frequencies, has been suggested for use in separate 1-30 Hz antennas to extend the LIGO frequency coverage.[3]

The main antenna parameters are summarized in Table I.

Antenna Parameters
Length - 5 km
Thermally Insulated End Stations
LIGO Optics
Temperature - 300° K
Suspended Mass - 1000 kg
Main Suspension:
    Resonant Frequency - 0.1 Hz
    Q - 3 x 10^8
Support Servo Isolation Factor - .001
Inertial Reference System:
    Two Stages
    Six Axes
    Resonant Frequency - 0.01 Hz

Table I. Parameters of a lunar antenna used in the evaluation of the noise budget.

SURVEY OF NOISE SOURCES

Seismic Noise

The seismicity of the moon has been measured by geophysical experiments placed on the lunar surface by the four Apollo landers. The results are summarized by Lammlein[5] with important additional information in Goins et al.[6] The Apollo data is dominated by discrete seismic events lasting 30 minutes to 2 hours. There were about 1000 events per year. The rate of events varied at the four Apollo stations by nearly a factor of 5. The Apollo 12 site was the quietest, probably because of its shallow regolith (2-4 m).

The lunar seismic events can be divided into four classes: (1) Small moonquakes, occuring at great depths, are triggered by tidal forces. Ninety percent of all deep seismic events had characteristic traces which identified their foci of origin. Most of the remainder were too weak to be traced. (2) Large moonquakes occur near the surface, but happen only a few times per year. (3) Meteroids impact on the surface. These rare events can be one hundred times larger than the average deep moonquake and can last up to 4-5 hours. (4) Small high frequency signals are attributed to thermoelastic stresses in equipment left at the site, small meteroid impacts with 10 km and micromoonquakes close to the seismograph stations. The micromoonquakes began two days after sunrise and decreased rapidly after sunset. They are thought to originate with thermally induced cracking or movement of rocks or with soil motion on slopes.

From this knowledge of lunar seismicity, several features desireable in a lunar interferometer site can be deduced. The site should be a distance comparable with the baseline length from acoustic noise sources, like the lunar base. Care must be taken in the design of equipment left at the site to reduce thermoelastic noise. And a site with a shallow regolith may prove more attractive.

The deep moonquakes, the surface moonquakes and the meteroid impacts all have signatures which allow them to be identified by seismometers at the site and vetoed out of the gravitational wave data. Even the weaker moonquakes could be traced with more sensitive seismographs, so long as they are separated in time. But the smallest moonquake signals become so numerous that they constitute a seismic background at an rms displacement spectral density of 1.0×10^{-9} cm/$\sqrt{Hz}$ (see Figure 30 in reference #5). Their spectrum is essentially flat over the bandwidth of the Apollo instrument, .25 - 2.0 Hz. By contrast, the terrestrial background is about 300 times greater and rising rapidly toward the low end of the band. The lunar seismic background equates to a strain spectral density of
1×10^{-14} /$\sqrt{Hz}$.

We expect that the steps outlined in the previous section will be taken to improve the seismic isolation. The final suspension for each 10^3 kg test mass will have a resonant frequency of 0.1 Hz and a Q of 3×10^8. Additional interferometers and servo systems will lock the motions of the main suspension support points together, achieving a common mode rejection of seismic background of 10^3. And the support points will be locked to an inertial

reference system, comprising two successive platforms stabilized in each of six axes with sensors on suspensions with effective resonant frequencies of 0.01 Hz. The lunar seismic noise level is assumed to be
1 x 10^{-7} cm/$\sqrt{Hz}$, a level which is 100 times the minimum level discussed above. This level is expected to be exceeded because of meteoroid impacts or large shallow moonquakes only about once per lunation. The resulting effect of the seismic noise can be seen in Figure 3.

Other Noise Sources

Thermal noise in the main suspension is a significant limiting factor. Inelastic dissipation in the flexing wire of the pendulum supporting the test masses causes kT noise. The rms displacement amplitude spectral density of the thermal noise for an oscillator with velocity damping is given by[7]:

$$\sqrt{\frac{2kT\omega_0}{mQ\omega^4}}$$

where k is the Boltzmann constant, T is the temperature, ω_0 is the natural resonant frequency of the suspension, m is the mass, Q is the quality factor of the material, and ω is the working frequency. The associated rms strain spectral density is shown in Figure 3. As can be seen from the figure, thermal noise in the suspension limits the sensitivity of a lunar antenna above 0.1 Hz.

Cosmic rays will impart spurious accelerations to the test masses of a lunar interferometer. For a mass density of 10 gm/cm^2 surrounding the test masses, the cutoff for cosmic ray penetration is about 100 Mev. Higher energy cosmic ray protons will typically deposit roughly this amount of energy in the test masses. The resulting white noise acceleration due to the galactic cosmic rays will produce the noise level shown in Figure 3. This source of noise, not found in terrestrial antennas, does not pose a significant problem. Major solar flares can cause much higher disturbance levels, but would not cause loss of data more than a few days per year.

The gravitational coupling of local mass motions to the test masses is called gravity gradient noise. Seismic waves, trapped in the thin surface layer, could be one source of this irreducible noise. However, it appears to be insignificant (Figure 3). Note, again, that a gravitational wave interferometer would have to be remote enough from the lunar base so that the activities there would not disturb the antenna.

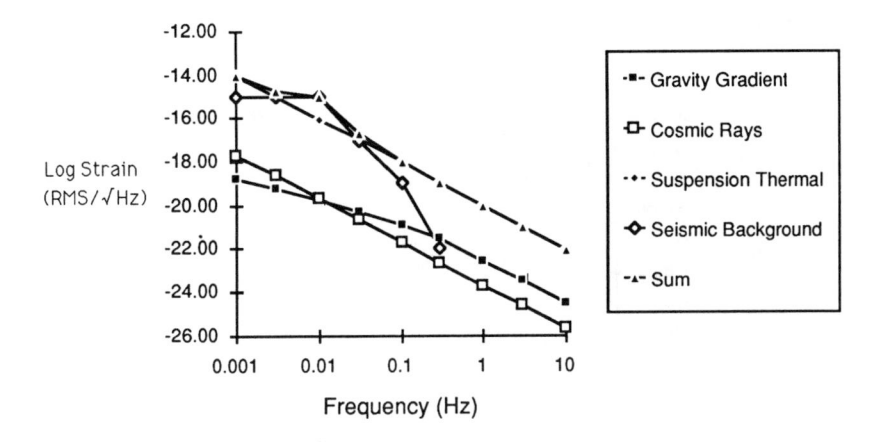

Figure 3. The noise budget for the lunar gravitational wave observatory.

COMPARISON OF SENSITIVITIES

Figure 4 compares the sensitivities of the terrestrial LIGO, the spaceborne LAGOS and the lunar base configuration analyzed in this paper. Above 3 Hz, both the terrestrial and lunar antennas have the same performance, governed by thermal noise in their suspensions. Although cooling would lower the strain sensitivity as $\sqrt{T}$, this strategy could be exercised on either antenna, giving neither an intrinsic advantage. Since both would use the same technology, a lunar antenna will only surpass LIGO when gravity gradient noise dominates.

Below 3 Hz, the lunar instrument would perform better than an earth-based one because of the quieter seismic environment. The lunar noise budget continues to be dominated by suspension thermal noise down to 0.1 Hz.

However, below 0.3 Hz, the spacecraft-based instrument has the advantage. Note that the present plans for LAGOS, on which Figure 4 is based, use a spacecraft separation of 10^7 km. The earlier version described in references #2 and 3 had a 10^6 km separation and assumed considerably higher accuracy in isolating the test masses from disturbances and in measuring their separation.

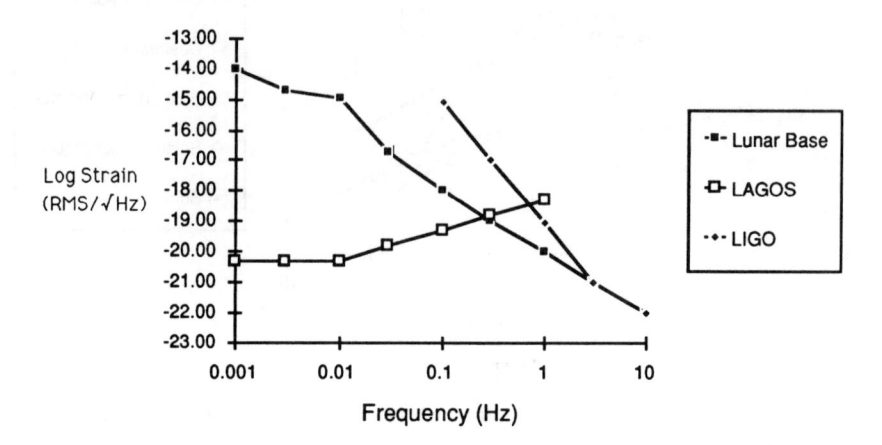

Figure 4. A comparison of sensitivities for an advanced terrestrial, spaceborne and lunar antennas.

Achieving the performance shown above for a lunar interferometer probably would require an advanced technology development program to develop high Q materials and to otherwise control thermal noise in inertial suspensions. A significant advance in the control of this noise source would affect the useful range of a lunar interferometer.

SUMMARY

A gravitational wave antenna at a lunar site may be capable of better sensitivity than other contemplated detectors between roughly 0.3 and 3 Hz. The sensitivity at 1 Hz may be an order of magnitude better. A lunar antenna would have the benefit of the design and operating experience of terrestrial antennas. It would need to be sited at some distance from the resources of a lunar base. Advances in the reduction of thermal noise in suspensions would improve the anticipated performance.

EPILOGUE

It was suggested by P. F. Michelson during the meeting and independently by J. W. Armstrong somewhat later that a lunar LIGO plus the planned detectors on the Earth would give valuable information on the location of pulsed gravitational wave sources. With this objective in mind, we have considered a simpler instrument which could be implemented soon after the establishment of a lunar base. The design bandwidth would be from 1 Hz to a few kHz, in order to match the range over which we believe useful measurements can be made on the Earth.

Reducing the mass of each test mass to about 30 kg would be an important simplification. The thermal noise in the final pendulum suspensions for the interferometer mirrors is proportional to $1/\sqrt{QM}$, where Q is the quality factor of the pendulum and M is the mass, so that using large test masses is usually considered desirable. However, larger masses require thicker wires or ribbons for their support, and the resulting Q is expected to decrease, possibly as $1/\sqrt{M}$. Thus a reduction in the mass from 1000 to 30 kg on the Earth would lead to a loss in sensitivity of only a factor 2.4. On the moon, the lower gravity means that somewhat thinner support wires can be used for a given mass, and in addition the resonance frequency for a given pendulum length will be lower. Thus the Earth-based LIGO thermal noise performance probably can be matched on the lunar surface with much smaller masses.

For vibration isolation, the system can have considerably less stringent requirements than discussed earlier if we limit the bandwidth to frequencies above 1 Hz for an initial antenna on the moon. As discussed earlier, a ground noise spectral amplitude of 1×10^{-7} cm/$\sqrt{Hz}$ at frequencies near 1 Hz probably would be exceeded only roughly once per lunation. There is very little information available about the noise level at frequencies higher than 10 Hz, where thermally induced noise in the surface layer or in the immediate surroundings of the apparatus may be the main problem.

The instrument design probably would be substantially different, depending on whether one settles initially for a lower cutoff frequency in the range of 30 to 100 Hz, as planned for the initial LIGO detectors on the Earth, or whether one includes frequencies down to 1 Hz. In the former case, passive isolation for the support points of the final pendulums may be sufficient.

However, active isolation at low frequencies of the kind we hope to develop in collaboration with the LIGO Group probably would be needed to get down to 1 Hz on the moon. The mass involved for such active isolation platforms does not appear to be a substantial limitation.

We believe that thermal problems may be one of the main issues in designing a lunar LIGO. The optical apparatus can be pictured as mounted in three to six units, with separate electronics packages. The optical units would be connected together as necessary and covered with insulating material, with care taken to avoid dust problems. The three sites then might have some kind of large sun shields or partial tents placed over them. It is very difficult to estimate the total mass and the power requirements for a lunar LIGO at present, and we will not attempt to do so. However, if the mass required to provide power at night turns out to be the largest item in the mass requirements, even operation during the lunar day only would be quite useful.

Acknowledgements

We would like to acknowledge helpful discussions with Jim Faller, and much useful information obtained from members of the LIGO group and other gravitational wave research groups.

References

1. K. S. Thorne, in *300 Years of Gravitation*, S. W. Hawking and W. Israel eds., Cambridge University Press, Cambridge, 364-399 (1989).
2. J. E. Faller, P. L. Bender, J. L. Hall, D. Hils, R. T. Stebbins and M. A. Vincent, *Adv. Space Res.* **9**, 107-111 (1989).
3. Peter L. Bender, in *Atomic Physics 11- Proc. Eleventh Int. Conf. on Atomic Physics*. S. Harroche, J. C. Gay, and G. Grynberg eds., World Scientific Pub. Co., Singapore, 567-588 (1989).
4. R. T. Stebbins, M. Ashby, M. H. Anderson, P. L. Bender and J. E. Faller, in *Proceedings of the Fifth Marcel Grossmann Conference on General Relativity*, (Cambridge University Press, Cambridge), in press (1989).
5. David R. Lammlein, *Physics of the Earth and Planetary Interiors*, **14**, 224-273 (1977).
6. Neal R. Goins, Anton M. Dainty and M. Nafi Toksöz, *J. Geophys. Res* **86**, B1, 378-388 (1981).

7. Paul Linsay, Peter Saulson and Ray Weiss, *A Study of a Long Baseline Gravitational Wave Antenna System*, unpublished report, Massachusetts Institute of Technology, p. V-21 (1983).

This view of the Moon after transearth injection shows the Mare Marginis (Border Sea) with craters Neper and Jansky at bottom, left of the photograph. Principal point is near 16^0 north, 94.5^0 east, adjacent to crater Al–Biruni. Diagonally, towards upper left-hand corner are Joliet, Vestine, and Fabry [43.0^0 north, 101.2^0 east:179 km] on top, left horizon. (AS15–95–12998).

Session On Cosmic Background Radiation Physics

The far-side highlands, viewed after transearth injection. The "wishbone" feature of King crater is clearly visible in lower, center of the photograph. (Apollo 16 metric frame 3005).

COSMIC BACKGROUND RADIATION PHYSICS

George F. Smoot

Space Sciences Laboratory and Lawrence Berkeley Laboratory
University of California, Berkeley CA 94720

INTRODUCTION

We are currently celebrating the 25th anniversary of the discovery of the Cosmic Background Radiation. As we look back over the history of the field, it is interesting to speculate on where the field will be in another 25 years and what we might do to see that it advances efficiently and effectively. It is well known that it is difficult to forecast the future, even the immediate future, much less 25 years; however, I am going to try basing my prediction by extrapolating trends and what we know to be physical principles.

TRENDS:
Immediately following the discovery of the CBR by Penzias and Wilson, it was recognized that the CBR was a unique and important tool for investigating the early and large scale universe. The understanding of its importance has increased over the years. The CBR traces the geometry of the universe and the evolution of matter and energy in the early universe. The CBR should contain evidence of the fossil progenitors of galaxies and evidence of the growth of structure in the universe. In addition, the CBR ties into particle physics. Its isotropy provides evidence for dark matter in that $\Delta T/T < 10^{-4}$ is too isotropic for galaxies and clusters of galaxies to condense under self-gravity of observed matter in the available time. Likewise the CMB isotropy over causally disconnected regions - the horizon problem - is difficult to explain. Both issues can be resolved by new physics at high energies - inflation and dark matter - and the spectrum and isotropy of the CMB and other cosmic backgrounds provide tests of these explanations.

25 years of effort and development in the field have lead to certain features that indicate trends. Features of the field that are easy to observe are:

1. CBR research uses complicated and specialized equipment (e.g. telescopes, gondolas, satellites) and much of this equipment is necessarily expensive to build and operate.

2. CBR research uses sophisticated techniques both in the experimental observations and in the data analysis. Also these techniques are becoming standardized as they have been around for a while and have been commented on by peers.

3. There is much theoretical development in prediction and interpretation. CBR related theory is a large and active area. We joke that one could develop an expert system for new theories and theoretical interpretation of new CBR observations - e.g.

When a new observations are published, then check against predictions from:
   a) cold dark matter theories
   b) cosmic strings (including superconducting) theories
   c) decaying particles (and vacuum) theories
   d) late phase transitions theories
   e) cosmic dust theories

using the proper formulas and write a paper with the blanks filled in with the proper new constraints and theory parameters.

4. There is now a growth in group size, collaborations, and longer term programs.

These trends signal a maturing field. No longer can a new person come in and very quickly join the forefront; instead, there is now a lore to learn or technology to develop.

MEASURING AND INTERPRETING THE CMB

For the Cosmic Microwave Background radiation everything else is foreground. That is to say, all currently observable objects in the universe lie between us as observers and interpreters and the source of the CMB. Not only are these objects in the foreground but they are also generally emitters at a level that is now

© 1990 American Institute of Physics

becoming significant (10^{-5} to 10^{-6}). This means that inevitably that CMB observation/interpretation is becoming a branch of astronomy requiring maps made at multi-wavelengths.

Figures 1, 2, and 3 show examples of the expected galactic emission, radio source count confusion, extra-galactic contamination and distortions and a map. These indicate to me that when observations reach the $\Delta T/T \leq 10^{-6}$ level, there will be structure observed in the sky maps. Thus the field will have to have models and data to allow the separation of foreground emission from the cosmic background.

Even when we do manage to find a way to separate the foreground from the cosmic background, we will (hopefully) have discovered regions of structure (anisotropies) and will be in the business of studying these structures - e.g. individual shapes and morphologies at that point we might get to produce a classification scheme like that for galaxies, the frequency of occurrences and distribution and so forth. This will hopefully be a rich and rewarding field and as you can tell much like extragalactic astronomy or x-ray astronomy.

STATUS of THEORY in 25 years
It is much too difficult to predict the theoretical developments of the next 25 years. We can predict that many of the theories that hold center stage today, will be no longer relevant are widely known - viz. the current question by new theorists "What is the Steady State Theory?" We can expect similar questions about theories such as T.O.E., inflation, cold dark matter, cosmic strings. As a close parallel to the Steady State Theory we have the new, updated, and revised Inflation theory which neatly explains the large scale isotropy, lack of monopoles, etc. but has little experimental evidence to solidify it place. A measurement of $\Omega=0.1$ or of not flatness could easily leave it in the library for elegant, beautiful, unused, and forgotten theories. In fact many of the theories and models that we will be testing in 25 years will be the products of people now in primary school (grades 1 through 6).

Thus researchers should design experiments as free as possible from model bias, as some of the models will no longer be around when the experiment is completed The field as a hold needs to emphasize experiments that measure the fundamental properties of the CBR and probe the available parameter space. Theories and models are good as examples but do not trust them implicitly but only as they are buttressed by observations.

Guiding Theoretical Ideas of CMB properties
Now having warned everyone not to trust to theory for experimental design, I review current thinking about the overall CMB properties as a rough guide. At the surface of last scattering there are two intrinsic scales affecting the CMB properties - the thickness of the surface of last scattering and the region of causal connectedness. The thickness of region of last scattering is estimated to encompass a range of redshift of about 80 at a redshift of about 1000. This corresponds to a length scale of about $4h^{-1}\Omega^{-1/2}$ Mpc and an angular extend of about $8\Omega^{1/2}$ arcmin. Structures of a size smaller than this will have their features washed out by the near isotropy of Thomson scattering as the CMB photons make their way out to us. This is not to say that one should not look for structure on a smaller scale but that it is likely that any structure on a smaller scale is due to interactions more recent than at a redshift of 1000. It is still worth checking that this picture is correct but as we shall see below making such fine scale maps will be a major undertaking.

The other angular scale of interest is the horizon scale at the surface of last scattering. In a universe without an inflationary epoch, the physical size of a causally connected region is about $3ctd \neq 200h^{-1}$ MPc or about 2 deg. In an inflationary universe the entire observable space can be causally connect.

CMB Measurement Goals:
If we want to make measurements a a sensitivity level of $\Delta T/T < 10^{-6}$, we then realize we must make sky maps with spectral sensitivity around the peak CMB signal. At minimum we will want 4 or more frequencies with very well defined bandwidths.
    Using a quantum limited receiver, a 1% bandwidth, and a nominal chopping or comparison observational scheme, it takes 3 hours of observation to reach the 10^{-6} level. Going to a larger bandwidth helps linearly for blip limited detectors like bolometers and with the square root of the bandwidth for coherent receivers. I estimate that a 10% bandwidth will be the ultimate limit if in 25 years we want to make multiwavelength full sky maps and that it will be extremely difficult to achieve such a wide bandwidth, quantum- limited device while still having good of beam rejection.
    The scales at the surface of last scattering gives us a natural angular resolution of 10 arcminutes on which to look for possible structures from the early universe. This implies a total to 1,500,00 fields of view (pixels) or about 500 years of observation. And we want to map selected areas with higher angular resolution. If we are going to measure at 4 or more frequencies the total observation time needed goes back to around 500 years even if we stretch to the larger bandwidths.
    Clearly, we are going to have to collaborate or build instruments with arrays of detectors and spectral response.

EXPERIMENT DESIGNS:

The constraints and factors discussed above point out that several features will be needed for an experimental program in place 25 years from now:
1. Instruments will feature receiver arrays. My predictions are that these arrays will take two primary forms:

    a) bolometer focal plane arrays with multichroic splitters A 20 by 25 element array working at quantum efficiency could achieve the necessary sensitivity to survey the full sky at the desired level and resolution in one year. In practice it is likely to take much longer but be within the scope of a graduate student's lifetime.

    b) coherent receiver arrays that synthesize the equivalent of dishes with focal plane arrays. There are serious fundamental problems with trying to put coherent receivers in a close-packed focal plane array with low sidelobes. However it is necessary to use an array of receivers in order to get the observation time to high multiples of the wall clock time. Thus we will need to create a focal plane array by using aperture synthesis and utilize them either directly observing the sky or as feeds for a large reflector.

2. Certain regions and objects will be studied at many wavelengths and even more importantly on various angular scales. Thus the specialized instruments described above will be charged with surveying the whole sky; while, others with higher resolution - OVRO and VLA types, will study certain structures in more detail. Likewise one can expect certain structures will be of enough interest to have other instruments with nominal properties like our survey instruments studying them to greater sensitivity or independently. Here we will be involved in the morphology of bumps and lumps and the independent study and verification of structures.

3. The experiments will be operated from carefully selected sites. We know know of certain problems that will not go away but have to be avoided by careful selection of site and experimental design:

    (a) There are places where the atmosphere is not a problem. The South Pole is a good possibility for the next generation or two of experiments however, in the long run getting above the atmosphere is the the only solution.

    (b) man-made RFI is a serious problem and a growing one at low frequencies and it is only a matter of time until the whole microwave and mm-wave region is contaminated. Near earth orbit is now already a problem with RF at the 1 to 10 volts/meter level common in the centimeter range. There are two sites that offer good promise for CMB astronomy for the foreseeable future, these are the Lagrange point L2 and the Lunar far side (either backside or over the polar horizon).

    (c) Objects in the beam and sidelobes. Both L2 and the lunar far side offer good opportunities to keep extraneous radiation out of the beam by careful antenna and shield designs. Very much care must be used but it is much less severe than for earth-based and near earth-based experiments. Removing the galaxy's emissions from the beam will have a long wait until we mount an out of the galaxy mission.

Clearly the sites suggested and the instruments outlined are major undertakings needing large resources. But then, just how many things are known to 10^{-6}. This is equivalent to mapping the surface of Venus and Mars to a precision of 20 feet. The people in the field will need to develop a consensus and coordination to be effective. For large projects to get started and sustain themselves, they need the support of the field. Everyone must believe them necessary, good concepts, well thought out, and well managed. Likewise the databases of experience, astronomical and galactic sources, interpretation - the infrastructure of the field is needed to support the overall undertaking.

PROPOSALS FOR NEW OPERATIONS AND PREPARATION FOR THE NEXT 25 YEARS

I. Hold workshops specifically on experimental designs
    (a) define atmosphere, galactic, etc. limits for various parameter regimes and experiments
    (b) develop designs and technologies such as nearly filled arrays
    (c) develop the design of next generations experiments

II Maintain and Improve Communication and Coordination System
    1. Exchange addresses including Electronic-mail
    2. Set up on-line system for CMB work
        (a) Names Addresses, and E-mail

(b) On-line list of articles - title, author, abstract, keys
(c) Bulletin board with proposed & work in progress
(d) Menus for adding names and articles and bulletin board
(e) Collect hard & electronic copies of articles/theses, etc.
   for CMB library with visitor/guest facilities
(f) Data summaries
   (1) CMB data - maps and tables of results
   (2) galactic models
   (3) radio source counts
   (4) software - i.e. experimental/theoretical programs
       with facility for comments and testing

As an example, I put forward the need for a study of good future sites- e.g. L2 and the moon far side base as well as developing the design of the hardware and techniques for doing the experiment. We could arrange a workshop on experiments designed for the short term at the South Pole and then longer term careful design of complementing experiments at lunar and L2-like sites.

SITE DISCUSSION

Both the Lagrange point L2 and the lunar far side (or lunar orbiting platform) offer good advantages for serious long term CBR astronomy. The moon provides shielding from the earth. Lagrange point L2 is located on the sun earth line opposite from the sun and at about 10^6 km from the earth. L2 is a good site because it is far enough from the earth-moo-sun system, which are thus all to one side so that the observing system can be designed to have little contamination from these sources and man-made interference. We can anticipate that most transmitters will be on the earth or in near earth orbit for the next half century and that care can be taken to avoid serious contamination of L2 or the lunar far side. If that cannot be achieved, we have to envision sending equipment to more remote locations or unusual orbits. Both L2 and the moon offer the advantage of relatively low cost of expendables for station keeping. Maintaining equipment at comparable of greater distances or low RFI environments will require substantially greater expenditure of fuels or more sophisticated control systems.

For full sky surveys, L2 offers the advantages of providing clear view access to the whole sky in that the earth-moon-sun system is in a small part of the sky and the orbit sweeps around to the opposite side of the sun in six months. L2's disadvantages compared to a lunar base is that it is difficult to:
   1) create large (high angular resolution) systems
   2) change observing strategies or up-grade the system, do repairs, refurbish
   3) to do more sophisticated systematic error checks after the data are being analyzed and the observers have learned from the data - e.g. change shields, frequencies, polarization

Thus instruments at L2 are likely to be used for full sky surveys at moderate angular resolution; while on the lunar sites one would set up observatories in concert with the other astronomical observatories. These facilities would be used to check selected regions of the full sky survey to see if it was correct, calibrated well, and free of serious systematic errors and to perform deeper studies or studies with higher angular resolution and detailed work on regions and structures of special interest. The lunar far side is likely to the be best location for long wavelength measurements of the spectrum of the CBR. while L2 may well turn out to be a superior location for shorter wavelength observations of the cosmic background radiation. We need careful studies and workshops to consider the lunar sites and the facilities likely to be available.

Session On Particle Astrophysics

The lunar far-side crater Van de Graaff [27.0° south, 172.0° east: 270 km long dimension] is the large, flat-floored double crater in this south-looking view. Adjoining it on the southeast is Birkeland [30.2° south, 173.9° east: 82 km diameter]. (AS17-150-22959).

High-oblique view looking north at the far-side crater Korolev [4.4° south, 157.4° west: 453 km diameter]. Doppler [12.8° south, 159.9° west: 100 km diameter] is in the foreground. (AS17-151-23112).

The Measurement of Elemental Abundances above 10^{15}eV at a Lunar Base

Simon P. Swordy

Enrico Fermi Institute, University of Chicago.

LASR, 933 E56th St, Chicago, Il 60637

At $\approx 10^{15}$ eV the slope of the energy spectrum of cosmic rays becomes significantly steeper than at lower energies. The measurement of relative elemental abundances at these energies is expected to provide a means to resolve the origin of this feature and greatly contribute to the understanding of the sources of cosmic rays. We describe a moon based detector for making well resolved elemental measurements at these energies using hadronic calorimetry. This detector is particularly well suited for a site on the lunar surface because there is no overlying layer of atmosphere and the large mass required can be provided by the lunar regolith.

Introduction

The Earth and its Moon are constantly bombarded with a wide range of particles, many of which have their origins outside the Solar System. The most energetic particles, the galactic cosmic rays, are the nuclei of atoms stripped bare of their electrons, ranging in atomic number over the entire periodic table of the elements. These may have travelled a distance of several hundred galactic disk diameters before arriving at Earth. The motion is diffusive because of the interaction of their electric charge with the galactic magnetic fields. These particles have energies which can far exceed those attainable with the largest accelerators on Earth. They are composed predominantly of Hydrogen nuclei, (protons), and Helium nuclei and have been observed up to energies of $\approx 10^{20}$eV. The precise origin of these particles remains unknown, but it seems likely that the bulk of them at lower energies are accelerated by the interactions with magnetic shock waves in the interstellar medium, generated by supernova explosions. Figure 1 shows the flux of cosmic ray particles as a function of energy, as measured at Earth. (The flux measurements have been multiplied by (Energy)$^{2.5}$ to compress the vertical scale). The measurements show that the total particle flux falls as a power law of

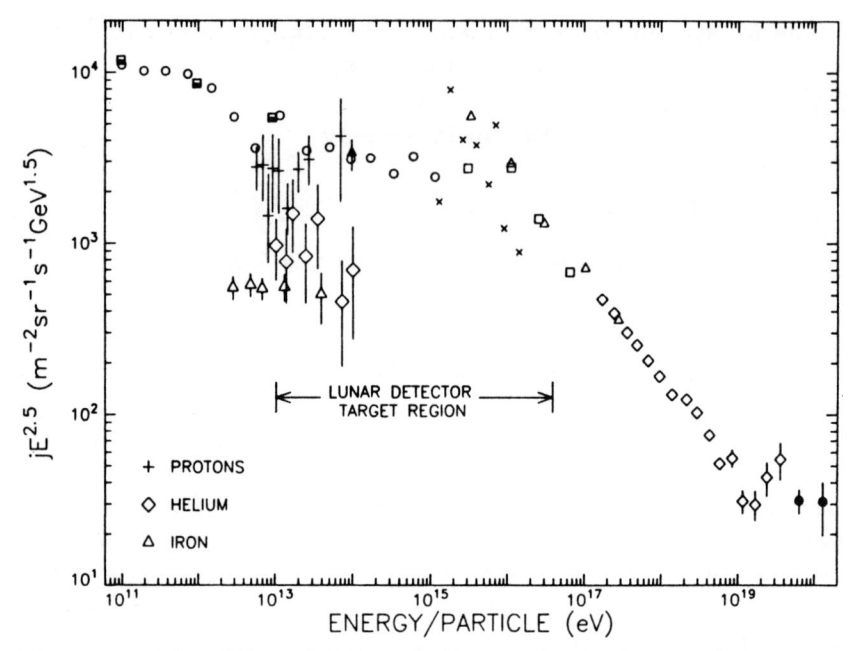

Figure 1. The differential flux of all particles in the cosmic rays as a function of energy, (note flux is multiplied by (Energy)$^{2.5}$), a compilation from Linsley 1983[3]. Also shown in larger symbols are measured elemental components H and He from Burnett et. al. 1983[4], and Fe from Grunsfeld et. al. 1988[5].

---

slope $\approx E^{-2.7}$ below 10^{15}eV, but at a steeper rate of $\approx E^{-3.2}$ at higher energy.

    The observation of this 'break' in the energy spectrum is tantilizing, since it is at an energy above which the mechanism of shock acceleration is predicted to become less effective. It is also possible that the 'break' is due to a sudden decrease in the efficiency of particle trapping in the galaxy, since the gyroradii of these particles in the galactic magnetic field is becoming significant compared to the thickness of the galactic disk. To further investigate these questions, we would like to know in detail how the abundances of various elements in the cosmic rays vary in this energy range. At present this is not known, with estimates ranging from pure protons to pure iron nuclei. The variations in elemental composition with energy can differentiate between rigidity dependent processes, such as trapping, and the effects of the momentum dependent spectra expected of shock acceleration. They may also reveal some new, unexpected phenomena, such as the dominance of an iron-rich source at high energies. The main obstacles to performing these measurements are the low fluxes of these nuclei and the presence of the Earth's atmosphere. Because of interactions in the ≈1000 g/cm^2 of overlying atmosphere the elemental character of a primary cosmic ray is lost and only a shower of secondary particles survives at ground level. This limits the measurements on the Earth's surface to a simple estimate of the total energy of an incident cosmic ray by observation of the secondary particles. Direct observation of the primary nuclei is limited to instruments exposed in space or on high altitude balloons near the top of the atmosphere. Since these remote detector systems cannot easily be made large, the direct elemental composition has only been determined up to energies of ≈10^{14}eV, the higher energy data coming from air shower experiments. Examples of the highest energy direct measurements and the contribution to the overall flux for protons, helium and iron nuclei are shown in Figure 1. The target region indicated for the measurements described here would explore new territory where the detailed elemental abundances are essentially unknown.

Why a Lunar Site ?

The only practical method to measure the energy of the particles at such large energies over a large area is to use a sampled hadronic calorimeter, such as has been proposed for the US Space Station[1]. In such a detector a hadronic shower develops in much the same way as in the Earth's atmosphere. The calorimeter measurement is preceeded by an ionization loss detector which measures the incoming particle charge before the first interaction in the calorimeter. For this type of experiment a lunar site is ideal. Firstly, there is no overlying atmosphere so direct observations of cosmic rays is possible with large, surface mounted detectors. Secondly the lunar regolith can be used as the bulk of the detector, eliminating the need to lift large masses of material into Earth orbit. The following sections discuss the detailed design and expected performance of such a detector.

A Detector for the Lunar Surface

The design of a suitable detector for these particles is shown in Figure 2. All signals are provided by thin, (≈1cm thick), plastic scintillation counter layers viewed by photomultiplier tubes. The uppermost layers of scintillator are used to determine the charge of the incident cosmic ray before interactions occur. These provide an output of light per unit pathlength which increases as Z^2 for relativistic nuclei of low atomic

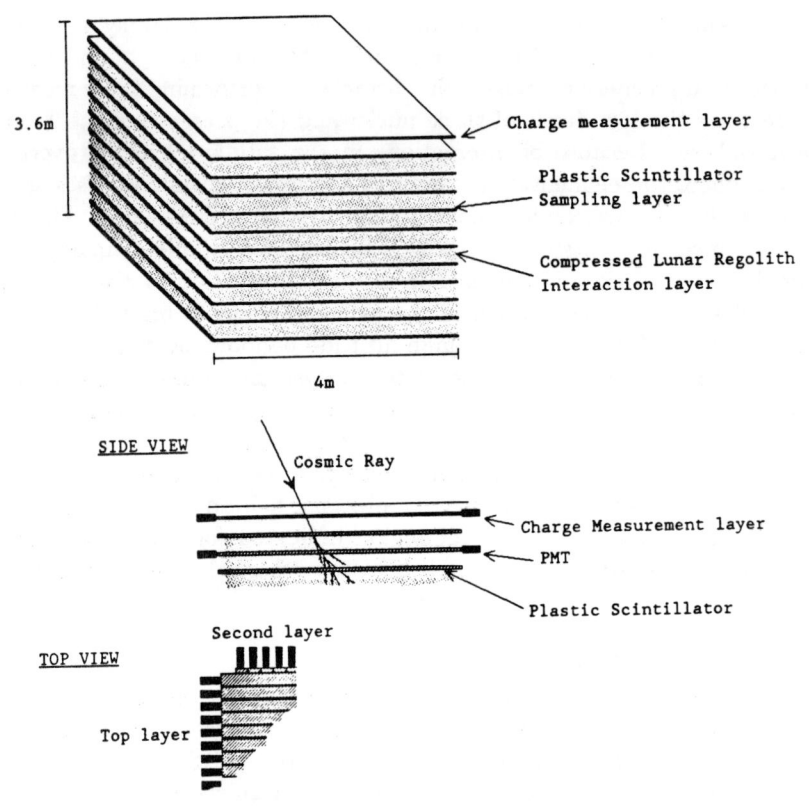

Figure 2. Schematic of a high energy cosmic ray detector for the lunar surface.

---

number, Z. For Z>10 there is some saturation of the yield, but the resolution is adequate even for iron nuclei, (Z=26). The charge resolution of plastic scintillators has been demonstrated[2] to be better than 0.4 charge units for iron nuclei of energy $\approx 3 \times 10^{13}$eV. The hadronic calorimeter is formed from layers of compressed lunar material separated by scintillator strips which provide the hadron shower samples. The hadronic interaction length, Λ, in the lunar material is ≈ 80 g/cm^2. Since the regolith is readily compressed to a density of 2.5 g/cm^3, each of the 10 layers in the detector is approximately an interaction length. The expected depth of hadronic shower maximum at these energies can be interpolated from accelerator measurements at lower energies and estimates from the Fly's Eye air shower experiment at $\approx 10^{18}$eV. At 10^{15}eV the shower maximum should occur at a depth of ≈5Λ. A significant part of the shower will be contained within the detector depth of ≈10Λ for a vertically incident particle. The expected energy resolution at this energy is $\Delta E/E \approx 15\%$. The profile of shower development measured with the detector will be extremely valuable for interpretation of air shower measurements made on Earth. To normalize the signals from the detector elements some determination of the cosmic ray trajectory is required. This is provided by segmentation of the scintillator layers into strips of width 20cm, as shown in the top view of the detector. Each strip is viewed by two photomultipliers. The orientation of these strips changes in each successive layer to provide views of the trajectory on orthogonal planes, these can be used to find the pathlength of the event in each detector element. The accuracy of the trajectory measurement with this device is ≈3° which will allow an investigation of the anisotropy in arrival direction of the particles in addition to the composition measurements. The triggering scheme for the detector can be extremely simple since all that is required is an event which crosses the charge determining scintillators and deposits a certain minimum energy in the calorimeter.

The aperture of the detector is estimated to be ≈24 m²str. If positioned near a lunar base it could easily collect data for 3 years or more since it only requires electrical power for operation. Table 1 lists the expected number of events for protons, helium, and iron nuclei predicted in the detector over this time period. These calculations assume a simple extrapolation of fluxes measured at lower energies.

It is estimated that the trigger rate, for a threshold energy in the calorimeter of 10^{13}eV, will be ≈10 per minute. The event information can be readily transferred to Earth for processing through a modest telemetry link which operates through the lunar base systems.

The execution of this experiment at a lunar base would be relatively straightforward. The scintillation counters would be preassembled and transferred to the lunar base complete with electronics. The counters would then be stacked to form the detector at a lunar site by interleaving the scintillator layers with trays of compressed lunar material. Physical and electronic characteristics of the detector are given in Table 2.

Conclusions

The idea of measuring the elemental abundance of cosmic rays above 10^{15}eV with a hadronic calorimeter in space is not new. However, the lunar base provides a unique opportunity for such an experiment because the extremely massive component of the detector is already available in the lunar regolith. It seems that a relatively simple detector can be designed and built with presently available technology. This instrument would finally resolve the long standing scientific question of the composition of particles at and above the 'break' in the cosmic ray spectrum at 10^{15}eV.

References
1. J. F. Ormes and R. E. Streitmatter, Proceedings of Workshop on cosmic ray and high energy gamma ray experiments for the Space Station era, Louisiana State Univ., 1985, pg 340.
2. J. L'Heureux, J. M. Grunsfeld, P. Meyer, D. Müller, and S. P. Swordy, Proc. 20th Int. Cosmic Ray Conf., vol 2, 1987, pg 366.
3. J. Linsley, Proc. 18th Int. Cosmic Ray Conf., vol 12, 1983, pg 135.
4. T. H. Burnett et. al. (JACEE collab.), Phys. Rev. Lett., vol 51, 1983, 11, pg 1010.
5. J. M. Grunsfeld, J L'Heureux, P. Meyer, D. Müller and S. P. Swordy, Ap. J. Lett., vol 327, 1988, L31

Table 1. Expected Number of Events in 3 Years

PARTICLE	$> 10^{13}$	$> 10^{15}$	$> 10^{16}$eV
Protons	3x10^6	10^3	25
Helium	1.5x10^6	600	12
Iron	8x10^5	300	6

Table 2. Detector Characteristics

MASSES	(kg)
Scintillator	3,500
PMTs/Electronics	1,500
Mechanical Support	1,000
Lunar Regoltih	160,000
POWER	1,000 W
TELEMETRY	1 kbit/s

Possibilities for Fundamental Particle/Astrophysics Experiments At a Lunar Base

Serge Rudaz

School of Physics and Astronomy, University of Minnesota,
Minneapolis, MN 55455

Abstract

It is suggested that some experiments such as the search for proton decay and for cosmic ray signatures of dark matter particles should be seriously considered in planning the scientific program of a Lunar Base Laboratory. Such experiments could elucidate many of the most fundamental unanswered questions of cosmology and particle physics, but face eventual basic limitations on Earth and in Earth orbit, respectively. Some of these limitations may be circumvented at a Lunar Base due to the lack of atmosphere and stable environment of the Moon.

1. Introduction

The past decade has seen an expansion in the traditional venues for experimental research in elementary particle physics, from high energy accelerators exclusively, to include large underground detectors dedicated to the search for rare events such as proton decay that would be indicative of a possible ultimate unification of all the forces (strong and weak nuclear, as well as electromagnetic and possibly even gravitational) acting on the fundamental contituents of matter. The quest for unification may be the single-most important goal of elementary particle physics, and the search for proton decay, its most fundamental experiment.

Another important development has been the realization that progress in elementary particle physics may have deep ramifications for cosmology and astrophysics, and vice versa (for a recent review, see Ref. 1). A very important case in point, of fundamental significance, is the cosmic dark matter puzzle: Cosmological studies indicate that as far as gravitational effects are concerned, the dominant component of the Universe on scales larger than say 10 kpc consists of non-luminous, "dark" matter, of unknown composition. This is altogether remarkable, as it amounts to an admission of our ignorance of the nature of what may comprise 90% or more of the mass present in the Universe! Particle physics may provide the most appealing and plausible explanation of the nature of dark matter as electrically neutral, stable particles of an unknown type, as yet unseen in terrestrial laboratories, that survived as relics of an earlier very hot and dense phase of the evolution of the Universe.

In what follows, we reiterate arguments that a Lunar Base, at a mature stage of its settlement, will provide an ideal venue for the staging of experiments that will further the search for proton decay and the elucidation of the nature of the cosmic dark matter, in all likelihood well beyond the possibilities of experiments set up on Earth or in Earth orbit, as a result of the most prominent characteristic of the lunar enviornment, its lack of atmosphere.

2. The Search for Proton Decay

As we enter the Nineties, elementary particle physicists face a paradoxical situation. On the one hand, all available experimental results are remarkably well accomodated in the framework of what has become known as the Standard Model of quarks and leptons (See Ref. 2 for a review directed at non-specialists, and Ref. 3 for a treatment at a slightly higher level and containing many further references). On the other hand, a number of important and fundamental questions remain unanswered: What accounts for the particular properties of the observed quark and leptons, for example their masses, and indeed, their very number? Do other, as yet unseen forces act upon them? Are the various known forces merely different manifestations of a single unified interaction acting in such a way that, at energies far removed from those attainable even at the projected Superconducting Super Collider, quarks and leptons themselves are revealed as different aspects of a single, fundamental type of matter particle?

This last possiblity is known as the Grand Unification hypothesis and almost invariably leads to the ultimate instability of all matter through its prediction of the phenomenon of proton decay. The experimental aspects of the search for proton decay have been ably reviewed by Perkins.[4] The reader can also trace the historical development of Grand Unification as well as find more recent updates of the search for proton decay by consulting the Proceedings of the annual Workshops on Grand Unification (WOGU's) held in the Eighties[5-14].

Briefly, early useful limits were obtained by nuclear and radiochemical methods based on the idea that if a proton decays (i.e., is removed from an otherwise apparently stable nucleus), its host nucleus is left in an excited state, leading for example to spontaneous fission or other decay to be detected. Proton lifetime limits of $\tau_p > 2 \times 10^{27}$ years have been obtained in this way[4] and do not depend on the particular nature of the decay products. The modern era of proton decay experiments involves direct searches for the decay products using very large dedicated detectors: Their size is dictated by the apparent very long lifetime of the proton. For example, a hundred-ton detector has roughly 10^{32} protons and neutrons. Typically, then, a a proton lifetime of 10^{31} years would lead to a few decays inside the detector per year.

Dedicated proton decay detectors (which in fact can also be used for the detection of astrophysical neutrinos) are of two general types. One is the large water Cerenkov detectors with the light produced by the decay products gathered by sensitive photomultiplier tubes, and the other is sampling calorimeters made of some dense material interspersed with devices sensitive to the ionization caused by the proton decay products. A crucial, common characteristic of both types of detectors (in addition to their enormous size) is their location at great depths. This is because of the need to suppress possible cosmic-ray induced background processes which would activate the detectors and fake a proton decay signature. The central problem[4] involves neutrinos and muons produced in cascades in the Earth's atmosphere that result from interactions of primary cosmic ray protons.

Suppose, for example, that one searches for the decay $p \rightarrow e^+ \pi^0$ of a proton into a positron and a neutral π-meson. This last particle immediately decays into two gamma rays, $\pi^0 \rightarrow \gamma\gamma$, and so effectively one searches for $p \rightarrow e^+ \gamma\gamma$, an ideal final state for water Cerenkov detectors. The present experimental limit on the partial lifetime into the particular mode is τ_p/branching fraction > a few $\times 10^{32}$ years. One searches for the equivalent energy deposition of the mass of the proton $E = m_p c^2 \simeq 10^9$ eV with a resolution of about 10% ($\Delta E/E$), consistent with momentum conservation in the observed (or inferred) tracks.

Now it could happen that atmospheric neutrinos with 10^9 eV = 1 GeV energy induce the process, say $\bar{\nu}_e p \rightarrow e^+ n\pi^0$ in the detector, where the final state neutron (n) goes unseen with very little energy. This would fake a proton decay signal, and constitute an irreducible background, even after additional constraints are imposed to reject those events that for example would not be consistent with momentum convervation if they were ascribed to proton decay. Let's estimate the lifetime limit τ_{max} obtainable in deep Earth detectors due to atmospheric neutrinos. The neutrino flux around E_ν ~ 1 GeV is roughly (there is some variation with neutrino type)

$$\phi_\nu^{Earth}(E) \sim 2 \times 10^{-3} \text{ cm}^{-2} \text{ sec}^{-1} \text{ sr}^{-1} (100 \text{ MeV})^{-1}$$

while the interaction cross-section for the relevant process is known to be (again, roughly)

$$\sigma_{\nu p} \simeq \tfrac{1}{2} \times 10^{-38} \text{ cm}^2 \quad .$$

Thus the number of fake proton decay events per second is estimated to be ($\Delta E \simeq 100$ MeV)

\# fake Proton Decay Events/sec

$$\simeq 4\pi \times 100 \text{ MeV} \times \sigma_{\nu p} \times \phi_\nu^{Earth}(E \simeq 1 \text{GeV})/\text{Rejection Factor}$$
$$\simeq [\ \mathcal{O}(10^{-41})/\text{Rejection Factor}]\ \text{sec}^{-1} \quad .$$

With a rejection factor of about 10, this leads to an irreducible proton partial lifetime limit into a particular decay mode of about[16]

$$\tau_{max} \simeq \mathcal{O}(10^{42}) \text{ sec} \simeq \mathcal{O}(10^{34}) \text{ years},$$

only about two orders of magnitude above present limits.

To understand the relevance of this number, consider the simplest theoretical estimate of the proton decay lifetime in Grand Unified Theories. This typically involves two parameters, namely the strength of the unified force between quarks and leptons, written as a dimensionless number $g^2/\hbar c$ analogous to the fine structure constant $\alpha = e^2/\hbar c = 1/137$ and the energy $M_x c^2$ at which quarks and leptons can no longer be distinguished in collision processes, written in terms of an equivalent Grand Unification mass scale M_x. By basic quantum mechanics, one finds

$$\tau_p \simeq (\hbar/m_p c^2)(g^2/\hbar c)^{-2}(M_x c^2/m_p c^2)^4 \quad .$$

With the scale set by the nuclear "unit of time" $\hbar/m_p c^2 \simeq \mathcal{O}(10^{-24})$ sec, we see that $M_x \gg m_p$ necessarily. In fact, early theoretical predictions took $g \simeq e$ and $M_x \simeq 10^{14-15} m_p$, leading to

$$\tau_p \simeq 10^{31 \pm 1} \text{ years} \quad .$$

This type of Grand Unified Theory is all but ruled out by presently available data. However, the search for proton decay must continue, with Earth-based detectors subject to an irreducible limit because of atmospheric neutrino backgrounds.

This is where the Moon's lack of atmosphere is crucial, a point noted by a number of people[15-18]. The shielding requirements there are different than on Earth, leading to a much larger limit on the ultimate sensitivity of proton decay detectors.

To obtain shielding from primary cosmic rays equivalent to the thickness of the Earth's atmosphere ($\simeq 10^3 \text{g/cm}^2$) requires a minimum of several meters of lunar regolith. Given this, what is the resulting neutrino flux $\phi_\nu^{Moon}(E \simeq 1 \text{GeV})$? Neutrinos and muons are produced in the decays of mesons resulting from primary interactions in the regolith shielding. A competing process is the prior nuclear absorption of these mesons, before they decay. All other things being equal, an estimate of Cherry and Lande[19] amounts to comparing the rate of absorption of mesons, that is

$$\phi_\nu^{Moon}(1 \text{GeV})/\phi_\nu^{Earth}(1 \text{GeV}) \simeq n^{\text{Earth atmosphere}}/n^{\text{Lunar regolith}} \simeq \mathcal{O}(10^{-3}) \quad ,$$

where n is a number density. This would suggest a lunar ultimate limit on proton decay searches[16,19] of

$$\tau_{max}(\text{Moon}) \simeq 10^3 \, \tau_{max}(\text{Earth}) \quad .$$

These crude and elementary considerations must be supplemented by further investigations. Are there other (astrophysical) sources of 1 GeV neutrinos? How much regolith shielding is actually required, especially as regards muon shielding? What is the natural radioactivity of Moon rock? Also, to take full advantage of τ_{max}(Moon) still requires say 10 kilotons of highly instrumented detector mass, with a sensor-to-detector mass ratio of hopefully 10^{-3} to 10^{-4}, as compared to present-day detectors with values of 10^{-1} to 10^{-2}. Novel detector concepts are clearly required.

Nevertheless, it is likely that the search for proton decay will remain an important goal of experiments far into the foreseable future, and should be part of the long-range basic science program of a permanent lunar laboratory.

Next, we turn to another type of experiment that could shed some light on fundamental questions of cosmology and particle physics, and for which the lunar environment is ideally suited, namely the search for cosmic ray signatures of dark matter.

3. High–Energy Gamma Ray Astronomy and the Search for Dark Matter in the Galactic Halo

Does non-luminous "dark matter," constitute 90% or more of the mass of the Universe? Studies of modern cosmology indicate that it is so[20], and that in fact as larger and larger distance scales are probed, there appears to be more and more non-luminous matter. On a galactic scale, a clear indication of the presence of dark matter comes from studies of rotation curves of spiral galaxies at distances beyond visible, luminous matter. These are measurements of the velocity of stars or gas an a function of distance to the Galactic center. Without dark matter, one expects velocities v to go as $v^2 \sim 1/R$ at distances R far beyond the visible Galactic radius. In fact, one repeatedly observes flat rotation curves, indicating the presence of unseen matter with density $\rho_{DM} \sim 1/R^2$, forming a "halo" of radius a and contributing a mass $M \sim R$ beyond the luminous regions.[20]

The precise nature of this dark matter is not known, and it is clearly a subject of fundamental importance. A number of arguments have been presented against the notion that it could be baryonic (See for example Ref. 21).

An exciting possibility is that the dark matter could consist of a new type of heavy, stable and electrically neutral particle (call it χ), that would remain as a relic of the Big Bang, in much the same way as the microwave background. One can imagine looking for cosmic ray signatures of the pair annihilation of χ particles in the halo of the Galaxy as a means of exploring this possibility. In this respect, the Earth's atmosphere is a considerable hindrance as it shields interesting and rare cosmic ray events. Detectors must be lofted in balloons or placed in orbit, where major size and weight limitations translate into relatively low expected collection rates for interesting cosmic ray events of exotic origin. A lunar base would afford a stable environment for larger scale cosmic ray detectors, unhindered by atmospheric absorption. We now outline a possibly promising method for the identification of dark matter as relic particles.[22,23]

The idea is to search for the nearly monoenergetic gamma rays that arise from the process $\chi\chi \rightarrow \gamma\gamma$ in the Galactic halo. The gamma rays have energy $E_\gamma \simeq Mc^2$ where M is the χ-particle mass, subject only to Doppler broadening by the typical virial velocities $v/c \simeq 10^{-3}$ of the annihilating particles. High resolution in energy is crucial in the detector design. The expected gamma ray line flux in the direction defined by Galactic latitude (θ, ϕ) is obtained from

$$F_\gamma(\theta,\phi) = \frac{1}{2\pi} (\sigma v_{rel})_{\gamma\gamma} \int_0^\infty dr(\theta,\phi) \, [\rho_\chi(r)/M]^2$$

where the line-of-sight integral is over the square of the number density of χ particles in the dark matter halo, and $(\sigma v_{rel})_{\gamma\gamma}$ is the cross-section times relative state velocity for the process $\chi\chi \rightarrow \gamma\gamma$, which must be estimated given a specific particle physics model for χ and its interactions.

A common parameterization of the halo dark matter density (assumed here to consist entirely of χ particles) is

$$\rho_{DM}(R) = \rho_0 \frac{R_0^2 + a^2}{R^2 + a^2} ,$$

where R is the galactocentric distance, with R_0 that of the solar system, ρ_0 is the local density and a the halo core radius. For $R \geq a$, this leads to flat rotation curves, while for $R \leq a$ it does not exceed the visible matter density, as required by observation. Typical values are $\rho_0 \simeq 0.4 \, m_p/cm^3$ and $R_0 \sim a \sim 8$ kpc with m_p the proton mass. With this, the line flux can be written as

$$F_\gamma(\theta,\phi) = (2\pi)^{-1} (\sigma v_{rel})_{\gamma\gamma} (\rho_0/M)^2 \, a \, L(\theta,\phi) \quad,$$

with $L(\theta,\phi)$ a dimensionless function of order unity for (θ,ϕ) in the direction of the Galactic pole. To get a very crude idea of the possible magnitude of the annihilation rate into two photons, we first need an estimate of the total annihilation rate of a $\chi\chi$ pair which can be obtained from a formula relating the present abundance of χ particles relative to the critical density representing the boundary between an eternally expanding (open) Universe and the alternate possibility of a closed Universe for which the expansion will eventually stop, leading to an era of contraction, and ultimately, a "Big Crunch." The result is roughly

$$(\sigma v_{rel})_{\chi\chi} \simeq \frac{10^{-26} \text{ cm}^3 \text{ sec}^{-1}}{\Omega_\chi h_o^2} \quad,$$

where $\Omega_\chi = \rho_\chi/\rho_c$ and the critical density ρ_c is expressed in terms of the Hubble parameter $H = 100 h_o$ km sec^{-1} Mpc^{-1} (with $\frac{1}{2} \leq h_o \leq 1$) and Newton's gravitational constant G as $\rho_c = 8\pi G/3H^2$. The annihilation cross-section is typically for processes $\chi\chi \to f\bar{f}$ where $f\bar{f}$ is an electrically charged particle–antiparticle pair. A resolution of the dark matter puzzle requires $\Omega_\chi h_o^2$ to be close to (in order of magnitude) unity. In general, $(\sigma v_{rel})_{\chi\chi}$ will be a function of the χ-particle mass, as well as of the interaction strength, and this formula correlates possible particle physics parameters with cosmological observations. An important contribution to the crosss-section for the reaction $\chi\chi \to \gamma\gamma$ is likely to be through the creation of an $f\bar{f}$ pair followed by its annihilation into two photons. As photons couple to charged particles with strength eQ_f where e is the electron charge and Q_f is the charge of the particle f in units of e, and two photons are involved, simple quantum mechanical rules lead to the estimate

$$(\sigma v_{rel})_{\chi\chi \to \gamma\gamma} \simeq (\sigma v_{rel})_{\chi\chi} (e^2/\hbar c)^2 Q_f^4 \quad,$$

which should be summed over the possible types of particles f. With $e^2/\hbar c \simeq 1/137$ and the rest of the factors of order unity, this suggests

$$(\sigma v^{rel})_{\chi\chi \to \gamma\gamma} \simeq 10^{-30} \text{ cm}^3 \text{ sec}^{-1} \quad,$$

in rough order of magnitude. This kind of number is in fact borne out by a detailed analysis in the framework of a particular type of particle physics model for χ known as a supersymmetric theory.[23] In this kind of model, all known fundamental particles have a so-called "superpartner" whose spin differs by one-half a unit, $\frac{1}{2}\hbar$. Thus, bosons have fermion superpartners, and vice versa. In general, superpartner particles have unknown masses, a situation known as broken supersymmetry. The idea is that in scattering experiments involving particles of energies far in excess of their rest masses, bosons cannot be distinguished from fermions. The lightest superpartner particle is stable, and is often the "photino," the spin one-half partner of the photon, denoted by $\tilde{\gamma}$. In Ref. 23, I calculate $(\sigma v_{rel})_{\tilde{\gamma}\tilde{\gamma}\to\gamma\gamma}$ subject to particular model assumptions, and find

$$(\sigma v_{rel})_{\tilde{\gamma}\tilde{\gamma}\to\gamma\gamma} = 4\times 10^{-31} \text{ cm}^3 \text{ sec}^{-1} (M_{\tilde{\gamma}} c^2/4 \text{ GeV})^2 \quad,$$

valid for $M_{\tilde{\gamma}} c^2 \geq 4$ GeV. By varying these assumptions, it is possible to increase this quantity by nearly an order of magnitude, but also, unfortunately, to make it much smaller! With this, an estimate of the line flux is, at high Galactic latitudes,

$$F_\gamma (\theta,\phi) \simeq 2\times 10^{-11} \text{ cm}^{-2} \text{ sec}^{-1} \text{ sr}^{-1} \quad,$$

for $E\gamma \geq 4$ GeV, independently of $M_{\tilde{\gamma}}$, to be compared with a diffuse background that falls steeply with $E\gamma$,[24]

$$dN/dE\gamma \text{ (pole)} \simeq 2\times 10^{-8} (E\gamma/4 \text{ GeV})^{-2.7} \text{ cm}^{-2} \text{ sec}^{-1} \text{ sr}^{-1} \text{ GeV}^{-1} \quad.$$

With a detector resolution $\Delta E\gamma/E\gamma \simeq 1\%$, the signal-to-background ratio is

$$\frac{F\gamma}{\Delta E\gamma (dN/dE\gamma)} \leq 2.5\times 10^{-2} (M_{\tilde{\gamma}} c^2/4 \text{ GeV})^{1.7} \quad.$$

It is clear that large collection areas and long exposure times are required. Typical designs for detectors in Earth orbit envisage areas of not more than a squre meter, leading to only of order 10 events per year.

It is important to note that no process other than dark matter annihilation could give rise to multi-GeV (or indeed, tens of GeV) photon lines in cosmic rays, and this makes such a process a remarkable signature for particle dark matter. A dedicated Lunar Gamma Ray Detector, with a large collection area and fine energy resolution will in all likelihood be required to continue the search for this particularly clear signature of dark matter as relic particles.

4. Conclusions

It is apparent that limitations in the performance of detectors deep underground or in Earth orbit are likely to result in the issues of the stability of the proton and the nature of the cosmic dark matter to remain unresolved into the long-term future. Under these conditions, it would seem that the exploration of these and other fundamental questions in particle physics and cosmology with experiments using the environment of the Moon to advantage, should not be overlooked in planning the scientific program of a Lunar Base in the early part of the 21st Century.

References

1. A. Salam, Int'l. J. Mod. Phys. **4A**, 583 (1989).
2. J. Bernstein, *The Tenth Dimension* (McGraw-Hill, New York, 1989).
3. L. B. Okun, *Particle Physics – The Quest for the Substance of Substance* (Harwood Academic Publishers, New York, 1985).
4. D. H. Perkins, Ann. Rev. Nucl. Part. Sci. **34**, 1 (1984).
5. *First Workshop on Grand Unification Proceedings* (New Hampshire, 1980), P. Frampton, S. L. Glashow, and A. Yildiz, eds. (Math. Sci. Press, Brookline MA, 1980).
6. *Second Workshop on Grand Unification Proceedings* (Michigan, 1981), J. Leveille, L. Sulak, and D. Unger, eds. (Birkhäuser, Boston, 1981).
7. *Third Workshop on Grand Unification Proceedings* (North Carolina, 1982), P. Frampton, S. L. Glashow, H. van Dam, eds. (Birkhäuser, Boston, 1982).
8. *Fourth Workshop on Grand Unification Proceedings* (Pennsylvania, 1983), P. Langacker, P. Steinhardt, and H. A. Weldon, eds. (Birkhäuser, Boston, 1983).
9. *Fifth Workshop on Grand Unification Proceedings* (Rhode Island, 1984), K. Kang, H. Fried, and P. Frampton, eds. (World Scientific, Singapore, 1984).
10. *Sixth Workshop on Grand Unification Proceedings* (Minnesota, 1985), S. Rudaz and T. Walsh, eds. (World Scientific, Singapore, 1985).
11. *Seventh Workshop on Grand Unification Proceedings* (Japan, 1986), J. Arafune, ed. (World Scientific, Singapore, 1986).
12. *Eighth Workshop on Grand Unification Proceedings* (Syracuse, 1987), K. Wali, ed. (World Scientific, Singapore, 1987).

References

13. *Ninth Workshop on Grand Unification Proceedings* (France, 1988), J. Barloutaud, ed. (World Scientific, Singapore, 1988).
14. *Tenth and Final Workshop on Grand Unification Proceedings* (North Carolina, 1989), P. Frampton, ed. (World Scientific, Singapore, to be published).
15. M. Goldhaber, P. Langacker, and R. Slansky, Science **210**, 851 (1980).
16. J. G. Learned, in Ref. 6, P. 125.
17. Report of the Lunar Base Working Group (April 23-27, 1984), compiled by M. B. Duke, W. W. Mendell, and P. W. Keaton, Los Alamos National Laboratory report LALP-84-43.
18. J. C. Pati, A. Salam, and B. V. Sreekantan, Int'l. J. Mod. Phys. **A1**, 147 (1986).
19. M. Cherry and K. Lande, in *Lunar Bases and Space Activities of the 21st Century*, W. W. Mendell, ed. (Lunar and Planetary Institute, Houston, 1985) P. 335.
20. V. Trimble, Ann. Rev. Astron. and Astrophys. **25**, 425 (1987).
21. D. Hegyi and K. Olive, Phys. Lett. **126B**, 28 (1983).
22. S. Rudaz, Nucl. Phys. B (Proc. Suppl.) **10B**, 114 (1989).
23. S. Rudaz, Phys. Rev. **D39**, 3549 (1989).
24. F. W. Stecker, in *The Large-Scale Characteristics of the Galaxy*, W. B. Burton ed. (Reidel, Dordrecht, 1979) p. 475; C. D. Dermer, Astron. Astrophys. **157**, 223 (1986).

Probing the Halo Dark Matter γ Ray Line from a Lunar Base [*]

Pierre Salati, [†] Alain Bouquet [‡] and Joseph Silk

*Center for Particle Astrophysics
Lawrence Berkeley Laboratory
and
Department of Astronomy and Physics
University of California
Berkeley CA 94720 U.S.A.*

Abstract

We study the possibility of detecting halo cold dark matter through the annihilation process $\chi\bar{\chi} \to \gamma\gamma$. This process produces monoenergetic γ rays, and may be a clear signature of particle dark matter. If there is a closure density of dark matter, we show that it will be very difficult to observe this annihilation line from a space station borne experiment. On the contrary, a large lunar based γ-ray telescope could detect hundreds of events per year.

1 - Introduction

There is a considerable interest in the possible detection of dark matter [1] through astrophysical observations as well as terrestrial observations [2,3]. If dark matter is made up of massive, weakly interacting particles, one of the best indirect signatures would be the detection of annihilation products of these particles [4,5,6,7]. It has been recently suggested [8,9,10,11,12,13,14,15] that if the halo of our Galaxy is made of particles in the mass range 1 GeV to 100 GeV, and if these particles self-annihilate, as most of the potential candidates do, then monoenergetic γ rays should be observed when the cold dark matter particles annihilate directly into two photons. The energy of these γ's is very nearly equal to the mass of the (non-relativistic : $v/c \sim 10^{-3}$) parent particles. Very narrow γ lines in the range 1 GeV to 100 GeV would provide a clear signature for dark matter annihilation, if they stand out above the background.

We evaluate the flux of such γ rays on a space detector, coming from the annihilation of neutral supersymmetric particles such as the photino $\tilde{\gamma}$ or the higgsino $\tilde{h}$, and from a heavy Dirac or Majorana neutrino. We then compare this flux to estimates of the γ background in the range 1 GeV to 100 GeV. We essentially conclude that observation of these γ lines will be very difficult in the case of satellite or space station borne γ ray observatories (GRO and

---

[*]Contribution to the NASA Workshop On Physics From A Lunar Base, held at Stanford University and The Sunnyvale Hilton, Stanford, California, on May 19-20 1989.
[†]Miller Research Fellow at the University of California at Berkeley. On leave of absence from LAPP, BP110, 74941 Annecy-le-Vieux Cedex, France and Université de Chambéry, 73000 Chambéry, France.
[‡]On leave of absence from LPTHE Université Paris VII, 75251 Paris Cedex 05, France.

ASTROGAM [16]) and advocate the possibility of a lunar based experiment (such as HR-GRAF [17]) but of considerably larger area.

2 - Flux from photino annihilation

The photino $\tilde{\gamma}$ is the supersymmetric partner of the photon, and one of the foremost candidates for dark matter because it naturally implies a near-critical density for the universe. We assume that the dark matter halo of our Galaxy is nearly spherical and nearly isothermal. Then, the dark matter density varies with the distance r to the galactic center as :

$$\rho(r) = \rho_\odot \frac{a^2 + r_\odot^2}{a^2 + r^2} , \qquad (1)$$

where $\rho_\odot$ is the dark matter density in the solar neighbourhood and a is the core radius. We consider the annihilation process $\tilde{\gamma}\tilde{\gamma} \to \gamma\gamma$, the cross-section for which is $\sigma_{2\gamma}$. The mass of the photino may range from 1 GeV to 100 GeV according to specific models, whereas their mean velocity in the halo is $v \sim 300\,\mathrm{km\,s^{-1}}$. The energy E_γ of the outgoing γ's is therefore practically equal to the photino mass $m_{\tilde{\gamma}}$. The number of photons received at a detector in the solar system, from photino pair annihilations taking place at a distance R and in the direction defined by galactic latitude b and longitude l is [18] :

$$F = \frac{\sigma_{2\gamma} v}{2\pi} \int_0^\infty \left[\frac{\rho(R,b,l)}{m_{\tilde{\gamma}}} \right]^2 dR \ \mathrm{cm^{-2}\,s^{-1}\,sr^{-1}} \qquad (2)$$

when summed over the distance R. This gives :

$$F = \frac{\sigma_{2\gamma} v}{4\pi} \left(\frac{\rho_\odot}{m_{\tilde{\gamma}}} \right)^2 a \frac{(1+A^2)^2}{B^3} \left[\frac{\pi}{2} + \frac{AB}{1+A^2} \cos b \cos l + \tan^{-1}\left(\frac{A \cos b \cos l}{B} \right) \right] \qquad (3)$$

where $A = r_\odot/a$ and $B = \sqrt{1 + A^2(1 - \cos^2 b \cos^2 l)}$. At high galactic latitudes ($b \sim 90°$), this reduces to :

$$F \sim \frac{\sigma_{2\gamma} v}{8} \left(\frac{\rho_\odot}{m_{\tilde{\gamma}}} \right)^2 a\sqrt{1+A^2} . \qquad (4)$$

The rate of annihilation of two photinos into two photons has been computed by Bergström and Snellman [8] and by Rudaz [13], whose estimates of $\sigma_{2\gamma}$ differ by nearly 2 orders of magnitude. We have recomputed the cross-section and agree with Rudaz's result :

$$\sigma_{2\gamma} v = \frac{16\alpha^4}{\pi} m_{\tilde{\gamma}}^2 \left| \sum_f \frac{Q_f^4}{m_{\tilde{f}}^2} I\left(\frac{m_f^2}{m_{\tilde{\gamma}}^2} \right) \right|^2 , \qquad (5)$$

where the sum runs over all the quarks and leptons f. Q_f is the electric charge of fermion f and $m_{\tilde{f}}$ is the mass of its scalar partners $\tilde{f}$. α is the fine structure constant. The complex valued function $I(x)$ is defined as :

$$I(x) = \frac{1}{2}\left(1 - x\left[\tan^{-1}\left(\frac{1}{\sqrt{x-1}} \right) \right]^2 \right) . \qquad (6)$$

One can assume that all the scalar masses $m_{\tilde{f}}$ are equal to some common value $\tilde{M}$. We have relaxed this (unrealistic) assumption without any major change in the results. Previous studies of the monochromatic gamma signature of dark matter annihilation took $\tilde{M}$ as fixed, leading to the conclusion that $\sigma_{2\gamma}v$ behaved as $m_{\tilde{\gamma}}^2$ (because $\lim_{x\to 0} I(x) = 0.5$), and that the flux F was therefore independent of the photino mass. As the background decreases at high γ energies (i.e. at high photino mass), the conclusion was that the signal-to-noise ratio would improve for heavier photinos [9,13]. However if one assumes that dark matter is made of photinos, consistency requires that the relic density be fixed, not the mass $\tilde{M}$. For a given photino relic density $\Omega_{\tilde{\gamma}}$, and a given photino mass $m_{\tilde{\gamma}}$, the mass $\tilde{M}$ is roughly [19] :

$$\tilde{M}^4 \sim (40\,GeV)^4 \left(\frac{m_{\tilde{\gamma}}}{1\,GeV}\right)^2 \Omega_{\tilde{\gamma}} h^2 \;, \tag{7}$$

except when the photino mass is just above one of the thresholds for annihilation into a quark or lepton pair. As usual, here h is the Hubble constant in units of 100 km/s/Mpc. From Equ. 5, the two-photon annihilation cross-section behaves as $1/\Omega_{\tilde{\gamma}} h^2$, and for a fixed value of $\Omega_{\tilde{\gamma}} h^2$, the annihilation rate into two photons is independent both of the photino mass and of the scalar quark and lepton mass $\tilde{M}$. Therefore the γ flux *decreases* with increasing photino mass, and the expected signal decreases. The two-photon annihilation cross-section $\sigma_{2\gamma}v$ is :

$$\sigma_{2\gamma}v \sim 3.1 \times 10^{-31} \text{cm}^3\text{s}^{-1} \left(\Omega_{\tilde{\gamma}} h^2\right)^{-1} \;. \tag{8}$$

At high galactic latitudes ($b = 90°$), the flux is then :

$$F \sim 2.2 \times 10^{-10} \text{ photons cm}^{-2}\text{s}^{-1}\text{sr}^{-1} \left(\frac{1\,GeV}{m_{\tilde{\gamma}}}\right)^2 \left(\Omega_{\tilde{\gamma}} h^2\right)^{-1} \;, \tag{9}$$

taking $\rho_\odot = 0.4\,\text{GeV cm}^{-3}$ and $a \sim r_\odot \sim 8\,\text{kpc}$. A detector of $1\,\text{m}^2\,\text{sr}$ collecting area, such as ASTROGAM [16] or HR-GRAF [17], will detect N_{line} photons of energy $E_\gamma = m_{\tilde{\gamma}}$ per year, where :

$$N_{line} \sim 70 \text{ photons} \left(\frac{1\,GeV}{E_\gamma}\right)^2 \left(\Omega_{\tilde{\gamma}} h^2\right)^{-1} \;. \tag{10}$$

Note that, if the photino mass is large, the number of photons detected becomes so low that there is no statistically significant signal. On the other hand, a lunar based γ-ray telescope with a large collecting area ($\sim 100\,\text{m}^2\,\text{sr}$), would receive hundreds of photons each year :

$$N_{line} \sim 7000 \text{ photons} \left(\frac{1\,GeV}{E_\gamma}\right)^2 \left(\Omega_{\tilde{\gamma}} h^2\right)^{-1} \;. \tag{11}$$

Actually, Equ. 10 and 11 are crude approximations to the signal that can be expected, since they do not take into account mass thresholds, nor the (very likely) possibility that different scalar fermions $\tilde{f}$ can have masses $m_{\tilde{f}}$ different from the common assumed value $\tilde{M}$. In addition, this estimate does not take into account the fact that the "freeze-out" temperature, where the photino density drops out of thermal equilibrium and stabilizes [20,19], depends in a complicated way on the photino mass, and thereby affects the resulting relic density. Therefore, we performed an exact computation of the "freeze-out" temperature as a function of the photino mass, taking into account the change in the number of degrees of freedom in the thermal radiation at the time of decoupling. We also computed the annihilation cross-section for the reaction $\tilde{\gamma}\tilde{\gamma} \to f\bar{f}$, taking into account the energy threshold for the opening of a new $f\bar{f}$ channel. We also took into account the possibility that some scalar partners could be anomalously light. In particular the scalar top quark is very often light in most supersymmetric models, but it must be heavier than the photino (otherwise the photino would be unstable). We studied the case of a scalar top quark almost degenerate in mass with the photino, and saw no noticeable difference. Since this is the most extreme situation, we do not think that the assumption of equal masses for the scalar quarks and leptons is a restrictive assumption (in *this* situation of course). We then obtained a more exact relation between the mass of the photino and the masses of the scalar quarks and leptons. This relation was used in the two photon annihilation cross-section (Equ. 5) to compute the expected γ flux and the expected signal. Figure 1 shows the result of this computation, for 3 different values of the relic density ($\Omega_{\tilde{\gamma}}h^2 = 1, 0.25 \text{ and } 0.025$). The effect of thresholds is seen as small dips in the curves (because the opening of a new $f\bar{f}$ threshold increases the total annihilation rate at the big-bang, which must be corrected by an increase in the scalar mass $\tilde{M}$, and therefore leads to a decrease of the 2γ annihilation rate).

3 - Background

There is no measurement of γ ray fluxes in the GeV-TeV energy range, so we must extrapolate from lower energy data. There are many different sources of background γ photons. Bremsstrahlung γ's from cosmic ray electrons should be negligible above 1 GeV, but the γ's from the decay of π^0's produced by collisions of cosmic ray protons with the interstellar medium are probably the dominant source of galactic background at energies larger than 1 GeV [21]. From the known distribution of cosmic ray protons, one expects a flux (Stecker 1978) :

$$\frac{dN_{galactic}}{dE_\gamma} \sim 8 \times 10^{-7} \text{ photons cm}^{-2} \text{s}^{-1} \text{sr}^{-1} \text{GeV}^{-1} \left(\frac{1\,GeV}{E_\gamma}\right)^{2.7\pm0.3} \quad (12)$$

at high galactic latitudes, and a correspondingly larger flux in the galactic plane. The extragalactic background is not well known, and different analyses do not agree on the extrapolation of COS-B and SAS-2 measurements to energies larger than 1 GeV [22]. Since it may drop sharply above 1 GeV, we do not consider it any further, but keep in mind the possibility that it could be of the same order of magnitude as the galactic background. Dark matter provides its own background when it annihilates into fermion-antifermion pairs, which give γ rays among their decay and annihilation end-products, but these γ's have an energy lower than the photino mass (for trivial kinematical reasons), and do not contribute to the line background. To summarize, we take Equ. 12 as representing the background, but we must keep in mind that the background could be an order of magnitude larger. Since the aim is to detect a narrow line, much narrower than the energy resolution of the detector, the interesting quantity is the number of photons received in one energy bin during the time of observation. For a $1\,\text{m}^2\,\text{sr}$ detector with a 1%

energy resolution (bin width $\Delta E_\gamma = 0.01\, E_\gamma$), one year of observation will give :

$$N_{background} \sim 2500\, \text{photons} \left(\frac{1\, GeV}{E_\gamma}\right)^{1.7\pm 0.3}. \tag{13}$$

If one compares the expected signal N_{line} (Equ. 10) to the background $N_{background}$ (Equ. 13), the situation seems hopeless. But this comparison is actually meaningless. We should *not* compare the expected signal to the background but to the *noise*, i.e. the error in measuring the background. This point was overlooked in previous works, but it does make a difference. The statistical uncertainty is the square root of the number of photons received :

$$Noise \sim 50\, \text{photons} \left(\frac{1\, GeV}{E_\gamma}\right)^{0.85\pm 0.15}. \tag{14}$$

The situation is then much better, although still not very bright for a satellite or space station borne detector for which the collecting area can hardly be larger than $\sim 1\, m^2\, sr$. For $\Omega_{\tilde{\gamma}} h^2 = 0.25$, very light photinos ($m_{\tilde{\gamma}} < 4\, GeV$) could be detected at the two standard deviations level (see Figure 1, and note that the scale is logarithmic).

On the contrary, a lunar based experiment could easily be designed to receive a very large γ-ray flux with, for instance, an aperture $\sim 100\, m^2\, sr$. Since the signal-to-noise ratio for the annihilation γ-ray line goes as :

$$(\text{collecting area} \times \text{angular aperture})^{1/2}$$

of the telescope, the lunar based option for detecting cold dark matter seems quite promising. For a lunar based detector, the background noise (dashed lines in Figures 1 and 2) is shifted downwards by an order of magnitude with respect to the various signals.

The observed diffuse galactic plane γ ray emissivity is found to correlate well with the total gas column density, confirming the interpretation of the observed diffuse flux as primarily being due to cosmic ray interactions (mainly π^0 production in the COS-B energy range) with interstellar atoms [23]. Analysis of the COS-B data yields an empirical diffuse galactic γ ray emissivity of $4 \times 10^{-27} s^{-1} sr^{-1} (H_{atom})^{-1}$ in the 300-800 MeV range, and $2 \times 10^{-27} s^{-1} sr^{-1} (H_{atom})^{-1}$ in the 800 MeV-6 GeV range. The inferred spectral dependence is approximately proportional to $E^{-2.7}$ [24]. It has been suggested [25] that one may examine high latitude holes in the galactic HI distribution for regions of expected low γ ray background. Specifically, we note that sensitive 21 cm surveys of high latitude HI regions [26] find that the HI consists of cold clouds (T ~ 100 K) with probability (at $|b| = 90°$) :

$$P(N > N_{cloud}) = 0.5 \left(\frac{N_{cloud}}{10^{20}\, cm^{-2}}\right)^{-0.8} \tag{15}$$

for $3 \times 10^{19} < N_{cloud} < 2 \times 10^{20}\, cm^{-2}$, together with an additional 50% of the HI in diffuse, warm (T ~ 5000 K) gas. In addition there is an ionized gas contribution that we take from the distribution of diffusion Hα emission and pulsar dispersion measures to be modelled by a layer of HII with scale height 1000 pc and electron density 0.03 cm^{-3} [27,28]. We infer a minimum column gas density towards HI "holes" (defined by $N_{cloud} < 3 \times 10^{19}$cm^{-2}) of about 6×10^{19}cm^{-2} in HI and 9×10^{19}cm^{-2} in HII, totalling 1.5×10^{20}cm^{-2}. For comparison, the lowest HI column density in the northern sky over a square degree or so amounts to 4.5×10^{19}cm^{-2} [29]. The inferred galactic γ ray background towards the holes is 6×10^{-7}cm^{-2}sr^{-1}s^{-1} in the 300-800 MeV range, and 3×10^{-7}cm^{-2}sr^{-1}s^{-1} in the 800 MeV-6 GeV range.

4 - Higgsinos and neutrinos

If the dark matter is made of Dirac or Majorana neutrinos, or of higgsinos, the situation is a little bit less satisfactory. The main change comes from the different dependences of the annihilation cross-sections. The exchange of scalar quarks and leptons contribute very little to the self-annihilation of the higgsinos, and not at all to the neutrino self- annihilations. The main contribution comes from the exchange of the Z^0 and gives [13,30] :

$$\sigma(\chi\chi \to \gamma\gamma)v \sim \frac{2\alpha^2 G_F^2 \cos^2 2\beta}{\pi^3} m_\chi^2 \left| \sum_f Q_f^2 T_f I\left(\frac{m_f^2}{m_\chi^2}\right) \right|^2 \qquad (16)$$

where χ is either a higgsino or a Majorana neutrino. G_F is the Fermi coupling constant, T_f is the third component of the weak isospin of fermion f, and β is an arbitrary parameter in the higgsino case (related to the difference of the vacuum expectation values of the two Higgs fields), and has value zero in the Majorana neutrino case. The annihilation cross-section is 4 times smaller for a Dirac neutrino than for a Majorana neutrino. The maximal effect is obtained for $\beta = 0$, and when the χ mass is much smaller than the top quark mass :

$$\sigma_{2\gamma} v \sim 6.2 \times 10^{-34} \text{ cm}^3\text{s}^{-1} \left(\frac{m_\chi}{1\,GeV}\right)^2. \qquad (17)$$

When the χ is much heavier than the top quark, the cross-section is zero as a consequence of the anomaly cancellation mechanism of the electroweak standard model [13]. We see immediately that the cross-section is much smaller than in the case of the photino, which can be traced to the higher mass of the exchanged Z^0 compared to the scalar quark and lepton mass $\tilde{M}$. It is not a surprise therefore to find a smaller flux and a smaller number of γ's at a detector :

$$F \sim 4.35 \times 10^{-13} \text{ photons cm}^{-2}\text{ s}^{-1} \text{ sr}^{-1} \qquad (18)$$

at high galactic latitudes. Note that the mass of the higgsino or neutrino cancels out in the computation of the flux. A detector of $1\,m^2$ sr collecting area will detect :

$$N_{line} \sim 0.14\,photon\,per\,year \qquad (19)$$

which is completely hopeless, *whatever the background*. Since the annihilation cross-section is 4 times smaller for a Dirac neutrino, its signal will be even less detectable.

We performed the same kind of numerical analysis for the neutrino and higgsino cases as for the photino cases, taking thresholds into account and differences in freeze-out temperature. As figure 2 shows, thresholds have an important effect, but in the Ωh^2 range of interest, they decrease the expected signal.

5 - Slightly more exotic models

In this section, we would like to explore the conditions under which a dark matter candidate could produce a very large γ-ray line signal. In order to achieve the cosmological relic density $\Omega_\chi h^2$, the total annihilation cross-section at freeze-out should behave like :

$$< \sigma_{ann} v >_{dec} \simeq \frac{5 \times 10^{-27} \, cm^3 \, s^{-1}}{\Omega_\chi h^2} . \qquad (20)$$

If $B_{2\gamma}$ denotes the cross-section ratio :

$$B_{2\gamma} = \frac{\sigma_{2\gamma} v}{< \sigma_{ann} v >_{dec}} , \qquad (21)$$

the expected γ-ray line flux reads :

$$F = 3.5 \times 10^{-6} \, cm^{-2} \, s^{-1} \, sr^{-1} \left(\frac{1 \, GeV}{M_\chi}\right)^2 \left(\frac{B_{2\gamma}}{\Omega_\chi h^2}\right) . \qquad (22)$$

A detector with a 1 m² sr effective area would therefore collect in one year :

$$N_{line} \simeq 1.1 \times 10^6 \, photons \left(\frac{1 \, GeV}{M_\chi}\right)^2 \left(\frac{B_{2\gamma}}{\Omega_\chi h^2}\right) , \qquad (23)$$

so that the signal-to-noise ratio would be :

$$\frac{N_{line}}{Noise} \simeq 2.2 \times 10^4 \left(\frac{1 \, GeV}{M_\chi}\right)^{1.15 \pm 0.15} \left(\frac{B_{2\gamma}}{\Omega_\chi h^2}\right) . \qquad (24)$$

The detection at the 10 σ level of a γ-ray line produced by the annihilations of neutralino particles inside the galactic halo requires therefore that the branching ratio $B_{2\gamma}$ fulfill the condition :

$$\left(\frac{B_{2\gamma}}{\Omega_\chi h^2}\right) > 4.5 \times 10^{-4} \left(\frac{M_\chi}{1 \, GeV}\right)^{1.15 \pm 0.15} . \qquad (25)$$

The larger the branching ratio $B_{2\gamma}$, the easier the detection. For instance, in order to detect at the 2 σ level a γ-ray line produced by a 100 GeV dark matter candidate, the ratio $B_{2\gamma}/\Omega_\chi h^2$ should exceed $\sim 2\%$. This fairly large value may be contrasted with the corresponding branching ratio in the photino case :

$$B_{2\gamma}(\text{photino}) = \left(\frac{1}{x_F}\right) \left(\frac{\alpha^2}{\pi^2}\right) \left|\sum_f Q_f^4 \, I\left(\frac{m_f^2}{M_{\tilde{\gamma}}^2}\right)\right|^2 \times$$

$$\times \left(\sum_f Q_f^4 \sqrt{1 - z_f^2} \left\{1 - \frac{7}{4}z_f^2 + \frac{3}{8}\frac{z_f^4}{(1 - z_f^2)}\right\}\right)^{-1} . \qquad (26)$$

which, assuming that the photino is well above any threshold, may be simplified further :

$$B_{2\gamma}(\text{photino}) \simeq 2.7 \times 10^{-5} \left(\sum_f Q_f^4 \right) \sim 1.3 \times 10^{-4} .\tag{27}$$

Condition (25) supplemented by relation (27) implies that the detection of a γ-ray line signal produced by hypothetical photinos from the galactic halo cannot be achieved with a large signal-to-noise ratio unless the relic density in these particles is unrealistically small, as already discussed in section 2, or unless the detector has a quite large aperture.

However, there should be models for which the branching ratio $B_{2\gamma}$ is large [14]. The most exotic possibility corresponds to $B_{2\gamma} = 1$, in which case the dominant channel for the χ's annihilations is the two-photon reaction. Such a dark matter candidate would have annihilated in the early universe predominantly into two photons. Our analysis of the Big-Bang production of dark matter particles may still be applied to this exotic particle. Even for a closure density in this species, i.e., $\Omega_\chi h^2 = 1$, the signal-to-noise ratio would be enormous up to a mass $M_\chi \sim$ 1 TeV.

Models for which $B_{2\gamma} = 1$ are still in their infancy, but we can sketch their main features. An interesting option for such an exotic candidate would be the neutral component L^0 of a weak SU(2) triplet. Since this neutral Majorana fermion has a weak isospin component $T_3^L = 0$, it does not couple to the Z^0 boson, so that one has to resort to non-standard interactions for its annihilations. Suppose, for instance, that L^0 couples only to a charged fermion L^+ and a charged Higgs boson T^+ with strength λ. Provided heavier masses for L^+ and T^+ are arranged, the L^0's annihilations only proceed through the two-photon reaction. If m_0, m_+ and μ_+ denote respectively the mass of L^0, L^+ and T^+, the annihilation cross-section for the exotic L^0 into two photons can be derived from the effective Lagrangian :

$$\mathcal{L}_{\text{eff}} = \frac{\lambda^2}{8\mu_+^2} \ \overline{L^0}\gamma_\mu\gamma_5 L^0 \ \times \ \overline{L^+}\gamma^\mu(1-\gamma_5)L^+ ,\tag{28}$$

and may be approximated by :

$$\sigma_{2\gamma}v \simeq \left(\frac{\alpha^2}{\pi}\right)\left(\frac{\lambda^2}{4\pi}\right)^2 \left(\frac{m_0^2}{\mu_+^4}\right) \mathcal{I}^2 ,\tag{29}$$

where the integral $\mathcal{I}$ is given by :

$$\mathcal{I} = I\left(\frac{m_+^2}{m_0^2}\right) .\tag{30}$$

As an example, let us consider the case for which $\lambda = 1$, mass $m_0 = 10$ GeV with L^0 slightly lighter than its charged partner L^+ (i.e., $\mathcal{I} \sim -0.73$). The annihilation cross-section into two photons simplifies into :

$$\sigma_{2\gamma}v \sim 6.7 \times 10^{-27} \text{cm}^3\text{s}^{-1} \left(\frac{10 \text{ GeV}}{\mu_+}\right)^4 .\tag{31}$$

By appropriately adjusting the mass μ_+ of the intermediate Higgs T^+ as well as the coupling constant λ, reasonable values for the relic density of the exotic candidate L^0 may be achieved. Since the two-photon reaction is their only annihilation channel, a closure density of L^0 would lead to a significant γ-ray line signal. This fairly simple analysis should motivate further investigation in the direction which we have just outlined.

6 - Conclusions

We have shown that the observation of the narrow γ ray line due to dark matter annihilation was barely possible if there is a closure density of dark matter ($\Omega h^2 > 0.25$). In the most favourable case, namely if the dark matter is made up of photinos with light masses ($m_{\tilde{\gamma}} < 5$ GeV), the signal-to-noise ratio is of order one for a detector of 1 m² sr acceptance, operating for one year. The detection of the γ ray line from Majorana or Dirac particle annihilations appears impossible. The signal is buried orders of magnitude below the noise of the background.

The best candidate for the γ ray line detection is the photino, provided its contribution to the dark matter density in the universe is small : $\Omega h^2 \approx 0.025$. Dark matter seems therefore to be elusive with respect to its γ ray line annihilation for a square meter class detector probing the galactic halo. There are at least three reasons for some degree of optimism, however.

- Suppose that our galactic nucleus, or some remote AGN, is found to be a source of GeV gamma rays. We propose that further scanning of the spectrum with better than one percent energy resolution is a worthwhile follow-up observation. Should a narrow line be found, this would be a "smoking gun" for a dense cloud of weakly annihilating photinos as an energy source for the galactic nucleus. Such dense clouds with gamma ray luminosities that could approach quasar-like luminosities are not unexpected in certain scenarios for galaxy formation [31,25].

- Use of high spatial resolution better than 1 degree will enable future γ ray telescopes to probe holes in the galactic halo gas distribution and thereby reduce the expected galactic gamma ray background and improve the signal to noise ratio by a factor of ~ 3 relative to the predictions given in the figures.

- Finally, this rapid investigation should motivate the design of a γ-ray telescope with a very large effective area (*i.e.*, a hundred m² sr), similar to the conclusion reached for X-ray astronomy [32]. Such a detector would be extremely useful for probing the structure and nature of the galactic halo and a lunar based experiment seems to be a very good option for such a device. A detailed analysis of the feasibility of this kind of project is therefore of considerable urgency.

Acknowledgments : It is a pleasure to thank H. Crawford and G. Smoot for explaining to us the ASTROGAM project. A.B. wishes to thank the Lawrence Berkeley Laboratory and the Astronomy Department of the University of California at Berkeley for their hospitality and support during the completion of this work. P.S. acknowledges a fellowship from the Miller

Institute for Basic Research in Science at the University of California at Berkeley. J.S. was supported in part by grants from NASA and DOE. We would like to express our gratitude towards Dr. T. Wilson for giving us the opportunity to contribute to the Proceedings of the NASA Workshop On Physics From A Lunar Base.

References

[1] Kormendy, J. & G.R. Knapp. 1987. *In* Dark matter in the Universe. IAU Symposium **117** (Reidel).

[2] Primack, J., D. Seckel & B. Sadoulet. 1988. Ann Rev. Nucl. Part. Phys. **38**:751.

[3] Smith, P.F. & J.D. Lewin. 1988. Report Rutherford RAL-88-045.

[4] Gunn, J.E., B.W. Lee, I. Lerche, D.N. Schramm & G. Steigman. 1978. Ap. J. **223**:1015.

[5] Silk, J. & M. Srednicki. 1984. Phys. Rev. Lett. **53**:624.

[6] Stecker, F.W., S. Rudaz & T. Walsh. 1985. Phys. Rev. Lett. **55**:2622.

[7] Hagelin, J. & G.L. Kane. 1986. Nucl. Phys. **B263**:399.

[8] Bergström, L. & H. Snellman. 1988. Phys. Rev. **D37**:3737.

[9] Bergström, L. 1988. Reports Stockholm USITP-88-12 and USITP-89-01.

[10] Bergström, L. 1988. Report Stockholm USITP-89-04.

[11] Stecker, F.W. & A.J. Tylka. 1988. Ap. J. **343**:169.

[12] Rudaz, S. 1988. *In* Proc. Workshop High Resolution Gamma Ray Cosmology, UCLA November 2-5, 1988, Nucl. Phys. B in press.

[13] Rudaz, S. 1989. Phys. Rev. **D39**:3549.

[14] G. F. Giudice and K. Griest; Report Fermilab-Pub-89/113A.

[15] Bouquet, A., P. Salati & J. Silk. 1989. Report LBL-27334. To be published in Phys. Rev. **D**.

[16] Adams, J.H. & al. 1989. Naval Research Laboratory proposal T-248-89.

[17] Fenyves, E.J. 1989. These workshop proceedings, and NASA Code E proposal UTD No. 890061.

[18] Turner, M.S. 1986. Phys. Rev. **D34**:1921.

[19] Goldberg, H. 1983. Phys. Rev. Lett. **50**:1419.

[20] Lee, B.W. & S. Weinberg. 1977. Phys. Rev. Lett. **39**:165.

[21] Bloemen, J.B.G.M. 1987. Ap. J. Lett. **317**:L15.

[22] Fichtel, C.E. & D.J. Thompson. 1982. Astron. Astroph. **109**:352.

[23] Bloemen, J.B.G.M. 1989. Ann. Rev. Ast. Ap., in press.

[24] Dermer, C.D. 1986. Astron. Ap. **157**:223.

[25] Silk, J. 1988. *In* Proc. Workshop High Resolution Gamma Ray Cosmology, UCLA November 2-5, 1988, Nucl. Phys. B in press.

[26] Payne, H.E., E.E. Salpeter & Y. Terzian. 1983. Ap. J. **272**:540.

[27] Vivekanand, M. & R. Narayan. 1982. Indian Ap. Ast. **3**:399.

[28] Reynolds, R. 1977. Ap. J. **216**:433.

[29] Jahoda, K., D. McCammon & F.J. Lockmar. 1986. *In* Gaseous Halos of Galaxies, Eds. J.N. Bregnon and F.J. Lockmar, NRAO:75.

[30] Kane, G.L. & I. Kani. 1986. Nucl. Phys. **B277**:525.

[31] Salati, P. & J. Silk. 1989. Ap. J. **338**:24.

[32] Wood, K.S. & P.F. Michelson. 1989. These workshop proceedings.

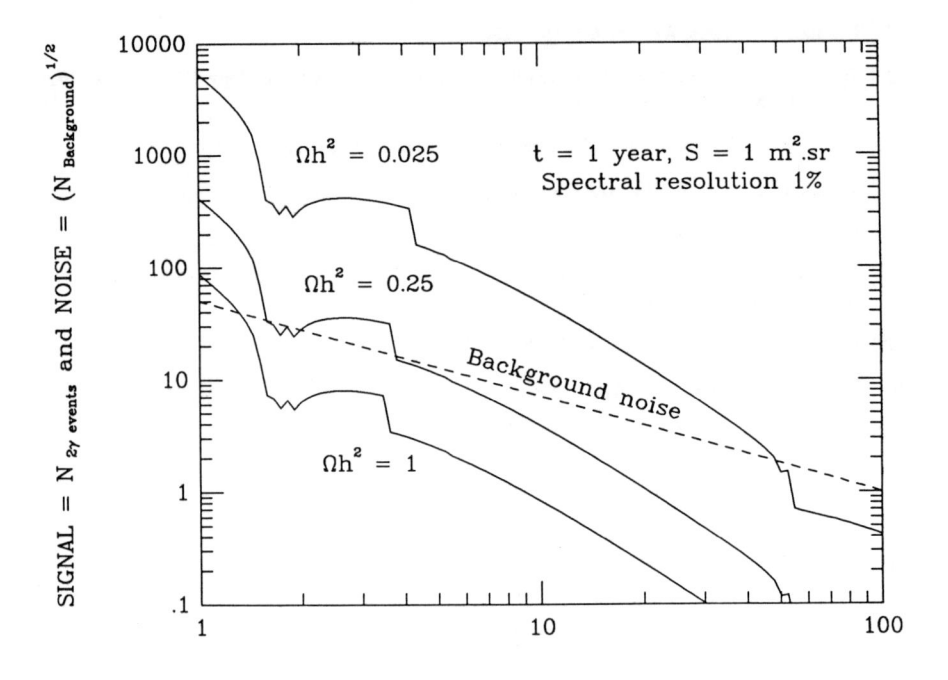

Figure 1 : Expected signal and noise for monoenergetic γ rays due to photino pair annihilation in the halo of our Galaxy. Full lines display the total number of annihilation γ's received on a detector like ASTROGAM (1 m^2 sr detector surface, 1% energy resolution, and one year of exposure), for 3 different relic photino densities $\Omega_{\tilde{\gamma}} h^2 = 1$, 0.25 and 0.025. The dotted line corresponds to the noise on the number of background photons received in *the same energy bin* as the annihilation line.

239

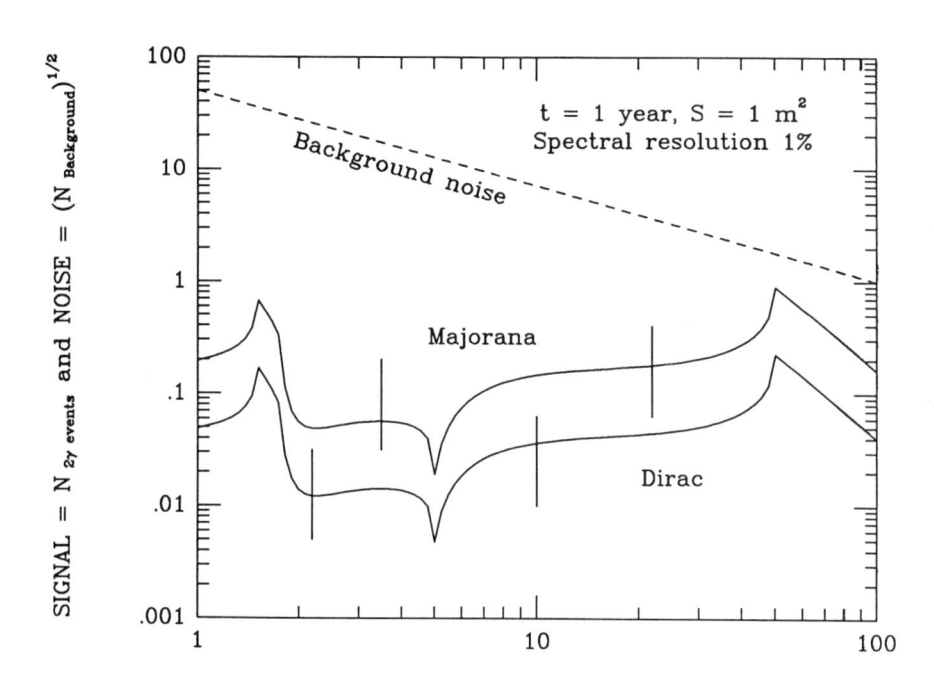

Figure 2: Same as Figure 1, but in the case of a Majorana ν_M and a Dirac neutrino ν_D. The vertical lines correspond to a relic density $\Omega h^2 = 1$ (left bars) and $\Omega h^2 = 0.025$ (right bars). The higgsino case would correspond to the Majorana line shifted downwards by a factor $cos^2 2\beta$ (see text and Equ. 16).

Session On
Gamma–Ray and X–Ray Physics

A near-vertical view of King crater [5.0° north, 120.5° east: 76 km diameter]. Principal point is 4.8° north latitude, 120.7° east longitude. (Apollo 16 metric camera frame 0891).

THE LARGE AREA HIGH RESOLUTION GAMMA RAY ASTROPHYSICS FACILITY

HR - GRAF

E. J. Fenyves, R. C. Chaney, J. H. Hoffman
University of Texas at Dallas

D. B. Cline, M. Atac, J. Park
University of California, Los Angeles

S. R. White, A. D. Zych, Q. T. Tumer
University of California, Riverside

E. B. Hughes
Hansen Laboratories of Physics, Stanford University

ABSTRACT

The main objective of this long term program is the development of the prototype of a space-based, large area High Resolution Gamma Ray Astrophysics Facility (HR-GRAF) capable of studying diffuse and point-like sources of gamma rays in the 1 MeV to 100 GeV energy range with a high angular and energy resolution, and very large sensitivity. The design of the HR-GRAF will be based on the High Resolution Gamma Ray Telescope (HRGT) developed by the University of Texas at Dallas and University of California, Los Angeles, collaboration for SDIO.

The long term studies planned for HR-GRAF to be flown first either in balloons or free fliers, and eventually in the Space Station are the following:

1) Map the galactic gamma radiation with better angular and energy resolution

2) Observation of characteristic angular variations in the diffuse gamma rays to identify the primary origin of this radiation as due to discrete structures

3) Study of point like sources of MeV and GeV gamma rays

4) Study the structure of the Galactic Center and test the hypothesis of the existence of a massive black hole there

5) Obtain some additional information on gamma ray bursters and on the possibility that gamma ray bursts may be originated by exotic objects

6) Search for gamma rays from the decay and annihilation of cold dark matter particles. Special emphasis will be given to the search for gamma ray lines in the 1 - 100 GeV range from the annihilation of weakly interacting, neutral, heavy Majorana fermions, such as photinos or higgsinos

7) Obtain some additional information on nuclear gamma ray lines.

The large area HR-GRAF will complement and extend the experimental studies to be carried out with GRO COMPTEL and GRO EGRET by introducing into gamma ray astrophysics high resolution detection technologies which were developed quite recently or are under development in high energy accelerator physics.

1. INTRODUCTION

The main objective of this program is the development of the prototype of a space based, large area High Resolution Gamma Ray Astrophysics Facility (HR-GRAF) capable of studying diffuse and point-like sources of gamma rays in the 1 MeV to 100 GeV energy range. A large fraction of this energy range from about 30-50 MeV up to several GeV has been studied by the SAS-2 and COS-B satellite experiments using spark chamber technique. They studied the galactic and extragalactic diffuse gamma ray emission[1-4], and produced a catalogue of high energy gamma ray sources[5].

Their data bases will be extended soon by the GAMMA-1 telescope which will study the energy range 50-500 MeV[6], and, particularly, by the launch of GRO in 1990, where EGRET a pair production telescope will detect and measure gamma rays in the range of 20 MeV to 30 GeV[7-10]. EGRET consisting of spark chambers detecting the electron-positron pairs, and a NaI(Tl) Total Absorption Shower Counter measuring the total energy of the incoming gamma, is characterized by a large field of view, with relatively good angular and energy resolution, low background and high sensitivity over a wide dynamic range[11].

GRO's Comptel detector, a Compton telescope using liquid scintillators and NaI(Tl) scintillators will measure gammas in the 1 - 30 MeV energy range, and is characterized by a relatively good angular resolution within a broad field of view[11].

Recently, The University of Texas at Dallas and the University of California, Los Angeles, have developed a space based, large area High Resolution Gamma Ray Telescope for studying point like astrophysical and man made sources of gamma rays in the 1 MeV to 20 GeV energy range[12-16] supported by SDIO Innovative Science and Technology Office.

The large area High Resolution Gamma Ray Telescope is a new concept in gamma ray telescope design employing a combination of liquid and gaseous drift detection techniques with scintillation fiber imaging systems and conventional scintillation counters for imaging and tracking the secondary particles generated by the gammas with a very high angular and energy resolution. This telescope is characterized by its event by event processing and decision making feature, which enables an extremely large flexibility in handling the data and obtaining the maximum possible information from the measurements.

The proposed large area HR-GRAF is a further development of the High Resolution Gamma Ray Telescope which will complement and extend the experimental studies to be carried out with GRO

COMPTEL and GRO EGRET by introducing into gamma ray astrophysics high resolution detection technologies which were developed quite recently or are under development in high energy accelerator physics. According to this HR-GRAF will have comparable or better angular and energy resolution than COMPTEL and EGRET, and its sensitivity will be also significantly larger.

The high energy part of the HR-GRAF spectral sensitivity will overlap with the low energy part of the proposed ASTROGAM[29] spectral sensitivity in the 1-100 GeV range, with comparable angular and energy resolution, and will be complementary to this facility in the energy range 1 MeV- 1 GeV.

2. LONG TERM SCIENTIFIC STUDIES

The long term studies planned for HR-GRAF to be flown first either in balloons or free fliers (shuttle or expendable space vehicles), and eventually in the Space Station are the following:

1) Map the galactic gamma radiation with better angular and energy resolution to help determine the origin and distribution of both the diffuse galactic gamma radiation itself and the primary cosmic radiation which produces it[17,18].

2) Observation of characteristic angular variations in the galactic and extragalactic diffuse gamma rays to identify the primary origin of this radiation as due to discrete structures such as point sources, active galaxies, or ridges of antimatter annihilation radiation. Special emphasis will be given to studying the "bump" in the diffuse gamma ray spectrum observed at MeV energies[19] with a much better energy and angular resolution than that of the previous experiments to answer the basic question whether the bump itself is real and of isotropic (and therefore cosmological) origin or not.

3) Study of point like sources of MeV and GeV gamma rays in order to
   (a) observe new, as yet unknown, localized gamma ray sources
   (b) identify at present unidentified sources of gamma radiation
   (c) resolve some controversies resulting from previous observations
   (d) improve the measurements of gamma ray spectra with a telescope larger in size, and better in angular

and energy resolution than those used in past experiments.

4) Study of the structure of the Galactic Center whether the gamma rays from the center come from a single object, such as a dark matter core of finite extent, or a complex of sources, and test the hypothesis of the existence of a massive black hole in the center[20].

5) Obtain some additional information on gamma ray bursters and on the possibility that gamma ray bursts may be originated by exotic objects such as superconducting cosmic strings and other topological defects[21,22].

6) Search for gamma rays from the decay and annihilation of cold dark matter particles. Special emphasis will be given to the search for gamma ray lines in the 1 - 100 GeV range from the annihilation of weakly interacting, neutral, heavy Majorana fermions such as photinos or higgsinos. The annihilation of these supersymmetric particles, which are prime candidates for the cold dark matter in the galactic halo, into two photons

$$\chi + \chi \rightarrow \gamma + \gamma$$

giving rise to discrete gamma lines may be observed experimentally for a broad range of M_χ the mass of χ with a telescope of sufficiently high energy resolution in the GeV range ($\Delta E/E \leqslant 1\%$)[23]. It was also shown[24] that (1) emission from a dark matter core at the galactic center may be the easiest source of dark matter annihilation gamma radiation to detect, (2) it may also be possible to detect radiation from a dark matter halo in the direction of bare patches in the interstellar medium at high galactic latitudes with a telescope of sufficiently high angular resolution, and (3) to measure the hardness of the gamma ray spectrum between 0.1 and 0.2 GeV may be a very promising way to study cosmologically significant neutral fermions of 5-25 GeV mass.

7) In addition to this the HR-GRAF will provide some additional information on nuclear gamma ray line spectra in the energy range 1 - 10 MeV.

3. THE HIGH RESOLUTION GAMMA RAY TELESCOPE

The large area High Resolution Gamma Ray Telescope (HRGT) consists of a Scintillation Fiber Converter and a Liquid Argon (or Xenon) Calorimeter separated by an Argon-Methane Gas Drift Chamber and operates as a

(1) Compton or Compton-Pair Telescope in the energy range 1-50 MeV, and as a
(2) Pair Conversion - Shower Telescope in the energy range 50 MeV-20 GeV (see Fig. 1)

with two different cryogenic liquid fillings of the Calorimeter: Argon or Xenon.

The entire telescope is surrounded with plastic scintillator plates to reject charged particles. Both the Liquid Argon Calorimeter and the Argon-Methane Gas Drift Chamber are Time Projection Chambers, and the Liquid Argon Calorimeter operates also as scintillation detector. It is used together with the Scintillation Fiber Converter for triggering the telescope, and for the time-of-flight discrimination to suppress the background of upward moving photons.

The Gas Drift Chamber helps to identify the electron pairs produced in the converter and thus to distinguish between Compton scattering and pair production events.

The HRGT has a surface area of 1 m^2, and the converter and the calorimeter are separated by about 70 cm. This improves the angular resolution of the telescope significantly. The thickness of the converter (20 cm) is chosen to be approximately one half of a radiation length for improving the efficiency for single Compton scattering events.

The converter consists of 1 x 1 mm^2 cross section plastic scintillation fibers made of polystyrene doped with butyl-PBD and POPOP (producing λ = 420 nm wave length photons), and clad with PMMA. The plastic fibers provide good energy and conversion point resolution due to their low density and low Z, and their attenuation length is over 2 meters.[25]

The scintillation fibers are stacked into 1 mm thick fiber planes U, V and W that are rotated by 60° angle relative to each other, and directly coupled to position sensitive Hamamatsu R2487 Series photomultiplier tubes[26].

This way a crude 3-dimensional image of the tracks of Compton electrons, electron pairs or showers produced by the gammas can be obtained. The number of photoelectrons per fiber per minimum ionizing track passing normal to the scintillating fiber is about 20. The photomultiplier tubes are equipped with 16 x 16 orthogonal wire anode planes which can provide a position resolution of $\sigma_{rms} \leq 1$ mm. The pulse length at the output of the photomultipliers is about 17 nsec. The fast rise time of the pulses provides a time resolution of about 2 nsec.

The 50 cm deep Argon-Methane Gas Drift Chamber indicates in the case of Compton events whether the scattered photon undergoes in the gas or the walls additional scattering or not, and images any electron track which comes out of the converter. In the case of pair production events it images the tracks of the electron pairs. The spatial resolution of the gas drift chamber is $\sigma_{rms} \leqslant 1$ mm.

The liquid drift chamber below the gas drift chamber serves as a calorimeter, and can be filled with liquid argon or xenon. The depth of the calorimeter is 60 cm (~4 radiation length for argon and ~22 radiation length for xenon). The 22 r.l. for xenon permits the energy measurement of showers properly located around the axis of the calorimeter up to several GeV with a $\leqslant 1\%$ energy resolution. The calorimeter is filled at present with liquid argon permitting the energy measurements up to about 50 MeV.

The position resolution of the Liquid Argon (or Xenon) Calorimeter is $\sigma_{rms} \leqslant 1$ mm, and the time resolution of the scintillation pulses in the chamber is approximately 2 nsec.

The overall energy resolution (FWHW) of the HRGT operated as a Compton Telescope is estimated to be varying under ideal conditions from 10% to 2.0% in the energy range 1 - 50 MeV, and its angular resolution (FWHM) varying from 1.5° to 10 arc min in the same energy range.

The angular resolution of the HRGT operated as a Pair Production Telescope is estimated to vary under ideal conditions from 5° to 5 arc min in the energy range 50 MeV to 20 GeV. The maximum effective geometric factor of the telescope is 1,800 cm^{2}sr.

4. DESIGN OF THE PROTOTYPE HR-GRAF

The design of HR-GRAF will be conceptually the same as the design of the High Resolution Gamma Ray Telescope (HRGT). However, several modifications and further improvements of telescope design will be carried out based on the data obtained in the extensive testings and measurements carried out on the already completed components of HRGT. Among others we plan to extend the energy range of HR-GRAF to 100 GeV.

The design study will concentrate on a very essential question: what is the best combination of liquid Argon or Xenon chambers, plastic or glass scintillation fibers, lead and scintillating fiber calorimeter systems, and gas drift chambers to achieve a relatively light, most reliable, and not too expensive telescope with the best angular and energy resolution parameters.

The design study will be carried out by detailed Monte Carlo simulation of the operation of the different possible combinations of the above detector components.

The most probable HR-GRAF designs are the following:

(1) Plastic Scintillating Converter and Scintillating Glass Fiber Calorimeter followed by a Lead and Scintillating Fiber Calorimeter, where the Converter and the Calorimeter are again separated by an Argon-Methane Gas Drift Chamber (Fig. 2).

The depth of the Plastic Scintillating Fiber Converter will be 20 cm (~1/2 radiation length), the depth of the Gas TPC 100 cm, the depth of the Scintillating Glass Fiber Calorimeter 40 cm (~4 radiation length), and the depth of the Lead and Scintillating Fiber Calorimeter 22 cm (~26 radiation length). The Lead and Scintillating Fiber Calorimeter will consist of alternating layers of 2 mm thick lead plates and 1 mm thick scintillator planes made up of 1 x 1 mm^2 fibers. The other dimensions of this HR-GRAF design will be similar to those of the HRGT.

The 4 radiation length deep Scintillating Glass Fiber Calorimeter will serve mainly as the calorimeter for the Compton and Compton-Pair Telescope operation, and the same Calorimeter together with the 26 radiation length deep Lead and Scintillating Fiber Calorimeter will be used mainly for the high-energy Pair Conversion - Shower Telescope operation.

The general design of the Plastic and Glass Scintillating Fiber units with the Position Sensitive Photomultipliers and connected electronics will be essentially the same as that of the Scintillating Fiber Converter of the HRGT.

The above telescope[15] is a fast detector with a time resolution better than 5 nsec. It is a self triggering system, can provide relatively good energy and angular resolution from 1 MeV to 100 GeV, does not require cryogenics and is easy to build and safe to use. The spatial resolution of the Compton electron conversion point in the improved converter, and the shower centroid in the calorimeter is estimated to $\sigma_{rms} \leq 0.4$ mm. The energy resolution of the Scintillating Fiber Calorimeter is, however, less good than that of the Liquid Xenon Calorimeter. The improvement of the energy resolution of the Scintillating Fiber Calorimeter requires further studies which will be carried out in the first year of the proposed project.

(2) Plastic Scintillating Converter and Liquid Xenon Calorimeter separated by an Argon-Methane Gas Drift Chamber. This design is almost identical to that of the HRGT with the only major change of increasing the depth of the Calorimeter to 83 cm (~30 r.l. for Xenon) to measure gamma ray energies up to 100 GeV with a good energy resolution, and the corresponding modification of the whole cryogenic detector system as well as the cryogenic liquid purification system from the present Argon system into a Xenon system.

The application of the liquid xenon which has a much shorter radiation length (2.7 cm) than liquid argon (14 cm) decreases the weight of the calorimeter of equal radiation length depth approximately by a factor of 2.4, and makes the whole telescope design much more compact.

The 83 cm deep Liquid Xenon Calorimeter assures the total (more than 99.5%) absorption of the energy of gammas up to about 100 GeV, as it was shown in Monte Carlo simulations carried out recently at UTD and Waseda University.

Using different sensing grid configurations than that used in the 60 cm Liquid Argon Calorimeter, the major characteristics of the Liquid Xenon Calorimeter have been calculated by T. Doke et al[27].

The energy resolution of this calorimeter was estimated by Doke under ideal conditions to be 0.44% at 3 GeV, and 0.24% at 10 GeV, and its position resolution was estimated to be 2.2 mm at 3 GeV, and 1.0 mm at 10 GeV. We intend to simulate the above calorimeter under less ideal conditions and try to estimate its energy and angular resolution in the first year.

The Liquid Xenon Calorimeter has the disadvantage of being a complex cryogenic system and requiring rather complex support systems, as compared to the simple operation of the Scintillation Fiber Calorimeter of the previous design.

On the other side the Xenon Calorimeter provides the ultimate energy resolution which may be most essential in some of the scientific studies, e.g. the annihilation of heavy Majorana fermions into photons.

The major parameters of the planned HR-GRAF are shown in Table I. These values are based on the measured and estimated HRGT parameters and could slightly vary for the different designs discussed above. We consider these parameters, however, as minimum requirements for the prototype HR-GRAF, and try to achieve better values wherever it is possible and seems to be necessary considering the changing requirements of the various scientific problems to be studied.

Some of the characteristic parameters of HR-GRAF are compared with those of COMPTEL in Table II, and with those of EGRET and Astrogam in Table III. The sensitivity limits and the centroid location precision for the above telescopes are compared in Figs. 3 and 4. It is important to emphasize that HR-GRAF is a modular unit, and the flux of gammas detected by a larger system consisting of many modular units is proportional to the number of the modular units, thus significantly increasing the sensitivity of the enlarged HR-GRAF system. According to this the improvement of the sensitivity is limited only by weight and financial limitations, and new modular units can always be added to the already operating facility when required.

The testing and calibration of the Prototype HR-GRAF will be carried out at an accelerator laboratory. The choice of this accelerator laboratory depends on the availability of the beam at the time of need. It is expected that a calibration beam suited to the needs of HR-GRAF can be found at an accelerator laboratory at a convenient time for this program.

5. INVESTIGATION APPROACH

The HR-GRAF will be flown first either in balloons or free flyers such as the shuttle or expendable space vehicles. Which one of these options will be chosen will depend on the availability of the above vehicles and technical considerations. The latter ones would definitely prefer the free flyers where one can test the behavior and operation of HR-GRAF in a real space environment.

The next step is the Space Station where the HR-GRAF could be flown as an Attached Payload. The Lunar HR-GRAF seems to be a natural extension of the Space Station HR-GRAF operation.

The combination of the Space Station and Lunar HR-GRAF operations will lead to very important astrophysical achievements, among others others to
 (a) better localization,
 (b) better identification, and
 (c) better mapping
of gamma ray emitting point sources.

REFERENCES

(1) Fichtel, C.E., et al. (1977). Ap. J. Lett. <u>217</u>, L9, and (1978) Ap. J. <u>222</u>, 833, and (1984) Astron. Astrophys. <u>134</u>, 13.

(2) Mayer-Hasselwander, H.A., et al. (1982) Astron. Astrophys. <u>105</u>, 164.

(3) Thompson, D.J. and Fichtel, C.E. (1982) Astron. Astrophys. <u>109</u>, 352.

(4) Fichtel, C.E. (1988) Workshop on High Resolution Gamma Ray Cosmology, UCLA (to be published in Nuclear Physics B).

(5) Bignami, G.F. and Hermsen, W. (1983) Ann. Rev. Astr. Astrophys. <u>21</u>, 67.

(6) Akimov, V.V., et al. (1987) Proc. 20th ICRC, Moscow, Vol. 2, 320.

(7) Hughes, E.B., et al. (1980) NASA TM 80590, and IEEE Trans. on Nucl. Sci., <u>NS-27</u>, 364.

(8) Fichtel, C.E., et al. (1983) Proc. 18th ICRC, Bangalore, Vol. 8, 19.

(9) Kanbach, G., et al. (1988) Space Science Reviews (to be published).

(10) Kanbach, G., et al. (1989) Gamma Ray Observatory Science Workshop, 10-12 April, 1989, Goddard Space Flight Center.

(11) Kniffen, D.A. (1988) 14th Texas Symposium on Relativistic Astrophysics, Dallas (to be published in the Annals of The New York Academy of Sciences).

(12) Fenyves, E.J. (1988) Proc. SPIE - The International Society for Optical Engineering, <u>879</u>, 29.

(13) Fenyves, E.J. (1988) Proc. Second Meeting on Low-Level Counting and Space-Based Use of Liquid Argon and Xenon Detectors, Waseda University and KEK, Japan.

(14) Fenyves, E.J. (1988) Workshop on High Resolution Gamma Ray Cosmology, UCLA (to be published in Nuclear Physics B).

(15) Atac, M., et al. (1988) Workshop on High Resolution Gamma Ray Cosmology, UCLA (to be published in Nuclear Physics B).

(16) Park, J. et al. (1988) Workshop on High Resolution Gamma Ray Cosmology, UCLA (to be published in Nuclear Physics B).

(17) Stecker, F.W. (1975) Phys. Rev. Lett. $\underline{35}$, 188.

(18) Stecker, F.W. (1988) Hyperfine Interactions, $\underline{44}$, 73.

(19) Schoenfelder, V., Graml, F. and Penningsfeld, F.P. (1980) Ap. J. $\underline{240}$, 350.

(20) Ramaty, R. and Lingenfelter, R.E. (1988) 14th Texas Symposium on Relativistic Astrophysics, Dallas (to be published in the Annals of the New York Academy of Sciences).

(21) Spergel, D.N. (1988) Workshop on High Resolution Gamma Ray Cosmology, UCLA (to be published in Nuclear Physics B).

(22) Schramm, D.N. (1988) Workshop on High Resolution Gamma Ray Cosmology, UCLA (to be published in Nuclear Physics B).

(23) Rudaz, S. (1988) Workshop on High Resolution Gamma Ray Cosmology, UCLA (to be published in Nuclear Physics B).

(24) Stecker, F.W. and Tylka, A.J. (1989) NASA/Goddard Preprint 89-008 (to be published in Ap. J., 1 August, 1989).

(25) Blumenfeld, H., et al. (1987) Nucl. Instr. and Meth. in Phys. Res., $\underline{A257}$, 603.

(26) Uchida, H., et al. (1986) IEEE Trans. on Nucl. Sci., $\underline{33}$, No. 1, Feb. 1986.

(27) Doke, T., et al. (1988) Workshop on High Resolution Gamma Ray Cosmology, UCLA (to be published in Nuclear Physics B).

(28) Mattox, J.R., et al. (1987) Nucl. Instr. and Meth. in Phys. Res., $\underline{B24/25}$, 888.

(29) Astrogam, Very Energetic Gamma-Ray Astronomy Experiment using the Astromag Facility, Investigation and Technical Plan, Naval Research Laboratory Preprint, 1989.

TABLE I

HR-GRAF Parameters

Energy Range	1 MeV to 100 GeV

Single Photon Angular Resolution (FWHM)

(a) Compton mode of operation $1.5°$ (at 1 MeV)
                                          10 arc min (at 50 MeV)

(b) Pair Production mode of operation $5°$ (at 50 MeV)
                                          5 arc min (at 20 GeV)

Energy Resolution (FWHM)   10%   (at 1 MeV)
                                          2%   (at 50 MeV)
                                          1%   (at $\geq$ 3 GeV)

Maximum Effective Area   1,000 cm^2 (at few MeV)
                                          3,000 cm^2 (E $\geq$ 50 MeV)

Maximum Effective Geometry Factor   600 cm^2-sr (at few MeV)
                                          1,800 cm^2-sr (E $\geq$ 50 MeV)

TABLE II

Comparison of COMPTEL and HR-GRAF

	COMPTEL	**HR - GRAF**
Energy Range	1 - 30 MeV	1 MeV to 100 GeV (1 - 50 MeV in Compton operation)
Energy Resolution	(5 - 8)% 8% (at 1 MeV)	10% (at 1 MeV) 2.0% (at 50 MeV)
Maximum Effective Area	50 cm^2	1,000 cm^2 (at few MeV)
Maximum Effective Geometry Factor	30 cm^2 - sr	600 cm^2 - sr (at few MeV)

TABLE III

Comparison of EGRET, Astrogam and HR-GRAF

	EGRET	ASTROGAM	HR-GRAF
Energy Range	20 MeV to 30 GeV	1 GeV to 1 TeV	1 MeV to 100 GeV
Energy Resolution	15% (100 MeV to 10 GeV)	1% (1 GeV to 100 GeV)	< 2% (at ≥ 50 MeV)
Maximum Effective Area	2,000 cm^2 (E > 200 MeV)	5,700 cm^2 (E > 2 GeV)	3,000 cm^2 (E ≥ 50 MeV)
Maximum Effective Geometry Factor	1,000 cm^2 - sr	7,000 cm^2 - sr	1,800 cm^2 - sr (E ≥ 50 MeV)

FIG. 1

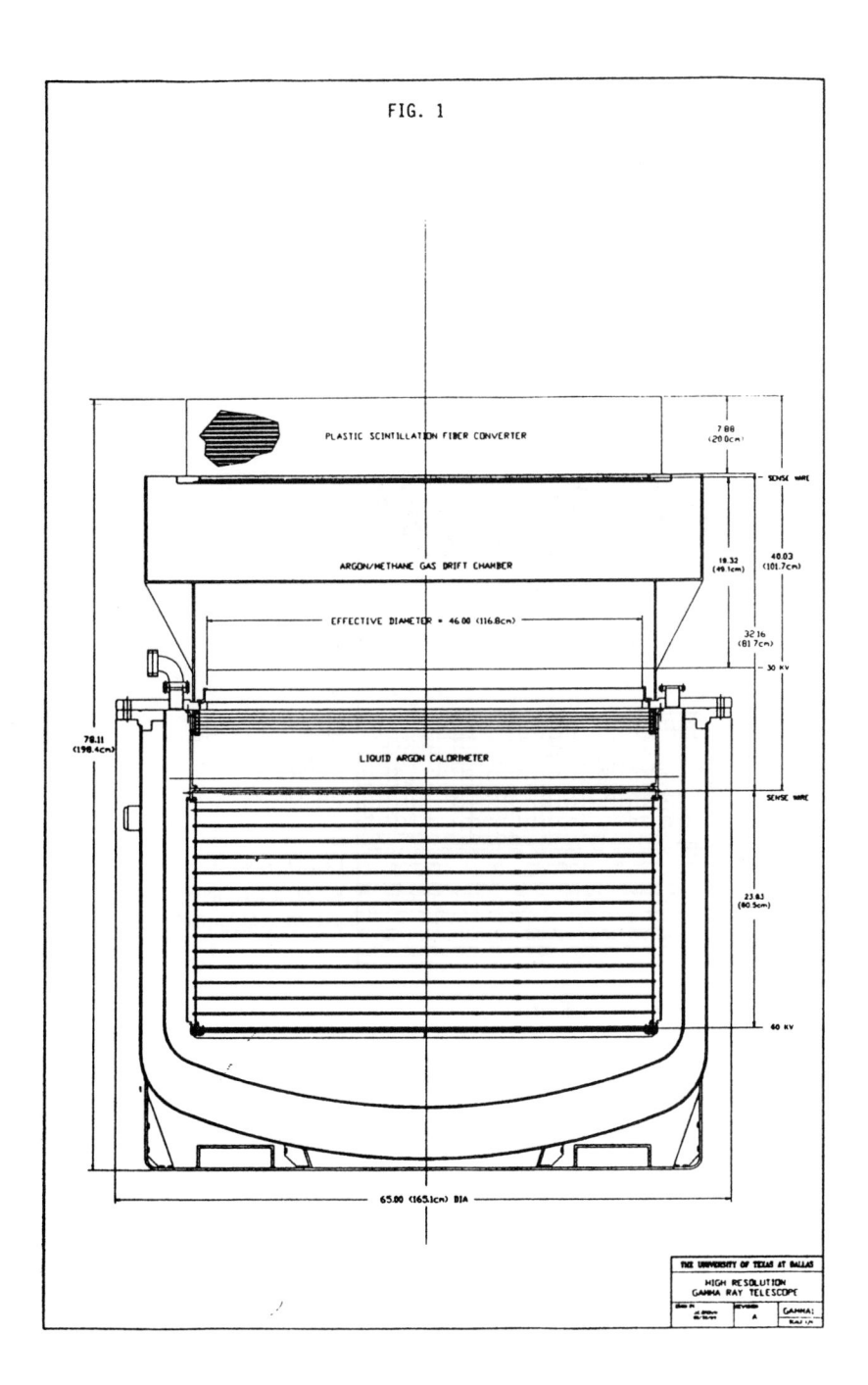

GAMMA RAY TELESCOPE

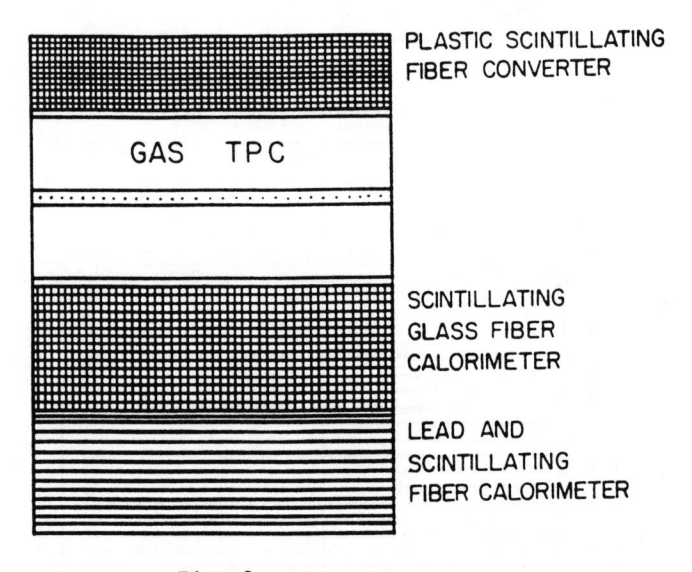

PLASTIC SCINTILLATING FIBER CONVERTER

GAS TPC

SCINTILLATING GLASS FIBER CALORIMETER

LEAD AND SCINTILLATING FIBER CALORIMETER

Fig. 2

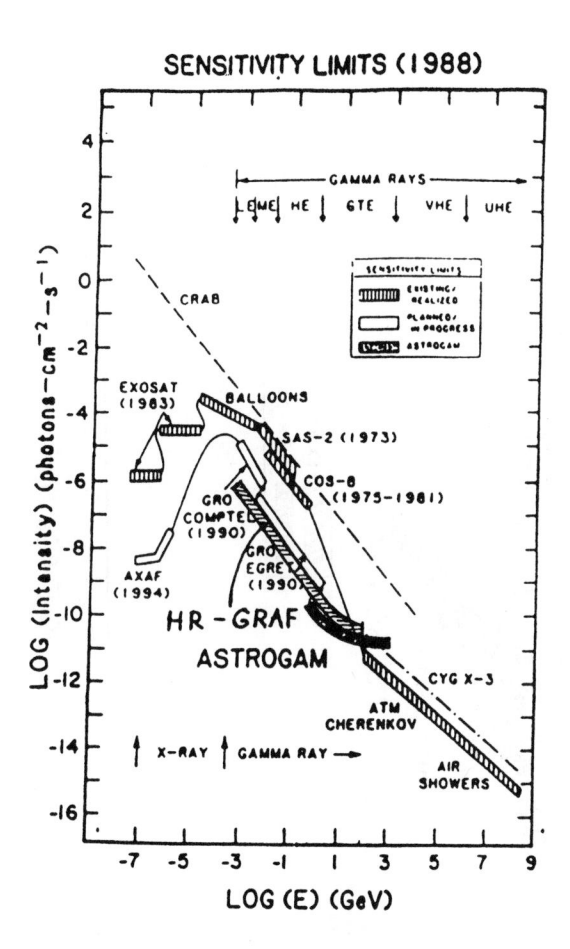

Figure 3: Representative sensitivity limits estimated for, or reached by, various x-ray and gamma ray astronomical instruments. Astrogam will fill the important gap in the GeV to TeV energy range (GTE). Astogram sensitivity assumes a two year mission and is based on a 5 sigma confidence limit for establishing an unknown source is a low background region. EGRET is expected to show an order of magnitude improvement over previous missions and has a limiting sensitivity of approximately 10^{-8} photons-cm^{-2}-sec^{-1} above 100 MeV (Ozel and Fichtel. 1988, Ap. J., in press). For both EGRET and Astrogam, increased background for observations in the galactic plane decreased sensitivity by factors of 3 to 10, depending upon the region of observation. Based on ref. 29.

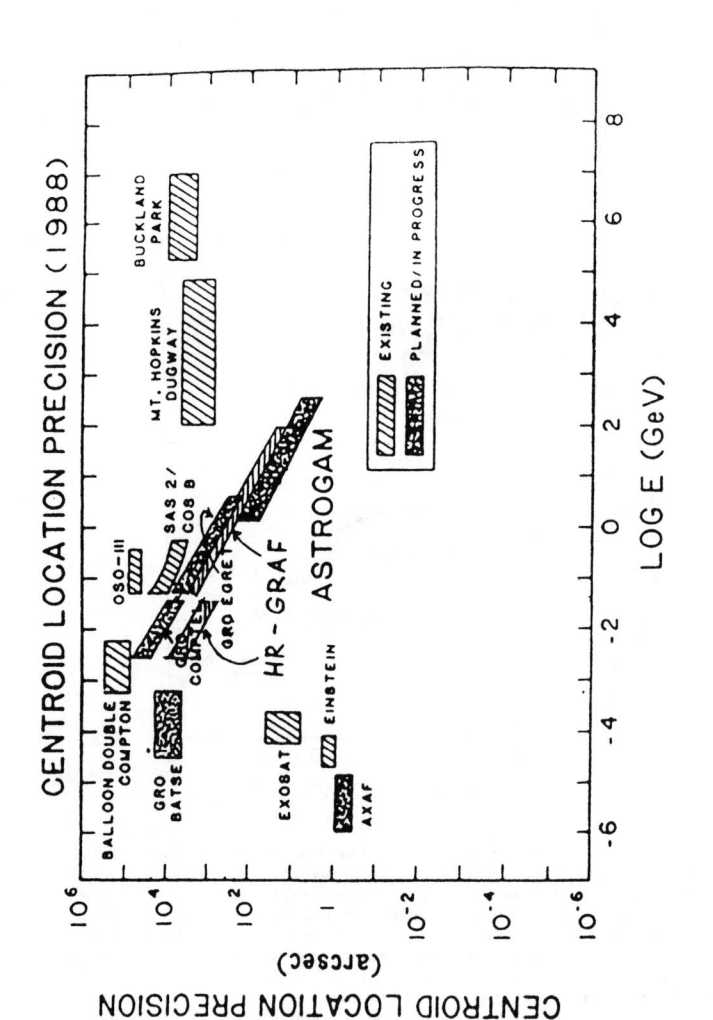

Figure 4: Source location precision reported by, or estimated for, various instruments used or planned in x-ray and gamma ray observations. EGRET values based on Kanbach et al. (1988, submitted to Space Science Reviews). Based on ref. 29.

LARGE X-RAY DETECTOR ARRAYS FOR PHYSICS EXPERIMENTS AT A LUNAR BASE

K.S. Wood
E.O. Hulburt Center for Space Research
Naval Research Laboratory
Washington, DC 20375

P.F. Michelson
Department of Physics
Stanford University
Stanford, CA 94305

Abstract

A large array of astronomical X-ray detectors can be constructed at a lunar base, and maintained there for a long time. Such an array permits development of a new astronomical subdiscipline based on collection of very large numbers of X-ray photons from bright X-ray sources. With such data qualitatively new experimental issues can be addressed, in gravitational physics, nuclear physics, and magnetospheric effects. Fundamental questions such as the existence of General Relativistic instabilities in rapidly rotating objects or the probing of orbits near a black hole can be investigated.

I. Introduction

This paper explores doing novel experimental physics from the moon by observing celestial X-ray sources. X-ray astronomy provides an established experimental approach, one that is comparatively low risk. The X-ray sky has been catalogued, sources have been identified, and physical conditions in those sources have been established to first approximation. From this work it is known that regions of strong gravity, high mass density (perhaps several times nuclear) and sometimes strong magnetic field strength are being observed. Interesting phenomena will be seen and major discoveries are likely.

What special advantages distinguish the moon? Two possibilities could be called "space" and time. "Space" refers to availability of enormous amounts of reasonably stable real estate, where one may place large instruments that view the universe with negligible intervening atmosphere. Time refers to cumulative observing time: a lunar base permits instruments to be active for decades. These features might be exploited in at least three classes of instruments: large arrays, long

© 1990 American Institute of Physics

baseline instruments, and all-sky monitors. Most of this discussion will concern large arrays operated for long spans of time. Radical increase in collecting aperture relative to past instruments shrinks integration times. Photometry and time-resolved spectroscopy benefit dramatically and angular resolution is also indirectly improved. If the array is operated for a long time further gains arise from monitoring long-term processes. Experiments done with large arrays range widely as to their physical content.

Skeptics might claim all further analysis could be circumvented right here by considering the array weight. The cost per unit weight delivered to the lunar surface is much higher than the corresponding number for low earth orbit (LEO). X-ray arrays that would be deemed "large" by comparison with past achievements would begin at apertures of ~10 m^2 and would weigh > 2,000 kg with current technology. Arrays > 100 m^2 would be highly desirable. If constructing the array amounts to achieving a certain total area of some design (held fixed for cost comparisons), then cost per unit weight times weight per unit area for the design gives cost per unit area. A disadvantage for the moon in cost per unit weight makes spurious any supposed lunar advantage for large arrays of all sizes. They belong wherever cost per unit area is least, presumably LEO. This amounts to saying the asset of "space" on the moon is deceptive: one cannot afford the cost of utilizing it and what one can afford would be unremarkable. Let us address this challenge before continuing.

It is premature to invoke transport cost in this way. Lunar base development will certainly bring economies in lunar transportation. The analysis assumes transport (not operation) dominates cost of an array, but the point should not be overlooked that in LEO it is necessary to build and maintain a suitable platform to house the array. Large platforms in LEO lead to nontrivial engineering requirements, including coping with gravity gradients and hazards of re-entry or collision damage. Total cost in LEO could grow nonlinearly with area, and the moon might win for sufficiently large areas. One could increase aperture indefinitely on the moon, with cost scaling perhaps less steeply than linearly with area, when economies of scale are included. Lunar arrays could be built in small increments (sending a module on each flight) and this could change the costing algorithm or spread cost so as to make it tolerable. The assumption that the array would be built with the same design on the moon as in LEO is plainly debatable.

The attempt to dismiss large X-ray arrays on the moon by transport cost involves too many assumptions to be rigorous. Accordingly, this concern will now be set aside, without denying its eventual importance. The new scientific rewards begin to be realized with arrays on the order of 10 m^2, which are completely reasonable with foreseen resources, while more ambitious arrays (>100 m^2) present a challenge that calls for further study.

The viewpoint taken here is that big arrays will become possible somewhere. When nations commit to building large space structures, there necessarily ensues parallel growth in the feasible size of space-based astronomical instruments. Otherwise the ratio of the size of the largest scientific payloads that are possible (in the engineering if not the political sense) to the size of other space hardware shrinks unreasonably as development proceeds. X-ray timing work needs exactly this growth in area to achieve a major advance. Payloads of order 1 m² have been flown for more than a decade. We shall inquire what physics a large array could do, irrespective of its site. Some ideas presented here have in fact been developed in connection with a concept for LEO, a 100 m² array for attachment to the Space Station, called the X-ray Large Array (XLA). This concept has been developed in studies, papers and proposals over the past few years (Wood et al. 1988; Dabbs et al. 1987,1988; Wood and Michelson, 1988; Michelson and Wood, 1989; Wood 1985).). Certainly it is viable as a Space Station payload, but it might also be appropriate for a lunar base. A lunar base might be the eventual site of an array even larger than XLA. We shall consider what such an array could do on the moon that it could not do in LEO and vice versa.

II. Photon-Rich X-ray Astrophysics

There is a venerable tradition of physics done by means of astronomical observation. The character of the physics depends on the source, which the observer cannot manipulate. Observational information about dynamical evolution proves invaluable in restricting physical interpretations. Two examples of past successes may clarify the role of large X-ray arrays. The first is that of the radio binary pulsar PSR 1913+16, whose orbit has been found to decay in a manner consistent with the emission of gravitational waves predicted by General Relativity (Taylor and Weisberg 1989). This was the first empirical evidence for gravitational waves. The need to find an astronomical test had to do with the weakness of the effect, which is proportional to the third time derivative of the mass quadrupole moment. PSR 1913+16 has proven useful for testing a specific, unified set of issues, which happen to be fundamental. The second example is solar physics. This is a field so diverse that journals are devoted to it and it comprises sub-disciplines that are almost independent of one another. One cannot easily sum up all the physics learned from the Sun, but its importance is beyond doubt.

The X-ray large array concept leads to an important new sub-discipline, one that we might give a name such as "photon rich X-ray astrophysics". It is what results from combining (i) ~1000 bright X-ray sources with (ii) observation by a very large X-ray collector, which is (iii) operated for long intervals of time. The X-ray photon yield is enormous and produces topical diversity like that of solar physics. At times it also takes on a fundamental quality, involving basic issues like those tested in PSR 1913+16, because it gives a probe of dynamical processes taking place in regions of unusual interest. Presenting this new territory calls for describing the sources that make it possible, the way large arrays change our ability to view the sources, some specific investigations, and the hardware approach. We take up these topics in that order.

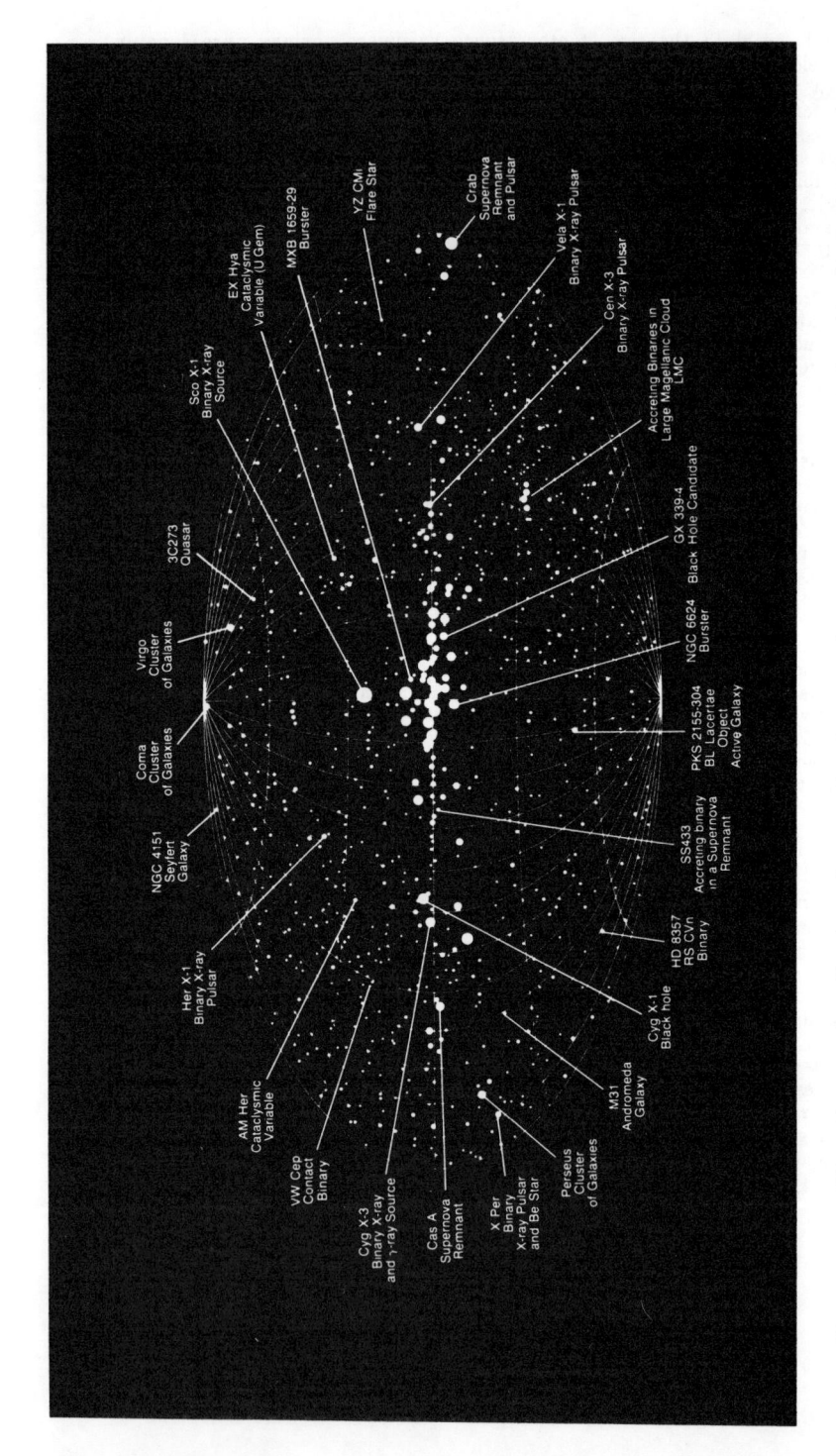

Figure 1. The X-ray sky displayed in galactic coordinates, as mapped by the HEAO A-1 experiment (Wood et al. 1984). The figure depicts 842 sources. The size of the dot is proportional to the logarithm of the source X-ray intensity. Prominent members of various source classes are indicated.

III. Appearance of the X-ray Sky: the Nature of the Bright Sources

We begin with the X-ray sky. Of particular importance is the variety of conditions encountered among the brightest X-ray sources, which are the primary observing targets.

The X-ray sky has been mapped with steadily-improving refinement over the past three decades. More than two thousand sources are now catalogued, the overwhelming majority found by X-ray satellites such as UHURU, ARIEL V, HEAO-1, the EINSTEIN Observatory, and EXOSAT. The Milky Way Galaxy stands out prominently; markers that delineate it are bright supernova remnants and accreting binary sources. In the binaries, a compact object (white dwarf, neutron star, or black hole) accretes material from a companion. A fainter population from our Galaxy consists of the coronae of nearby normal stars. Beyond the Galaxy one sees active galactic nuclei such as Seyfert galaxies, N galaxies, quasars, and BL Lacertae objects. These are probably accreting systems on a much larger scale, possibly accreting black holes with masses exceeding 10^6 solar masses. There is also emission associated with entire clusters of galaxies, where gas trapped in cluster potential wells reaches temperatures of several keV.

The X-ray sky is thus quite unlike the sky in visible wavelengths, where the brightest sources are normal stars. (The photospheres of normal stars, with roughly blackbody spectra at temperatures of order of a few eV, are not seen at all in X-rays.) Whereas in the visible sky the main features are point sources that remain nearly constant, bright sources in the X-ray sky are all either extended or variable, hence timing and mapping are very important. The prevalence of gravitational energy release (rather than nuclear) and of major inflows and outflows on various scales characterizes the X-ray sky. The bright X-ray sources provide testbeds for ideas about which empirical data cannot otherwise be gathered. Specific examples we shall describe will be drawn mainly from the compact stars in binary systems. Densities in neutron stars are ~10^{15} g cm^{-3} and magnetic fields reach 10^{13} gauss. Conditions in neutron star binaries have been exploited to study plasma physics and nuclear physics. Issues in relativistic gravitation are ripe for study.

IV. How Large Arrays Alter Modes of Observation

The path from large arrays to new physics was sketched earlier and will now be developed in greater detail. The timescale, δt, in which a detection can be made at a particular level of significance (designated by σ, i.e., $\sigma = 5$ corresponds to a 5σ detection) scales inversely with detector area because of Poisson photon counting statistics, i.e.,

$$\delta t = \sigma^2(S+B)/(S^2 A), \qquad (1)$$

where S is source intensity (in photons cm^{-2} s^{-1}), B is background in the same units and A is the effective aperture of the array. There are roughly a dozen sources in our galaxy for which S is of order
1 photon cm^{-2} s^{-1} or greater, the brightest objects in Figure 1. A large increase in photon-collecting power brings orders of magnitude of quantitative improvement in certain kinds of measurements, permitting qualitatively new problems to be studied. In photometry and time-

resolved spectroscopy, an aperture of 100 m² can bring 3 to 5 orders of magnitude improvement over the typical performance of past instruments. There is also 1 to 2 orders of magnitude improvement in the sensitivity to fractional flux variations, relative to what has been achieved to date. These breakthroughs can be applied to the brightest sources. (Another analogy with solar physics comes up here, in that the closest, brightest members of a class yields detailed physical information that cannot be obtained on more distant members of that class.)

Photometry refers to the accumulation of a series of counts as a function of time, in some chosen energy band. The shortest integration time is set by source brightness and detector aperture. With a 100 m² array a source such as the Crab Nebula gives 4 million photons per second, enough in one microsecond to constitute a detection and enough in 20 microseconds to determine source brightness to < 15% uncertainty. Photometric measurements are applicable to periodic, aperiodic and quasiperiodic phenomena. They can be used to study millisecond pulsars, black holes, X-ray bursts, eclipses and many other phenomena. It is not necessarily true that the level of modulation will be anywhere near 100%. Sometimes a basic issue hinges on observing subtle modulations that are small fractions of the total flux. With suitable summations of data, a large array can work at the level of 10^{-4} or even 10^{-5} of the total flux from the source, particularly at higher frequencies. This can be applied to such work as the study of quasiperiodic oscillations, searches for vibrations or gravitational lens effects in neutron stars, and accretion shock phenomena in AM Herculis stars and other sources.

Fast time-resolved spectroscopy is the process of making many (millions) of determinations of the X-ray spectrum of a source in rapid succession, with a short integration time for each spectrum. Spectral resolution depends upon the collector. If we assume a proportional counter then the resolution is about 18% at 6 keV. One can detect line or edge features as they appear and vanish, or determine rise and fall times of X-ray spectral components.

A by-product of having very large area and fast timing is that very fine angular resolution can be achieved. The idea is simple in principle. The large aperture permits a source such as an active galaxy or a cluster of galaxies to be seen in milliseconds. If an occulting edge can be made to sweep slowly across the field, sources can be resolved to a limit which is the angle scanned in the limiting integration time. Under favorable conditions this can be better than any other current means of achieving high resolution in X-rays.

A large array thus permits measurement of variability phenomena and determination of rapidly varying spectra on timescales of milliseconds or shorter and can be made to provide source mapping on very fine scales. Large arrays are complementary to mirror instruments, such as the Advanced X-ray Astrophysics Facility (AXAF). The primary strength of such instruments is in imaging, spectroscopy, and detection of faint sources. They do not produce photon-rich astronomy on millisecond timescales. Future X-ray missions now planned overwhelmingly consist of such focusing instruments, and there is no mission now planned with an area exceeding 1 m². This imbalance between imaging and timing should be corrected by inclusion of a large array in long range plans.

V. Examples of Observable Physical Effects

Representative experiments will now be described. The first examples will be chosen from the field of relativistic gravitation, a field where large arrays can open truly new territory for astrophysics. Regions of strong gravity exist in binary systems containing neutron stars or black holes. The gravitational potential ϕ/c^2 can reach levels of 0.1 - 1. Such objects stand out conspicuously in X-rays when they accrete matter, and characteristic variability timescales can be milliseconds or perhaps shorter.

Rapidly Rotating Compact Objects

The first example concerns effects of extremely rapid rotation. The misstatement is sometimes encountered that the sole limitation on the spinup of a weakly magnetized neutron star by accretion is the centrifugal breakup limit, i.e., the mass shedding limit for rapid rotation. It has been recognized since a pioneering paper by Chandrasekhar (1970) that rapidly rotating neutron stars are subject to general relativistic instabilities associated with gravitational radiation reaction. This process can become dominant at frequencies **below** the mass shedding limit. Growth of nonaxisymmetric modes of deformation -- with consequent gravitational radiation emission -- permits the star to reach lower energy states. The idea was first applied as a possible effect to be sought in young neutron stars in the seconds following core collapse. Additional modes and secular instability criteria were identified by Friedman and Schutz(1978).

The discovery of the 1.5 ms radio pulsar (PSR 1937+214; Backer et al. 1982) motivated Wagoner (1984) to consider how this instability (called the CFS instability) could work in a very old neutron star with a weak magnetic field accreting from a binary companion, as in a LMXB source. Because the nonaxisymmetric deformation modes that grow by gravitational radiation are damped by viscosity, these modes do not grow until the star is spinning very rapidly. At a critical rotation frequency the general relativistic instability growth time becomes shorter than the viscous decay time; this onset comes for spin periods on the order of a millisecond. A steady state can be reached in which angular momentum added to the star by accretion is carried away by gravitational radiation. The metric perturbation, h, observed at earth is proportional to the mean X-ray flux, S_x, hence the brightest accreting X-ray sources such as Sco X-1 are prime targets for detecting the phenomenon. The same stellar distortion that produces gravity waves can produce low amplitude, but perhaps detectable, modulation of the X-ray emission at this same frequency.

The amplitude and frequency are sensitive to details of theoretical modeling. A significant fraction of the energy released by accretion occurs within the amplitude δR of the nonaxisymmetric distortion. Under some circumstances the resulting modulation can be concentrated in a particular part of the spectrum. The frequency of the gravitational wave should be in the range 200 < f < 800 Hz, below the rotation frequency of the star. When accretion eventually turns off, the neutron star quickly sheds angular momentum by means of the CFS modes until it is at a period near the neutral stability point, a period on the order of a millisecond. It then enters an evolutionary phase like that seen in PSR 1937+214. The CFS gravitational wave luminosity becomes negligible.

It is important to try to detect CFS-unstable stars while the nonaxisymmetric distortion is still present, both to verify the CFS mechanism and to achieve simultaneous detection in X-rays and gravitational waves. One could discover the pulsar in X-rays and use the frequency found in X-rays to search for the gravity wave signal. Detection of the X-ray modulation alone would constrain the neutron star equation of state and viscosity.

Observationally this entails looking for (i) a millisecond X-ray coherent pulsation, which is (ii) at a rather low relative amplitude (in a source exhibiting other temporal structure) and (iii) is further disguised by the frequency modulation (FM) associated with binary orbital motion. Data analysis task requires first finding an optimal search strategy for such signals. Methods have been developed for searching for faint millisecond pulsations from sources in binary systems. The method is computationally intensive and involves searching a grid of trial binary orbits in order to find the right one (Wood et al. 1986, 1988; Norris and Wood, 1987)

Accreting Black Hole Candidates

The accreting black hole is among the central problems of observational astrophysics. If active galactic nuclei are accreting black holes, then much of the observable energy release in the universe involves such objects. Black holes having masses greater than a few M_o are permitted by General Relativity and conventional theories of neutron star structure. Examples are thought to have been identified in binary systems, the best known being Cyg X-1. Criteria on which past claims for black holes have been based include (i) determination of the mass of the compact object, (ii) observation of rapid aperiodic variability in X-rays, and (iii) distinctive spectral signatures. Cyg X-1 satisfies all three criteria, but each of them involves difficulties.

Consider first the mass limit, on which principal reliance has been placed in recent work (McClintock 1986). The maximum mass of a neutron star (rotating or nonrotating) becomes crucial. Such predictions involve the theory of nuclear matter. In view of uncertainties about the equation of state of nuclear matter theorists have attempted to place bounds on the maximum neutron star mass by means of the following assumptions: (i) the matter is a perfect fluid, (ii) the energy density is positive (i.e., gravity is attractive and is the only force relevant on large distances), (iii) causality holds (v<c), (iv) matter is microscopically stable (dP/dρ>0) and finally (v) the equation of state (EOS) is known below some fiducial density ρ_o, typically between 10^{14} and 10^{15} g cm^{-3}. For a wide range of proposed EOS it is found that (Hartle 1978) $M_{bound} = 6.8\ M_o\ \rho_{14}^{1/2}$, where M is the stellar mass and ρ_{14} is ρ_o scaled to 10^{14} g cm^{-3}. Thus for $\rho_o = 10^{14} - 10^{15}$ the bound ranges from 6.8 to 2.1 M_o.

The last of these assumptions is the least well founded. The equation of state of nuclear matter is not well-constrained by either laboratory or astrophysical data. For example, the possible detection this year of a very fast pulsar in the supernova 1987A (Kristian et al. 1989) underscores our ignorance. If this is a spin frequency, then it excludes nearly all otherwise-viable equations of state, although not exotic possibilities involving support against collapse by rapid rotation are not ruled out (Wood and Michelson 1988; Michelson and Wood 1989; Friedman et al. 1986). On the theoretical side, the extrapolation from models that fit laboratory data based on systems with few nucleons to systems with large numbers of nucleons is uncertain. For example, Bahcall, Lynn and Selipsky (1989) have described a new class of hadronic field theories in which stable configurations with masses > 3 solar masses are permitted.

Maximum neutron star mass "theorems" are only a useful guide to identifying candidates for further study, and not rigorous proofs of the nature of the compact object. Candidates identified this way include Cyg X-1, LMC X-1, LMC X-3, A0620-00, and, by somewhat different reasoning, SS433. There remains a need for identifying observational signatures of material that is near or crossing the event horizon near a black hole.

Aperiodic variability has been observed in Cygnus X-1 and in at least three other binary systems in our galaxy, although only in Cyg X-1 has it been seen to timescales as short as 3 ms (Meekins et al. 1984). For a time it appeared that rapid variability was a unique signature of a black hole. However, from EXOSAT observations, Stella et al. (1985) found that the transient V0332+53 showed aperiodic variability but was also a 4.4 sec pulsar. Rapid X-ray variability should only be thought of as a signature of an accretion disk around a central object that is either a black hole or a neutron star. Combined with constraints on the compact object mass, such variability can be reinstated as a useful probe of the inner regions of the disk. Much remains to be done with this approach. The existence of 3-ms bursts and their power are inferred from the study of long strings of data. The bursts are not seen individually but are studied in a statistical aggregate, because past instruments have been too small to see them in the presence of Poisson noise. Individual bursts would be well isolated with a large array and their time profiles and spectra could be determined.

Other Applications

The preceding effects, both relevant to relativistic gravitational physics, do not begin to capture the phenomenological diversity accessible to a large array. There are many other applications in gravitational physics, which might include searches for low-modulation gravitational lensing effects in neutron stars and the mapping of extended X-ray components of clusters of galaxies which can be used to infer presence of dark matter.

Nuclear physics could benefit from observations relating to the equation of state for neutron stars, measurement of neutron star radii, the study of X-ray bursts (thought to be nuclear explosions occurring when material accreted on neutron star surfaces undergoes a deflagration), and potentially from the detection of neutron star vibrational modes, which would be sensitive to details of the stellar structure dependent on nuclear physics. The study of the magnetospheres around neutron stars and white dwarfs could be furthered in many ways, including the observation of quasiperiodic oscillations (already seen in 1 m^2 instruments at the level of a few percent, but with many open questions calling for improved photon collection). Accretion shocks in AM Herculis white dwarf systems or the study of accretion torques in binary pulsars would provide other kinds of magnetospheric probes.

Long life of a lunar facility would lend itself to further studies, those that require a long time baseline in addition to a large area. One example would be observation of decay of binary orbits, a physical process that involves gravitational radiation and other physical mechanisms.

Other scientific investigations with a large array result from capabilities for angular resolution. These uses are sensitive to where the array is placed. In LEO, the occulting edge that one would use is the limb of the moon, which can move < 1 arsec/second in favorable circumstances. Combined with the ability of a large array to see, for example, bright active galaxies in 1 millisecond, this would translate into milliarcsecond angular resolution, although limitations of knowledge of lunar terrain would degrade it by about an order of magnitude. The lunar limb observed from LEO eventually gives access to about 10% of the sky for occultation purposes. This 10% happens to include many interesting sources including the Galactic Center, the Crab Nebula, and the brightest X-ray quasar, 3C273. Terrain and sky access limitations can be overcome by placing an orbiting artificial occulter in an orbit having the radius of the lunar orbit, but perpendicular to it. (Wood and Breakwell, 1987). This occulter could be steered to intersect the line of sight to targets of interest over the full sky.

If the array were placed instead at a lunar base, the limb of the moon could not be used in quite the same way, because it becomes the horizon. It would now give full-sky access at coarser angular resolution. To supplement horizon crossings, artificial occulters might again be considered in connection with an array at a lunar base, for example around the outer Lagrangian point, requiring a far-side location for the array. (For angular resolution of order 0.1 arcsec it would be possible to utilize coded masks in front of the detectors. This hardware approach is discussed in section VII.)

VI. Concerning Hardware

Substantial study has been devoted to hardware appropriate for observing bright sources. Gas-filled proportional counters are currently an attractive technology, since resulting weight per unit area is low, and the energy range can extend from below 1 keV (depending on choice of front window material) to above 20 keV. Units for space flight have been built for as little as $100 per cm^2 of effective area. In order to isolate and study individual X-ray sources, a proportional counter detector must be combined with collimators to restrict the field of view (typically to about 1 square degree, which is sufficient to isolate roughly the 2,000 brightest sources in the sky). There must also be a pointing system to acquire the target. Proportional counters sensitive from 0.5 to 25 keV with effective areas up to 2000 cm^2 have been flown in space. Detector area as well as energy range might increase somewhat in future development. Very large areas are naturally achieved by assembling many individual detector modules into an array. For the X-ray Large Array on the Space Station, an array with 512 individual detectors that achieved an effective area of 100 m^2 was considered in a Pre-Phase A study conduced at the NASA Marshall Space flight Center (MSFC). Assembly options that were examined ranged from emphasizing on-orbit self-deployment to extensive use of astronaut extravehicular activity. Much of the XLA study is relevant to the hardware issues associated with deploying a large array at a lunar base.

The cost of transporting the array to its intended site is influenced by the stowed configuration. For example, the detector weight and detector effective depth (volume per unit area) govern how many modules can be packed into the Shuttle. The Pre Phase A study for XLA found that without technology improvements one could pack hardware for an array approaching 100 m^2 into the Shuttle. Whether weight or volume is the limitation depends on the exact design.

In the Introduction we noted that, whenever transport dominates cost, the achievable size for fixed financial resource varies as the product of two factors, the weight transportable for given cost and the effective area per unit weight. Both should improve as technology advances. The former factor affects all payloads equally but the latter is specific to the X-ray detector design. Re-examination of designs for lunar use is warranted because of at least three key factors that may lead to radical change. First, the availability of "space" at the destination means that the usual premium on area efficiency (high ratio of effective to gross frontal area) does not apply; second, any creative use of material from the destination reduces transported weight per unit area, hence indirectly increases achievable size; third, structural members assembled at the destination never have to encounter either accelerations or gravitational fields exceeding ~1/6 g.

Formally one could begin with the irreducible requirements. These are that the detector must have good quantum efficiency for X-ray photons incident from a certain restricted direction, which is determined by a collimator. Nothing in this specification strictly requires use of large quantities of machined metal. Metal parts are traditionally used to give a conducting outer boundary to the proportional counter gas volume, to provide strength for containing the gas at its operating pressure, and for mechanical stability and alignment in collimator structures. On the moon one is free to operate a larger detector at lower pressure. Options for using lunar material include extracting counter gases from the lunar outgassing or using fine-grained lunar surface material as a passive, X-ray absorbing filler that could be stuffed into light, hollow, machined structures brought from earth. Much weight in traditional designs is functionally associated with passive absorption of unwanted photons. That purpose does not demand compositional purity in the absorber nor does it otherwise require technology available only on earth. Achieving minimal transported weight per unit area in a detector design is a well-posed problem that merits further study; a breakthrough could have far-reaching consequences. The advances in composite materials over the past two decades would justify such a study. Certainly, the scientific rewards that would result from large arrays should serve as a powerful incentive for exploring all ways to get them constructed.

Data handling requirements call for peak rates of ~100 Mbps, arising only when the array views the brightest celestial objects. In the case of XLA it was clear that preliminary processing of data at the Space Station was advantageous, while most data would be archived for later analysis on the ground. For a lunar array, substantial on-site computational power would be advantageous, and ease of data transfer might favor a near-side site.

VII. Other Instrumentation Options for a Lunar Base

Large arrays are not the only way to take advantage of the moon as a platform. An all sky monitor (ASM) exploits the temporal advantage of a lunar base with an instrument that provides continuous or suitably frequent recurring coverage of a large solid angle. There are several design options that fulfill this function; the exact variants need not concern us. The scientific importance comes from the frequent changes in levels of activity that are characteristic of accreting X-ray sources and also from the bright transients that appear at rates of a few per year. Ordinarily the total exposure to a source (the product of instrumental aperture and observing time) is not large in an ASM but important long timescales are covered for hundreds of sources. An ASM is almost a required adjunct to a large array, to assist it in finding targets of opportunity. Acting by itself the ASM can do valuable work on source parameters such as orbital periods. The quality of information an ASM provides depends on its timescale coverage. A lunar base maintained for many years would permit excellent long-timescale coverage to be obtained. A lunar ASM would not suffer from coverage gaps near the LEO orbital period, ~ 90 min, which sometimes prove inconvenient. An ASM on the moon would have a gap near 28 days and possibly a 50% duty cycle at that period, not a serious handicap since it is phased with an identical gap for the large array. (Note also that if the large array is segmented into independently-pointed modules it can itself carry out some monitoring functions.)

Another hardware option is use of a half-transparent, half-opaque aperture in front of the detector, to provide imaging capability. Such apertures are essentially pinhole cameras. Two broad classes that have received much study are the coded aperture imager (CAI) and the Fourier Transform Telescope (FTT). The CAI has a single mask placed some distance L in front of a position-sensitive X-ray detector. The size of holes in this mask roughly matches the detector position resolution. The FTT employs two masks, one displaced some distance L in front of the detector and the other immediately in front. In the FTT the mask opening is not set by detector resolution hence it can be made finer, to the diffraction limit. Masks usually have ~50% open area, hence the CAI cuts the effective area of the detector in half and the FTT reduces it by a factor of 4, because the two masks are in series. The angular resolution of these systems is the mask hole size divided by L.

Part of the scientific appeal of CAI and FTT designs, lies in the promise they offer for achieving high angular resolution, particularly in hard X-rays. The finest resolution attainable is set by a combination of the technical limitation on the maximum instrument length, L_{max}, and the diffraction limit for that L_{max} at the energy of interest. In LEO L_{max} is on the order of 50 m, but on the moon it could increase to ~5 km, simply by setting up the detector and mask platforms at two points on the surface that far apart (the "space" asset). Angular resolution improves as $L^{1/2}$ once the diffraction limit is reached, hence the hundredfold increase in L translates into a tenfold improvement in resolution over LEO instruments such as the proposed Pinhole/Occulter Facility (POF). Either the mask or the detector must be translated to maintain coalignment of the system with the line of sight to a celestial target, as the moon rotates. To preserve alignment for about an hour calls for translating the mask along an arc ~45 m in length, and he direction of motion is roughly perpendicular to the surface, except very near the poles. Integration time is directly proportional to the size of the structure that can be built for this purpose. A strategy to achieve good time utilization could run thus: begin by selecting several interesting targets in a small declination band. The first target could be tracked for ~1 hour, then the mask apparatus could be reset to be ready for the next target a few hours later. Over a few months each target in the declination band would accumulate several hours of data, and then the mask apparatus could be moved to a new position corresponding to a different declination band. Whether this is a reasonable lunar application of CAI/FTT methods depends on difficulty of engineering the mask-translation apparatus.

There are two further points to mention regarding CAI or FTT on the moon, both directly concerning large arrays. First, there is a "zoom" aspect to these techniques: if L is large they provide high-resolution imaging over a small field, but if L is small one has instead a wide-field camera. Such a camera is a strong candidate for inclusion in a large array. Regarding total array area, A, as fixed, one can ask what fraction of A is best devoted to such an imaging capability, to give enhanced performance on low-frequency variations and steady faint sources in the field. If all the array were covered with such masks, net throughput would suffer considerably. A 100 m^2 array with a FTT mask would have an effective area of only 25 m^2; with a CAI mask it would be about 50 m^2. It would also increase the weight and the cost, not only because of the masks but also because of additional complexity in the detectors. Optimizing for timing work, the best value of the fraction is probably not in excess of 10% of the total array area. A lower bound might be about 1 square meter, to provide ability to image to low flux levels. Thus one could well envision devoting 1-10% of an array to position-sensitive detectors suitable for use in conjunction with coded masks. The second point, then, is that this same position-sensitive section of the large array could also function as the detector element of the high-resolution long baseline system described previously. Conversion from one to the other would be a matter of removing the nearby mask and using the distant one in its place, with the same caveats about the need to maneuver it as were noted earlier. The translation structure for the distant mask could represent a second phase of array development. Thus both all sky monitors and coded apertures can be incorporated naturally into the concept of an advanced lunar X-ray facility for timing work.

VIII. Summary

The main points we have made are summarized as follows. (i) A lunar base offers advantages of extensive real estate where large instruments might be constructed with the possibility of maintaining them for long times. (ii) A large X-ray array, which is one utilization of those assets, will bring about photon-rich X-ray astronomy. (iii) This would be a powerful instrument for extracting information about physical conditions and processes in bright X-ray sources. It would address a diversity of topics, and would not be a single-purpose instrument. The sources in question are known to exist and observing them at their natural dynamical timescales will bring rich rewards. (iv) The challenges of building such an array anywhere include processing the data and realizing economies of scale in fabricating hardware. Building such an array at the moon additionally involves the challenge of minimizing the transportation cost. This might mean a general solution (reducing the cost per unit weight delivered) or one specific to an X-ray array (changing the design to reduce transported weight). (v) All sky monitors and incorporation of hard X-ray imaging through addition of masks are important options for increasing the power of such a facility.

There are several important items that require further study. (i) A size versus science tradeoff study should explore what can be done with arrays of 10 - 100 m^2, as a function of area. The realistic range for fabrication cost per unit area, weight per unit area, and volume per unit area should be examined. (ii) The optimum lunar layout should be studied. Should the array be composed of a number of independently-pointable modules, and if so, how many? (iii) Detailed simulations of the instrument performance for various observing programs should be carried out. This should include a study of what fraction of the array is optimally devoted to a coded aperture or Fourier Transform Telescope. (iv) Lunar operations, including data handling, need to be studied. (v) The optimum site should receive attention, a comparative study of the scientific programs achievable from various lunar sites and from LEO.

IX. References

Backer, D.C., et al. 1982, Nature, **300**, 465.
Bahcall, S., Lynn, B.W., and Selipsky, S.B. 1989, submitted, Nuclear Physics B.
Chandrasekhar, S. 1970, Phys. Rev. Lett., **24**, 611.
Dabbs, J. et al., 1987, Report on Pre-Phase A Feasibiliy Study for XLA, NASA/MSFC.
Dabbs, J., Davis, B., and Davis, J. 1988, Proceedings of the AIAA 26th Aerospace Sciences Meeting, Paper AIAA-88-0654, Reno, Nevada, January 1988.
Friedman, J.L., and Schutz, B.F. 1978, Ap. J., **222**, 281.
Friedman, J.L., Ipser, J.R., and Parker, L. 1986, Ap. J., **304**, 115.
Hartle, J.B. 1978, Physics Reports, **46**, 201.
Kluzniak, W., and Wagoner, R.V. 1985, Ap. J., **297**, 548.

Kluzniak, W., and Wilson, J. 1986, talk presented at 13th Texas Symposium on Relativistic Astrophysics, Chicago, IL, Dec. 1986.
Kristian, J., et al. 1989, Nature, **338**, 234.
McClintock, J.E. in The Physics of Accretion onto Compact Objects, p. 211, K.O. Mason, M.G. Watson, and N.E. White Eds. (Springer Verlag, Berlin, 1986).
Meekins, J.F., et al. 1984, Ap. J., **278**, 288.
Michelson, P.F., Price, J.C., and Taber, R.C. 1987, Science, **237**, 150.
Michelson, P.F., and Wood, K.S. 1989, MNRAS, in press.
Norris, J. and Wood, K.S, 1987, Ap. J., **312**, 732.
Stella, L., et al. 1985, Ap. J. Lett., **288**, L45.
Taylor, J.H., and Weisberg, J.M. 1989 "Further Experimental Tests of Relativistic Gravity Using the Binary Pulsar PSR 1913+16", preprint.
Wagoner, R.V. 1984, Ap. J., **278**, 345.
Wood, K.S., et al., 1984, Ap. J.Suppl., **56**, 507.
Wood, K.S., 1985, in The Fourth Marcel Grossman Meeting on General Relativity, ed. R. Ruffini (Elsevier Science Publishers), p. 783
Wood, K.S., and Michelson, P.F. 1988 in International Symposium on Experimental Gravitational Physics, eds. P.F. Michelson, Hu Enke, and G. Pizella (World Scientific Publishing: Singapore), p. 475.
Wood, K.S., Norris, J.P., Hertz, P., and Michelson, P.F. 1986, in Variability of Galactic and Extragalactic X-ray Sources, ed. A. Treves (Bologna: Associazione per L'Avanzamento dell'Astronomia), p. 213.
Wood, K.S. and Breakwell, J.V. 1987, Acta Astronautica, **15**, 9.
Wood, K.S., and Michelson, P.F. 1988, MNRAS, **232**, 225.
Wood, K.S., et al. 1988 "A Concept Study of the X-ray Large Array for the NASA Space Station".

ARCSEC SOURCE LOCATION MEASUREMENTS IN GAMMA-RAY ASTRONOMY
FROM A LUNAR OBSERVATORY

David G. Koch
NASA Ames Research Center
Moffett Field, California 94035

E. Barrie Hughes
Hansen Laboratories of Physics, Stanford University
Stanford, California 94305

ABSTRACT

The physical processes typically used in the detection of high energy γ rays do not permit good angular resolution, which makes difficult the unambiguous association of discrete γ-ray sources with objects emitting at other wavelengths. This problem can be overcome by placing γ-ray detectors on the Moon and using the horizon as an occulting edge to achieve arcsec resolution. For the purposes of discussion, this concept is examined for γ rays above about 20 MeV for which pair production dominates the detection process and locally-generated nuclear γ rays do not contribute to the background.

I. INTRODUCTION

After the flight of future γ-ray telescopes, such as the GRITS telescope or one with similar capabilities, it is conceivable that the positions of the point sources first detected by SAS-2, COS B and EGRET will be known to about 1 arcmin. Even this degree of precision is unlikely to permit the unambiguous association of all the sources with objects emitting at other wavelengths. In addition, limited angular resolution also inhibits the separation of the diffuse galactic emission from embedded discrete sources. These goals will not be achievable until the positions of the γ-ray objects can be measured to the level of several arcsec. To achieve this goal will require a quantum step forward in the position measurement of high energy γ-ray sources. A potential solution to this problem is to place one or more γ-ray detectors on the surface of the Moon and to use the horizon of the Moon or the edge of a crater as an occulting edge. In this note, the utility of this method will be outlined and the items identified that will need further study.

II. CONCEPT

The proposed unique objective of lunar-based γ-ray astronomy is to make a quantum step forward in point-source position measurement. The principle involves the placement of a large area detector on the Moon with a clear view of the horizon or of the edge of a large surrounding crater. The horizon or crater edge is used as an occulting edge to achieve, over time, arcsec resolution. The γ-ray detector need not itself possess good angular resolution, but should have a small field-of-view to reduce the acceptance of background events. For a detector placed at or near the lunar equator, the horizon defines a plane that sweeps the whole sky in one lunar month. The detector would require an azimuth-elevation mount so that it could accommodate an occultation occurring at any place on the horizon and thereby achieve all-sky coverage and also compensate for irregularities in the horizon at the arcsec level.

Take as a goal an angular resolution of two arcsec for a weak source. The rotational velocity of the Moon is 33 arcsec per minute, or roughly two arcsec in four seconds of time. For sources off the lunar equator, the apparent velocity decreases as the secant of latitude. In five years of observing, 134 risings and settings of each object of interest can be measured. This integrates to 536 seconds of observing time per 2 arcsec bin. The brightest γ-ray source (Vela) has a flux above 25 MeV of 5×10^{-5} photons $cm^{-2}s^{-1}$. The goal is to make measurements for sources with flux 5×10^{-7} photons $cm^{-2}s^{-1}$, or in five years to accumulate 2.7 photons m^{-2} of detector area. Typical detection efficiencies are about 20%, so about 0.54 counts m^{-2} can be detected in five years. The cosmic background seen in the same period (taking the diffuse flux from the galactic plane near the galactic center to be 2.2×10^{-3} photons $cm^{-2}s^{-1}sr^{-1}$) for a detector having a field-of-view and resolution element of one degree would be 0.58 counts m^{-2}. Background will also occur due to cosmic ray interactions in the lunar surface, producing γ rays that can also appear in the field-of-view. If the celestial background is arbitrarily doubled, to account for this locally-generated background, the background rate is 1.2 counts m^{-2} in five years. Given a detector of fifty square meters, the background count is 60 on the average when the source is below the horizon and the source plus background count is 87 when the source is above the horizon. This corresponds to a 3.5 sigma detection with two arcsec position resolution.

It is a reasonable expectation that GRO will generate a catalog of from 500 - 1000 sources. The best positions, which will be for the brightest objects, may be on the order of a few arcmin for the centroid uncertainty. Most of the objects, however, will be near the detection limit of 10^{-7} photons $cm^{-2}s^{-1}$ and have a positional uncertainty on the order of one degree. Assuming a goal of high-precision position determination on about one-thousand sources, one occultation will be measured about every half-hour by a lunar-based detector. This appears to be a reasonable observing program.

The detector must be designed uniquely for the lunar application. Specifically, it will require the smallest possible field-of-view, which would also be its resolution element. This requirement could make it very different from the traditional large field-of-view spark chamber design. Considering that the transportation of mass to the Moon is very costly, the detector should also be designed with this factor in mind. These requirements may generate a novel detector design. One concept might be to use gas discharge devices with the wires aligned along the pointing direction. The electron-positron pair, resulting from a γ-ray conversion traveling nearly parallel to the axis, produces a large fast pulse compared with any other direction. Lightweight plastic scintillator would be used to surround the detector to reject charged primary particles and also provide time-of-flight information to determine direction. A 50 m² detector would best be configured as 2 m horizontally x 25 m vertically to minimize the impact of irregularities in the profile of the occulting edge. Two bore-sighted star trackers could be used to measure the absolute location of the horizon and a zenith tracker for those times when the horizon is illuminated by the Sun. The irregularities in the edge across the two-meter width of the detector and the distance to the horizon, along with the vertical resolution of the detector, determines the angular resolution of the system. Ignoring for now the detector spatial resolution and requiring an angular resolution of 10^{-5} (2 arcsec), the horizon would have to be at least 10 km away if the irregularities in the horizon (assuming them to be unmapped) were as rough as 10 cm. The detector would have to be maximally 30 meters above the surface, the exact height depending upon the mean elevation of the horizon. An angular resolution of 2 arcsec will also permit a mapping of the galactic arm structure in the nearby galaxy M31 and the identification of discrete sources within that galaxy.

The measured source positions for a single detector on the equator would have an uncertainty of 2 arcsec by about one degree, i.e., a line of position. If two detectors were used, at say 20° degrees above and below the equator, the lines of position would intersect by 40°, giving an error box (diamond) of about 2 x 3 arcsec. The detectors could not be much more than 20° from the equator, since the galactic plane, where many of the sources of interest will be located, is inclined to the lunar orbital plane by about 64°.

III. DEVELOPMENT

Development of the concept of source-position measurement by occultation from the lunar surface could proceed as follows. A principal objective will be to design an appropriate detector with the properties of large area (about 50 m²), light weight and narrow field-of-view (about one degree). Calculations will also be necessary to estimate the background to be expected from cosmic ray interactions with the lunar surface. Application of the Monte Carlo codes FLUKA and HETC will provide one way of estimating this background. This will lead to improved estimates for the angular sensitivity for discrete sources and to the development of methods for the high resolution mapping of extended emissions. Both the scientific goal of arcsec position measurements and the need for an observatory lifetime of 5 years or more are factors compatible with the effort of establishing a facility for high energy γ-ray astronomy on the Moon.

A simple and yet scientifically useful proof of concept would be to place a single 2 x 1 m^2 detector on the Moon to obtain data from Geminga, the second brightest γ-ray source in the sky, whose flux above 25 MeV is 1.8 x 10^{-5}. This object is brighter than the Crab and yet its position at γ-ray energies is uncertain to about one degree. After one year of integration for E > 25 MeV, about 8 counts per 2 arcsec of rotation of the Moon could be measured with essentially no background counts (0.5 per 2 arcsec). Even after two occultations, using larger angular bins, the position could be obtained to within 30 arcsec. Thus, within just a few occultations, the concept can be proven as well as a significant astronomical measurement made.

IV. SUMMARY

Occultation measurements appear to be the only solution to the long-standing limitation on angular resolution in γ-ray astronomy. Lunar-based detectors will provide an opportunity to make angular resolution measurements with a precision which cannot otherwise be achieved. Such measurements could lead to unique identifications for many detectable γ-ray sources and greatly improve the separation of discrete sources embedded within the diffuse galactic emission.

A radioactive thermoelectric generator (RTG) is shown on the Lunar surface in the center foreground, producing electrical energy for the central scientific station beyond. Although expedient for free-flying interplanetary probes and the tight schedules for the original manned Lunar landings, they pollute the environment with spurious γ-rays and antineutrinos. They could have been used to normalize the orbital γ-ray spectrometer measurements, but were not. They symbolize the basic problem that human exploration also brings with it human encroachment, often spoiling the natural environment. The Space Physics Group of this Workshop emphasized a need to perform measurements of the Moon before a permanent manned or man-tended presence was established. A need for a strong precursor mission phase (and not simply a re-flight of 25-year-old Apollo instrumentation) was stressed. A point of view was also stated that any exploration group which takes RTG-type technology into the lunar environment should later be required to remove it. (AS14-67-9366).

Session On Surface Physics

View from the Lunar surface at Station 6 of Apollo 17's landing site [20^0 9' 55" north latitude, 30^0 45' 57" east longitude] near the center of Taurus–Littrow valley. These boulders are comprised of blue–gray breccia. (AS17-140-21497).

SURFACE PHYSICS - MATERIALS SCIENCE RESEARCH POSSIBILITIES ON A LUNAR BASE

Alex Ignatiev
Department of Physics
and
Space Vacuum Epitaxy Center
University of Houston
Houston, TX 77204-5507

ABSTRACT

The possibilities of return to the Moon and resultant setup of research laboratories at a permanent Lunar Base will allow for the undertaking of experiments in surface physics and materials science that are expected to benefit from the unique environment of the Moon. These researches include studies of lunar atmospheric contamination profiles as pertaining to the use of the vacuum environment of the moon for surface physics and thin film growth research, studies of radiation damage (cosmic, solar, etc.) of the surface and bulk properties of materials, the study of microclusters (charging, agglomeration, sticking probabilities, affects on mechanical systems operation), solar ultraviolet effects on materials surface properties, the growth of thin epitaxial films in the lunar vacuum principally for application to solar cells for lunar use, and dark matter studies (identification, characterization, utilization).

INTRODUCTION

Surface physics and thin film materials research has progressed since the early 1960's mainly as a result of the introduction of ultra-high vacuum technology. The attainment of vacuum levels of the order of 10^{-10} torr allowed for the study of atomic surface properties of materials on the hour timescale without significant contamination of the surface from the ambient vacuum[1-3]. Thin film growth with unparalleled atomic perfection followed the ability to identify and reproduce growth surfaces of known atomic order[4-6]. The contribution of ultra-high vacuum to progress in the field of surface and thin film physics cannot be understated.

Based on the above comments, it is expected that the vacuum environment of the moon (~10^{-11} torr) would be highly beneficial to continued research in the surface physics arena as well as a vehicle for extended thin film research and development.

DISCUSSION

Of specific interest in Lunar Base surface physics studies will be the study of the interaction of ultra-violet radiation with surfaces of materials. It has been previously shown that UV/near UV radiation in the range of 100 to 400 nm can significantly affect the chemical state of the surface of a material[7,8]. The nature of this UV interaction is not well documented, however, it is known that it is not a direct classical photon bond breaking process[9]. To add to interest of the studying this effect is the fact that the effect can be very detrimental to the chemical stability of the surface of a material[10]. Detailed studies of this interaction must be coupled with an ultra-vacuum environment and a continuous broad band soft UV radiation source, and a large working volume to house the variety of characterization tools required for critical analysis of the effect.

The interaction of charged particle radiation with surfaces is another area of interest in surface physics. A large amount of data is available for the interaction of high energy ions (30 keV to 100 MeV) with materials. However, the interaction of the incident ions is principally described in terms of nuclear interactions with atoms of the bulk[11,12]. The surface character of the interaction is little known, and yet that interaction is critical in determining the stability and reactivity of the surface. Study of the basis of the surface character of the high energy ion interaction with materials, as well as the study of the low energy ion (100 eV to 30 keV) interactions with surfaces is required[13]. Such information, in light of the long term charged particle doses expected for materials on the moon, will also be prominently required in the application of stable materials to lunar facilities.

Of recent interest is the study of the interaction of small clusters of atoms. Such macro-atomic physics is finding application in catalysis[14] and magnetic studies[15]. The surface of the moon is covered with a fine micro dust, and as a result the study of the behavior of small particulates (several 10's to several 100's Angstroms in diameter) could be undertaken there. This can only be readily done in a ultra-vacuum environment so as to reduce the effects of adsorbed contaminants on the particle dynamics. The reduced gravitational acceleration levels on the moon will also add to the ability to study coalescence of small particles. The cosmic radiation present will serve to charge the particles and as a result the particle dynamics can be investigated under electrostatic conditions.

An overriding concern in the use of the vacuum and radiation environment of the moon for surface and thin film research is the lunar atmospheric contamination distribution. Little is known of the contamination 'plume' or local atmosphere about a lunar facility and its variation with time. It has been identified that MIR generated and dragged around the cosmos its own contamination cloud. Such an occurrence would be unacceptable for a lunar base. Studies would have to be undertaken, therefore, for the determination of and the prediction of the dynamics of base contaminants. The 'rarefied' lunar atmosphere and the 1/6 g gravity level make for dynamics predictions significantly different from terrestrial prediction algorithms. Therefore, data acquisition and model generation would be required to allow prediction and control of base contaminants.

An additional possibility for the use of a lunar base for thin film technology is the ability to produce high quality thin films in the lunar vacuum environment. Thin film growth requires ultra-high vacuum conditions. For large area thin film growth, as would be required for the generation of solar cells, a large vacuum volume is also required. Such conditions could be met on the surface of the moon and as a result solar cells could be produced at a lunar base by using raw materials from the moon. Silicon-based solar cells would be the most direct to produce due to the prevalence of silica on the moon, however, *GaAs* or *InP* cells could be realized with added lunar materials processing. The solar cells would clearly be used for power generation for lunar base and other lunar as well as interplanetary purposes.

Finally, an area which overlaps materials science and nuclear physics is the search for 'dark matter' in the universe. A recent proposal[16] has identified a form of dark matter which is singly charged (+ or -) with a mass of $\sim 3 \times 10^5$ amu. The proposal projects that the dark matter should be prevalent in the moons surface at a concentration of 0.1% and that in the nucleosynthesis only the negative particle reacts with a limit at He^* (3 protons and 1 negative dark matter particle). The positive dark matter particle remains as H^* (hydrogen-like). Such particles could readily be identified within lunar material, and if existent would be mined for future energy production. Although this is very speculative, the search for the super heavy dark matter in the lunar soil could be undertaken using materials analysis techniques available on a lunar base.

SUMMARY

A proposed Lunar Base would be of use to scientist and engineers addressing surface physics and materials science problems. Not only would data become available to identify basic understanding of questions such as the interaction of low energy and high energy charged particles and solar UV with surfaces of materials, and the dynamics of interaction of (charged and uncharged) small particles, but also materials applications in the lunar environment could bring such things as solar cells to production possibilities on the lunar base.

REFERENCES

1. M. Hablanian, in Advances in Vacuum Science and Technology, ed. E. Thomas (Pergamon, N.Y. 1960).

2. S. Dushman, Scientific Foundation of Vacuum (Wiley, N.Y. 1962).

3. L. Holland, W. Steckelmacher and J. Yarwood, Vacuum Manual (E & F.N. Spon, London 1974).

4. A.Y. Cho and J.R. Arthur, Prog. Sol. State Chem. 10, 157 (1975).

5. V. Narayanamundi, Physics Today 37, 24 (1984).

6. A.Y. Cho, Thin Sol. Films 100, 241 (1983).

7. E. Ekwelundu and A. Ignatiev, Surface Sci. 179, 119 (1987).

8. A. Mesarwi, A. Ignatiev and J.S. Liu, Sol. State Comm. 65, 319 (1988).

9. A.Z. Moshfegh and A. Ignatiev, Energy 12, 277 (1987).

10. A. Mesarwi, Y. Sun and A. Ignatiev, Energy 3, 269 (1987).

11. J. Lindhard, M. Scharff, H. Schiott, Mat. Fys. Medd. Dan Vidensk. Selsk 33 14 (1963).

12. K. Bruce Winterbon, Ion Implantation Range and Energy Deposition Distributions, (Plenum, N.Y. 1975).

13. A. Zomorradian, J. Tougaard and A. Ignatiev, Phys. Review B30, 3124 (1984).

14. C.L. Pettiette, S. Young, M. Craycraft, J. Conceicaa, R. Laaksonen, O. Cheshnovsky and R. Smalley, J. Chem. Phys. 88, 5377 (1988).

15. T. Arnoldussen and E-M Rossi, Ann. Rev. Mater. Sci. 15, 379 (1985).

16. S. Glashow (Private Communication).

288

The possibility of an active Moon, indicated by volcanic activity or atmospheric venting, is important. Below is the crater Aitken [16.5° south, 173.1° east: 131 km diameter] whose dimpled clusters to the right of its central peak may be lava domes. (Apollo 17 metric frame 0484). Above is a possible volcanic extrusion in a small crater west of Aitken. (AS17 metric frame 0481). A standard reference is the Basaltic Volcanism Study Project: *Basaltic Volcanism of the Terrestrial Planets*, 1286 pages (Pergamon, New York, 1981), available from the LPI.

Group Leader Summaries

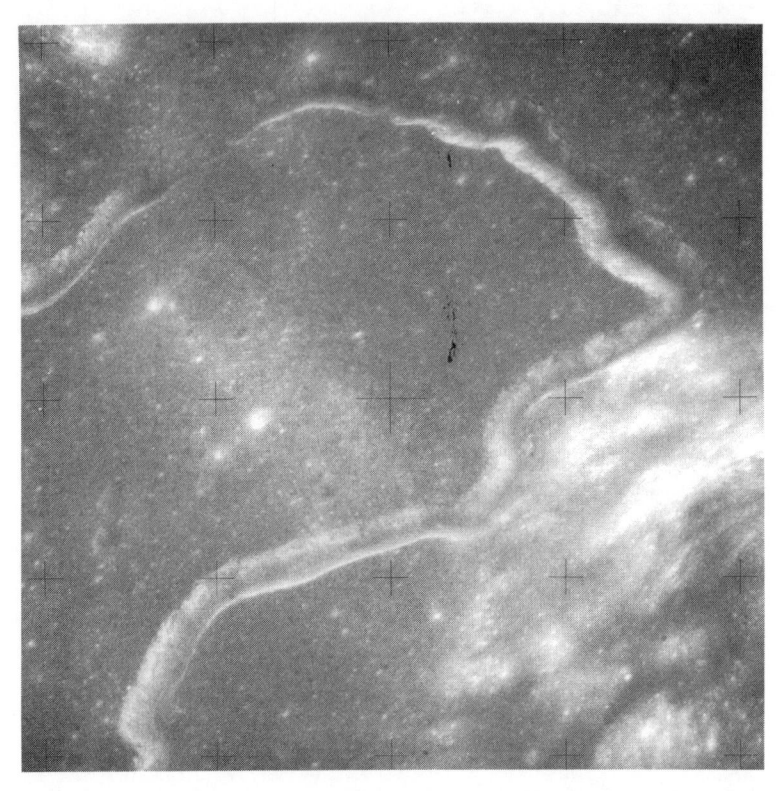

Lunar rilles as well as craters represent alternatives for large-scale instrumentation of particle detectors. The idea is either to excavate a tunnel at the crater or rille bottom for the instrumentation, or to erect it directly in the open or in an inflatable habitat (providing a shirt-sleeves environment) and subsequestly to cover it with dust and regolith. Lunar rock and regolith would then be used to pack the detector array with the desirable mass as a particle target. Liquid scintillators could also be used, particularly when and if the liquid substance is produced *in situ* on the Moon. View above is the Hadley Rille landing site from 113 km altitude. Principal point is 26.0^0 north, 3.5^0 east. (AS15-81-10887).

Possible Cosmic Ray Experiments on the Lunar Base
Discussion in Stanford on May 20

The main advantages of performing low energy (∼GeV) cosmic ray experiments on the Moon is the lack of atmosphere and the fact that Moon's orbit is well outside the radiation belts. Two types of experiments seem to be of particular importance now: a measurement of the cosmic ray composition beyond the iron group (M.Wiedenbeck) and a measurement of the antiproton fluxes up to 10-15 GeV(S. Rudaz). The knowledge of the composition up to high Z values will vastly improve the ratio of primary to secondary nuclei in cosmic rays and allow for much deeper understanding of the comic ray propagation, i.e. galactic structure and magnetic fields strength. The measurement of the antiproton flux will serve the same purpose and in addition will help setting much stricter limits on (or even discover) dark matter particles and scenarios.

At high energies, however, the atmosphere actually provides a target for interactions and helps the detection of the low magnitude cosmic ray fluxes. Thus the detection at a lunar base will require the replacement of the atmosphere with a target of similar thickness. The advantage than will be that individual partcles will be detected instead of atmospheric showers, where the properties of the primary cosmic ray are hidden behind numerous secondary partciles and interactions. S. Swordy suggested to build a several tens of m^2 scintillator calorimeter usind lunar dust as absorber. In combination with plastic scintillators this device will be able to measure the composition of cosmic rays for energies up to 10^{17} eV with a charge resolution of about 0.5 and energy resolution of ∼10%. Such a measurement will be very sensitive to the origin of the cosmic rays in this range as well as to galactic structure.

An experiment which will bring a lot of knowledge on the acceleration of cosmic rays at point sources, such as binary X-ray systems and young supernova remnants, is the monitoring of such sources in the TeV range. Producing radiation in this energy region requires extreme conditions at the source, thus it is even more sensitive to the conditions and dynamics of the sources and might be crucial for understanding the acceleration processes. The fluxes are again very low (∼10^{-11} photons.cm−2.s−1 above 1 TeV and require large area detectors.

© 1990 American Institute of Physics

J. Learned suggested the observation of Earths atmosphere for Cherenkov flashes produced by giant ($>10^{18}$ eV) air showers. The technique requires development, but it might be able to extend the measurement of the cosmic ray spectrum by orders of magnitude in a region which is extremely important from astrophysical and cosmological point of view.

Generally most of the experiments suggested can be as well performed on a Space Station and are certainly only suited for a "mature" Lunar Base. The scientific topics deserve serious consideration and I hope that better justification for putting these experiments on the Moon will be given in the written versions of the talks.

Todor Stanev

Comments on Particle Astrophysics at a Lunar Base

Simon P. Swordy

Enrico Fermi Institute, University of Chicago.
LASR, 933 E56th St, Chicago, Il 60637

The Moon has no atmosphere and a low intrinsic magnetic field. As such it is an ideal platform for observations of those particles from space which are disturbed by passage through the Earth's atmosphere or magnetic field. Although the same can be said for an artificial Earth satellite in general, the Moon has additional practical advantages. It is able to physically support extremely large experiments without requiring additional structure. It is also capable of providing raw materials for use in the experiments themselves, this can significantly reduce the mass required to be lifted from Earth. The suitability of a lunar base for observing various types of particles is considered under the following headings. Some critical comparison is made with potential satellite experiments having similar objectives.

High Energy Photons ($> 10^8\, eV$)
The lunar base would be highly appropriate for this type of observation, particularly at high energy,($> 10^{11}$eV), where extremely large area, (≈ 1000 m^2), detectors are necessary. Such large areas would be almost impossible to acheive on a satellite.

Antiparticles ($\bar{p}$, e^+, $\bar{\alpha}$, ...)
The Moon has a distinct advantage over presently planned experiments in this area for the US Space Station. The lack of magnetic field on the Moon will allow measurements of $\bar{p}$ and $\bar{\alpha}$ to well below 1 GeV, which is not possible on the Space Station because of its low inclination orbit. This will investigate the energy region where kinematic limits prevent the production of these particles by collisions in the interstellar medium. A lunar base could therefore provide the first detection of primary antimatter.

High Energy Cosmic Rays ($> 10^{15}\, eV$)
The lunar base could provide an ideal site for an investigation of the composition of these particles by hadronic calorimetry, with $\approx$ 10m^2 of detector area. The provision of the bulk of the calorimeter ($\approx$100 tons) by lunar regolith represents a major advantage over an equivalent satellite experiment.

Weakly Interacting Particles
The Moon possesses no intrinsic advantage as a detector site for such particles over an equivalent Earth-based detector. It might however be useful as a screen against noise of terrestrial origin for a sensitive detector placed on the dark side.

© 1990 American Institute of Physics

View northwestward in Mare Imbrium (Sea of Showers or Rains). Crater Timocharis [26.7° north, 13.1° west: 33 km diameter] is at the edge of the termintor in the upper left-hand corner, and Archimedes A [28.1° north, 6.4° west: 12 km diameter] is on the right-hand edge. Beer [27.1° north, 9.1° west: 9 km diameter] and Feuillée [27.4° north, 9.4° west: 9 km diameter] are the pair just left of and above center. Rilles are clearly visible. (AS15-94-12776).

Afterword

View looking north, with crater Triesnecker [4.2° north, 3.6° east: 26 km diameter] near the center in Sinus Medii. Hyginus [7.8° north, 6.3° east: 9 km diameter] is the crater on top of the rille, where the latter bends. Ukert [7.8° north, 1.4° east: 23 km diameter] is just above left, center. (AS10-32-4813).

THE MOON AS THE SEARCHING GROUND FOR PROTON DECAY*

J. C. PATI

Department of Physics and Astronomy
University of Maryland, College Park, MD 20742, USA

ABDUS SALAM

International Centre for Theoretical Physics
34100 Trieste, Italy

and

B. V. SREEKANTAN[†]

Department of Physics and Astronomy
University of Maryland, College Park, MD 20742, USA

Received 7 November 1985

The distinct advantage which a lunar detector would have over a terrestrial one in searching for proton decays corresponding to rather long nucleon lifetimes $\gtrsim 6 \times 10^{32}$ yrs is noted.

Proton-decay searches at various laboratories of the world have so far led to results which are either negative or whose interpretations are subject to some ambiguity. The IMB group[1] working with the largest detector, particularly well-designed for detecting the $e^+\pi^0$-mode, have as yet reported no candidates for this mode, setting a lower limit $[\tau_p/B(p \to e^+\pi^0)] \gtrsim 2 \times 10^{32}$ yrs. On the other hand, several groups[2-5] have reported contained events which could arise from genuine proton decays (in the experimenters' views) of the type $p \to \mu^+ K^0$, $p \to \mu^+ \eta$, etc. Unfortunately the interpretations of these events appear to be subject to ambiguity, especially if one demands near consistency between the observations of the different groups. The sole cause of this ambiguity is the expected background owing to atmospheric neutrino-induced reactions, whose rate in the relevant energy-range roughly appears to coincide with that of the observed events.

Since proton decay bears on fundamental issues,[6] one would, of course, like to see that the efforts in its search be strengthened in *all directions* so that one may finally ascertain beyond a shadow of doubt whether or not it decays with observable lifetimes $\lesssim 10^{34}$ yrs (say).[a]

If the background-to-signal ratio in the search for proton decay continues to be a problem even after the present detectors have run for the next three to four years (whether or not this will happen largely depends, of course, upon the proton lifetime and the nature of its dominant decay modes), it seems to us that one must then consider

---

[†] Permanent address: Tata Institute of Fundamental Research, Bombay, India
[a] From a theoretical standpoint, within grand unified symmetries like SO(10) and SU(16) which unify members of one family in contrast to SU(5) and which permit a *two-stage-descent* into $SU(2)_L \times U(1) \times SU(3)^c$, proton lifetime is expected to be of order 10^{31}–10^{34} yrs for such a two-step descent with $\sin^2 \theta_W \simeq 0.22$–$0.23$.

seriously the possibility of using the moon as the searching ground for proton decay. The major advantage of the moon in this regard is the fact that it has no atmosphere and, therefore, no atmospheric neutrinos or atmospheric muons. The background of neutrinos, muons and neutrons arising from interactions of primary cosmic rays with the moon rock and/or the detector materials can be made sufficiently small by a suitable design and a choice of the location of the detector, as we discuss, so that even a few events of the type observed on the earth (e.g., the $\mu^+ K^0$-type), if seen on the moon, should constitute evidence for proton decay. As we shall see, the galactic and extragalactic neutrino backgrounds do not pose a serious problem on the moon either.

The purpose of this note is to stimulate thinking in this regard. Conscious of our lack of expertise in rocket technology and its limitations and also of our knowledge of detectors which may be available, say five to ten years from now, and the lack of detailed information about the surface and the underground structures of the moon, we still wish to make a start here by spelling out what appear to us to be some of the relevant issues.

We are encouraged to entertain the idea of moon experiments in the first place because several space planners are now enthusiastic to revive trips to the moon and have suggested that few people may even live there, mining raw materials, growing their own food and using solar energy.[b] This could provide a very useful experience which is needed for exploration and possible colonization of space in the future. If such an exploration of the moon materializes, it would become much easier to entertain the idea of setting up the proton-decay experiment on the moon as a parasitic endeavor to the main program at a very small fraction of the total cost. We therefore proceed with the presumption that the moon exploration will resume in full force in the not too distant future.

Since we wish to have a source which effectively provides at least (few) $\times$ (10^{33} to 10^{34}) nucleons, which would need at least (few) $\times$ (10,000 tons) of matter,[c] it is clear that we must utilize the moon rock (or dust) itself to provide the bulk of this. In other words, we should plan to carry from the earth only those components and systems which are needed to detect the decay products, and to record and transmit the information.

Since the flux of primary cosmic rays unhindered by the atmosphere and magnetic field would be rather enormous at the surface of the moon (e.g., the flux of 10 GeV protons is $\simeq 1$ per cm^2 per sec $\sim 10^{14}$ per 1000 m^2 per year) we must reduce this and also the secondaries produced in these interactions, especially the neutrons, by at least 15 to 16 orders of magnitude (say) to help isolate proton-decay signals which may be of the order of 10 to 5 per year for a 20 kiloton detector with $\tau_p \sim (1 \text{ to } 2) \times 10^{33}$ yrs (say). To reduce the flux by this extent, we must shield the detector by placing it, for example, in a tunnel with, say, 100 meters of rock above it (see Fig. 1).[d] One may, for

---

[b] For a descriptive report on this, see e.g. J. Bigger, Washington Post Magazine Section, p. 8, September 1, 1985.

[c] 10,000 tons of matter yields about 6×10^{33} nucleons.

[d] By having at least 100 meters of rock above the detector, the flux of primary cosmic rays would be cut down by a factor of at least 10^{20}. See discussions later about possible background due to secondary particles.

Fig. 1

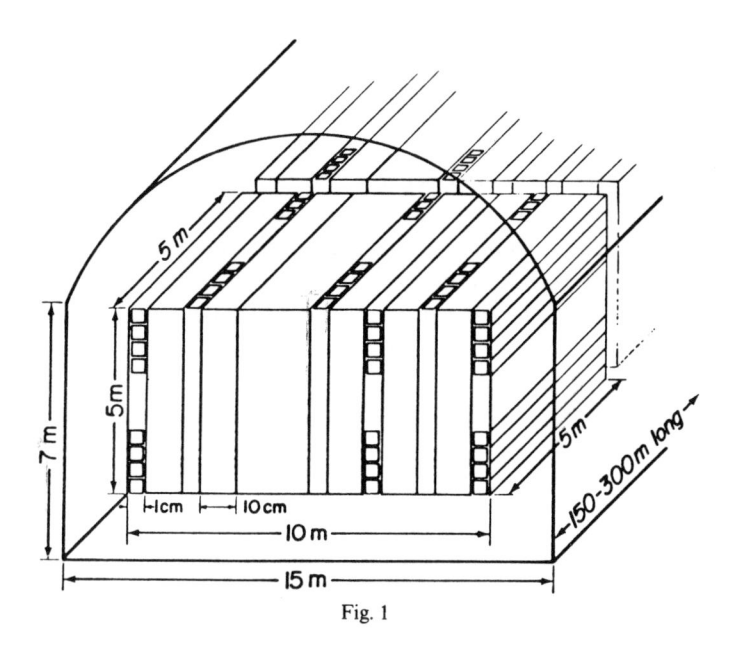

example, blast a tunnel on one side at the bottom of a crater which is fairly deep[e] (depth $\gtrsim$ 100 meters, say) but not too wide at the top.

While the exact details of the nucleon decay detector to be installed on the moon will naturally depend on the state-of-the-art of detector technology, the limitations on the size and weight of individual packages and the restrictions on the type of gases that can be carried to the moon, etc., we may for purposes of illustration envisage systems that are similar to the ones operating in the terrestrial underground laboratories. Since there appear to be no water or any other transparent liquid readily available on the moon, we have to rule out water Cerenkov type of detectors and plan in terms of the fine grain calorimeters. In line with the currently operating systems the detector could be modular in construction, *each module* having the following features: overall dimensions $\approx$ 5 m (height) $\times$ 5 m (length) $\times$ 10 m (width) (see Fig. 1), composed of 80 vertical layers of moon rock packed in suitable lightweight containers—each 5 m $\times$ 5 m $\times$ 10 cm, interspersed with 80 crossed layers of gaseous discharge counters made of lightweight plastic walls. Each counter in the layer may be 5 meters long with a cross-sectional area of 1 cm $\times$ 1 cm. The lunar rock (with a packing density[f] of about 2 g/cm^3) in such a module would weigh approximately 400 tons and the plastic containers nearly 2 tons.

For a 10–20 kiloton nucleon detector, we require 25–50 such modules and the length

---

[e] If the idea of mining raw materials on the moon materializes, then blasting a tunnel may not need new equipment.
[f] The average density of the moon is $\simeq$ 3.39 g/cm^3. That at the surface is $\approx$ 2.2–2.8 g/cm^3, corresponding to rocks like basalt.

of the tunnel for the detector system alone would be 125–250 m—the overall dimensions of the tunnel would have to be approximately 150–300 m long, 15 m wide and 7 m high. From the mouth side of the tunnel there would be a background of neutrons produced by the bombardment of the unhindered cosmic rays on the surface of the crater. To reduce this background, the assembling of detector modules should start sufficiently deep inside the tunnel—i.e. about 30–40 meters from the mouth—and this space could be utilized for all the electronic and other equipment.

We see that for a 10–20 kiloton detector, the total earth-weight of the detector equipment which needs to be carried from the earth to the moon for the sake of the proton-decay experiment is around 25–50 modules × 2 tons = 50–100 tons. The recording and transmitting equipment may weigh at least half as much. Once the moon exploration project is established, the cost of carrying this weight from the earth to the moon might be manageable.[g]

Expected Background

On the moon one does not have to worry about atmospheric neutrinos or muons. But secondary hadrons like $\pi^\pm$ and kaons produced by interactions of the primary cosmic rays with the moon rock or dust can decay into muons and neutrinos. (Neutrinos can also arise from the secondary muon decays.) These will provide a source of background for nucleon decays, provided $E_\nu \sim 1$ GeV $\pm$ 200 MeV. Fortunately, owing to the intervening moon rock or dust, which is much more dense ($\rho \gtrsim 2$ g/cm^3) than the earth's atmosphere, the secondary hadrons will most of the time interact strongly with the lunar rock, lose energy and be absorbed before having a chance to decay. For example, for a 2 GeV pion, the effective decay length is about 110 meters, whereas the interaction mean free path is less than 1/2 m. For this reason, the neutrino background ($E_\nu \sim 1$ to 2 GeV) arising from secondary decays in the moon would be reduced by at least a factor of 200 compared to that on the earth. This is the major advantage of the moon experiment as stressed before.

The production of energetic muons ($E_\mu \gtrsim 2$ GeV) arising from $\pi^\pm$ (or $K^\pm$) decays would also be considerably reduced compared to that in the earth's atmosphere for the same reason as above—i.e., most of the correspondingly energetic $\pi^\pm$ (or $K^\pm$) will not have a chance to decay. The higher the energy of pions (or kaons), the lower is the relative probability for their decays. Those muons which would still arise from pion (or kaon) decays with energies ~ 1 to 5 GeV (say) would rapidly lose energy in the rock due to ionization loss; the range of a 5 GeV muon in the moon rock with a density $\simeq 2.5$ g/cm^3 is about 10 meters, while the decay length is nearly 9 km. Thus the relevant muon flux arriving at the nucleon decay detector which is shielded by moon rock ($\gtrsim 100$ m) on all sides would be reduced by at least five to six orders of magnitudes compared to that on the earth. For this reason, there would be no

---

[g] Any cost estimate would, of course, be significantly lowered beyond the initial stages of the exploration of the moon.

equilibrium neutron background, either. Furthermore, the component of the neutrino background arising from muon decays would be essentially eliminated. This would further reduce the net neutrino background on the moon relative to the earth. Conservatively, we will still take the relative reduction factor to be $\gtrsim 200$.

So far, we have not considered the contribution from galactic neutrinos which are predominantly produced in the interaction of cosmic rays with the interstellar gas and dust. Due to the lens shape of the galaxy and due to the fact that the solar system is located off the center, this flux is highly anisotropic as seen on the earth and on the moon. The maximum flux will naturally be from the direction of the galactic center and the minimum from a direction perpendicular to the galactic plane. Detailed calculations[7,8,9] show that (i) the neutrino flux from the direction of the galactic center (i.e., from a band distributed along galactic longitudes of -60 to $+60°$ and latitudes of $\pm 6°$) is lower by a factor of nearly one hundred compared to the terrestrial atmospheric neutrino flux; (ii) it is further reduced by a factor of 5–10 from other directions in the galactic plane and by a factor of about 100 in a direction perpendicular to the galactic plane compared to that from the galactic center. Therefore it is clear that the galactic neutrino flux in the lunar underground laboratory would be comparable to the locally produced neutrino flux discussed above only in a restricted solid angle in the direction of the galactic center, or at the worst in all azimuths within the galactic belt of say $\pm 20°$. However, if we restrict ourselves to events away from this solid angle, the galactic neutrinos are again down by at least three orders of magnitudes[h] compared to the atmospheric neutrino background on the earth.

To summarize, the backgrounds due to all sources—i.e., neutrinos, muons and neutrons—can be reduced by at least a factor of two hundred and perhaps even better on the moon compared to those on the earth.

In Table 1, we present a comparative summary that exhibits the distinct advantage which a lunar detector would have over a terrestrial one. For example, if $\tau_p \gtrsim 10^{33}$ yrs, the expected ν background event rate on the earth, even after the relevant reduction due to topological cuts, would be equal to or *exceed* the rate of proton decay events.[i] On the other hand, the background-to-signal ratio on the moon, with the same topological reduction factor as on the earth, is at best only 0.5% for nucleon lifetimes (τ_N) of nearly 10^{33} yrs and 5% for $\tau_N \simeq 10^{34}$ yrs.

---

[h] The galactic and extragalactic neutrino flux calculations depend on (i) the assumed distribution of matter and (ii) distribution of cosmic rays in the galaxy and outside. If these are radically different from what is assumed in the calculations, then the same experiment on the moon will serve to open up new vistas in our understanding of the structure of matter and radiation in the galaxy.

[i] For this reason, 10^{33} yrs seems to be about the limit which proton-decay searches on the earth can optimistically hope to reach as regards proton lifetime, at least under the present state-of-the-art. Since the raw neutrino background cannot be reduced, the only way to reduce the effective background is to improve the reduction factor beyond that of 100 shown in Table 1. If detectors are designed so as to improve this factor significantly *and such efforts should by all means be undertaken*, it is clear that such improved detectors transplanted to the moon would, in principle, permit searches for proton-decays with lifetimes approaching even 10^{35} yrs.

Table 1

	Nucleon Decays Expected in 20 KTY	ν-Induced Events for Earth Detector $E_\nu = 1000 \pm 200$ MeV		ν-Induced Events for Moon Detector $E_\nu = 1000 \pm 200$ MeV	
τ_N		Without topological cut	With topological cut	Without topological cut	With topological cut
10^{32} years	120	1000	10	<5	<0.05
10^{33} years	12	1000	10	<5	<0.05
10^{34} years	1.2	1000	10	<5	<0.05

A comparative summary of earth-versus-moon nucleon decay detectors. The fourth and the sixth columns show the expected number of events due to neutrino interactions which would simulate nucleon-decay events in the earth and the moon detectors respectively for 20 KTY exposure in each case. Based on observations and theoretical calculations, we have assumed that for a detector on the earth, there will effectively be a total of nearly 50 neutrino induced events with $E_\nu \simeq (1000 \pm 200)$ MeV per KTY-exposure and that this background can be further reduced by about a factor of a hundred for high resolution fine grain detectors by imposing topological cuts (see e.g., Ref. 10). The raw neutrino background on the moon is expected to be lower by at least a factor of 200 than that on the earth (see text). It is assumed that the background reduction factor due to topological cuts is the same for the moon as it is for the earth detector.

Thus, if the technical and financial aspects involving transportation of detector materials weighing about 100 to 200 tons from the earth to the moon and setting up the detector on the moon can be managed, proton decays with lifetimes $\lesssim 10^{34}$ yrs can be measured with confidence on the moon. This would be a significant improvement over terrestrial experiments. Needless to say, the proton-decay detector set up on the moon will also be most useful as a detector of neutrinos from astronomical sources because of the significant reduction of the background due to local sources (like the atmosphere).[j] Furthermore, it would permit one to search for neutron-antineutron oscillations with far greater sensitivity than that available on the earth.

We believe that many physicists must have already given thought to the idea of searching for proton decay on the moon.[k] We, therefore, hope that this note will help focus on the idea and generate further thinking leading to an optimum design of a proton-decay detector specially suited for the moon.

Acknowledgments

We thank S. Miyake, George A. Snow, John Strathdee and G. B. Yodh for many helpful discussions. The research of JCP is supported in part by the National Science Foundation.

[j] This has been noted e.g. by M. Shapiro and R. Silberberg, M. Cherry and K. Lande and A. G. Petschek in abstracts of the *Symposium on Lunar Bases and Space Activities in the 21st Century*, held Oct. 29–31, 1984, published by L. B. Johnson Space Center, Houston, Texas.
[k] In the process of writing this note, we came across an abstract by A. G. Petschek (see footnote j) mentioning this point. We thank Dr. C. Alley for kindly drawing our attention to the abstracts. For a short remark, see also M. Goldhaber, P. Langacker and R. Slansky, *Science* **210** (1980) 851. We have subsequently learned that F. Reines has also been considering this idea (private communication).

References

1. R. Bionta et al., *Phys. Rev. Lett.* **51** (1983) 27; *Proc. 22nd Int. Conf. on High Energy Phys.*, Vol. I Leipzig (July 1984) p. 244.
2. M. R. Krishnaswamy et al., *Pramana* **19** (1982) 525; *Proc. 22nd Int. Conf. on High Energy Phys.*, Vol. I Leipzig (July 1984) p. 244.
3. G. Battistoni et al., *Phys. Lett.* **118B** (1982) 461; *Proc. 22nd Int. Conf. on High Energy Phys.*, Vol. I Leipzig (July 1984) p. 246.
4. M. Koshiba et al., *Proc. 22nd Int. Conf. on High Energy Phys.*, Vol. I Leipzig (July 1984) p. 250.
5. E. Aprile-Giboni et al., *Proc. 22nd Int. Conf. on High Energy Phys.*, Vol. I Leipzig (July 1984) p. 252.
6. J. C. Pati and A. Salam, *Phys. Rev.* **D8** (1973) 1240; *Phys. Rev. Lett.* **31** (1973) 661; H. Georgi and S. L. Glashow, *Phys. Rev. Lett.* **32** (1974) 438.
7. V. S. Berezinsky, *Proc. 19th Int. Conf. on Cosmic Rays*, Vol. **6** Budapest (1977) p. 231.
8. R. Silberberg and M. M. Shapiro, *ibid.*, (1977) p. 237.
9. A. Dar, *Proc. 4th Workshop on Grand Unification*, p. 101 (1983).
10. A. Grant, *Proc. 1982 Summer Workshop on Proton-Decay Experiments*, Argonne National Lab., ANL-HEP-PR-82-24, ed. by D. S. Ayres, p. 203.

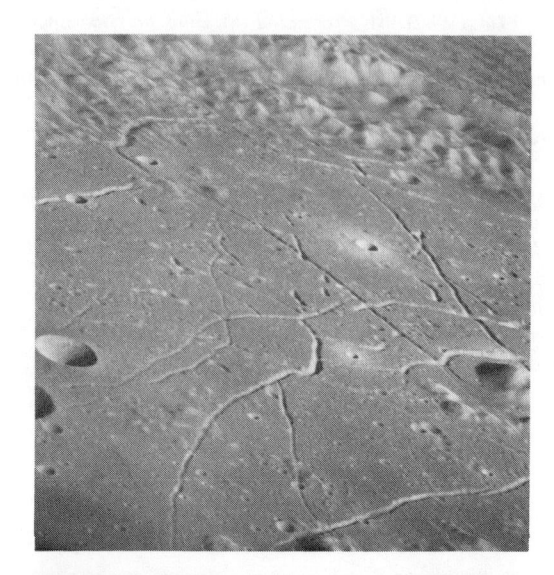

Eastern floor of crater Humboldt [27.2^0 south, 80.9^0 east: 207 km diameter]. Radial cracks and concentric rilles are visible in the mare-type material on its floor. The "doughnut" crater at left is one of its unique features.
(AS15-93-12641).

View southwestward across the bright albedo of crater Aristarchus [23.7^0 north, 47.4^0 west: 40 km diameter] in the upper left, and half-open crater Prinz [25.5^0 north, 44.1^0 west: 46 km diameter] at lower left. Aristarchus rilles flow toward the camera.
(AS15-93-12602).

Participants
NASA Workshop on Physics from a Lunar Base
May 19–20, 1989

Gautam Badhwar
NASA
Johnson Space Center
Houston, TX 77058

Paul E. Boynton
University of Washington
Department of Physics &
 Astronomy
Seattle, WA 98195

David B. Cline
University of California, Los Angeles
Department of Physics
Los Angeles, CA 90024–1562

C. W. Francis Everitt
Stanford University
High Energy Physics
 Laboratory
Stanford, CA 94305–4085

James E. Faller
University of Colorado
Joint Institute for Laboratory
 Astrophysics
Boulder, CO 80309–0440

Ervin J. Fenyves
University of Texas at Dallas
Department of Physics
P. O. Box 830688 4085
Richardson, TX 75083

John W. Freeman, Jr.
Rice University
Department of Space Physics &
 Astronomy
P. O. Box 1892
Houston, TX 77251

Ronald W. Hellings
Jet Propulsion Laboratory
4800 Oak Grove Drive
Pasadena, CA 91103

Edgar Barrie Hughes
Stanford University
High Energy Physics Laboratory
Stanford, CA 94305–4085

Alex Ignatiev
University of Houston
Department of Physics
Houston, TX 77004

Warren Johnson
Louisiana State University
Department of Physics & Astronomy
Baton Rouge, LA 70803-4001

John G. Learned
University of Hawaii at Manoa
Department of Physics & Astronomy
2505 Correa Road
Honolulu, HI 96822

F. Curtis Michel
Rice University
Department of Space Physics &
   Astronomy
P. O. Box 1892
Houston, TX 77251

Carl B. Pilcher
NASA
Headquarters, Mail Code Z
Washington, D. C. 20546

Arnold Rosenblum
Utah State University
Department of Physics
Logan, UT 84322-4415

Michael Salamon
University of Utah
Department of Physics
Cosmic Ray Physics, 201 JFB
Salt Lake City, UT 84112

Paul W. Keaton
Los Alamos National
   Laboratory
Los Alamos, NM 87545

Alfred K. Mann
University of Pennsylvania
Department of Physics
209 South 33rd Street
Philadelphia, PA 19104-6396

Peter Michelson
Stanford University
Department of Physics
Stanford, CA 94305-4060

Andrew E. Potter
NASA
Johnson Space Center
Houston, TX 77058

Serge Rudaz
University of Minnesota
School of Physics &
   Astronomy
Minneapolis, MN 55455

George F. Smoot, III
Lawrence Berkeley
   Laboratory
1 Cyclotron Road
Berkeley, CA 94720

Todor Stanev
University of Delaware
Bartol Research Institute
Newark, DE 19716

Victor J. Stenger
University of Hawaii at Manoa
2505 Correa Road
Honolulu, HI 96822

Richard R. Vondrak
Lockheed Palo Alto Laboratories
B/255, 091-20
3251 Hanover St.
Palo Alto, CA 94304

Mark E. Wiedenbeck
University of Chicago
Enrico Fermi Institute
5720 Ellis Avenue
Chicago, IL 60637

Kent S. Wood
Naval Research Laboratory
Space Science Division,
Code 4121
Washington, D. C. 20375-5000

R. T. Stebbins
University of Colorado
Joint Institute for
    Laboratory Astrophysics
Boulder, CO 80309-0440

Simon P. Swordy
University of Chicago
Enrico Fermi Institute
5640 Ellis Avenue
Chicago, IL 60637

Joseph Weber
University of Maryland
Department of Physics &
    Astronomy
College Park, MD 20742

Thomas L. Wilson
NASA
Johnson Space Center
Houston, TX 77058

Mason R. Yearian
Stanford University
High Energy Physics
    Laboratory
Stanford, CA 94305-2184

View of lunar crater Letronne [10.6⁰ south, 42.4⁰ west: 119 km diameter] on the southern edge of Oceanus Procellarum (Ocean of Storms). North is at the top of the photograph, and the principal point is 10.3⁰ south latitude, 41.5⁰ west longitude. (Apollo 16 metric frame 2994).

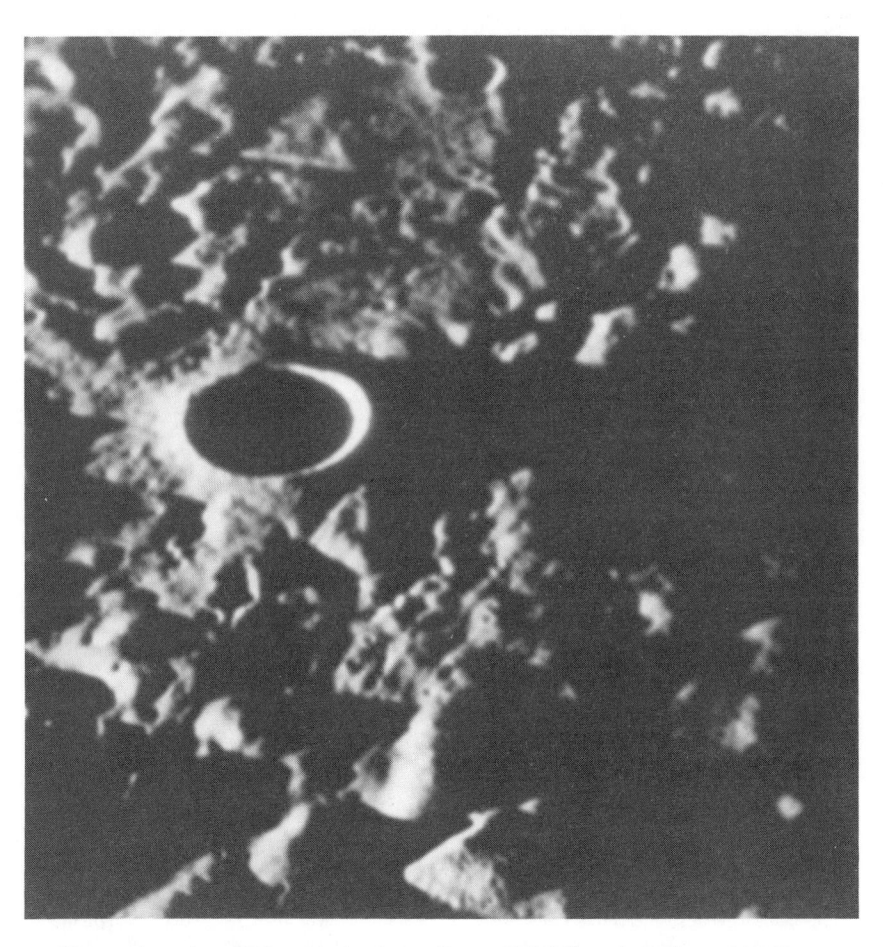

View southeast from 122 km altitude of crater Gassendi F [15.0° south, 44.9° west: 9 km diameter], east of Billy [13.8° south, 50.1° west] and south-southwest of Letronne [10.6° south, 42.4° west]. (AS16-120-19344).

Oblique view westward across Lansberg crater [0.3⁰ south, 26.6⁰ west: 38 km diameter] on the earthside Lunar equator, southwest of Copernicus and northwest of Frau Mauro. (AS14-70-9825).

Lunar Photography Credits and Maps

The editors wish to thank Mary Ann Hager of the Lunar and Planetary Institute (LPI) for her assistance in providing the photographs selected for the manuscript volume.* As Manager of the LPI's Regional Planetary Image Facility (RPIF), she has helped in identifying the images and data products** which were processed at the Johnson Space Center. RPIF's are located regionally (U.S.) and in Europe, in order to aid in the identification of photographic data and maps. These can then be obtained through the National Space Science Data Center (NSSDC), which is the repository for all lunar and planetary photographic products (available in almost any form). RPIF locations are the following: ***

U.S. Facilities:

University of Arizona
RPIF, Lunar & Planetary Lab
Tucson, AZ 85721
(602) 626-4861

Brown University
RPIF, Box 1846
Dept. of Geological Sciences
Providence, RI 02912
(401) 863-2526

Cornell University
RPIF, Center for Radiophysics
and Space Research
Ithaca, NY 14885
(607) 256-3833

University of Hawii
RPIF, Hawaii Insitute of Geophysics
Planetary Geoscience Division
Honolulu, HI 96822
(808) 948-6588 or 6320

Jet Propulsion Laboratory
RPIF, MS 264-115
4800 Oak Grove Drive
Pasadena, CA 91109
(818) 354-3343

Lunar and Planetary Institute
RPIF, Ctr for Inform. & Res. Services
3303 NASA Road One
Houston, TX 77058
(713) 486-2136 or 2172

Smithsonian Institution
RPIF, Room 3101
National Air and Space Museum
Washington, DC 20560

U.S. Geological Survey
RPIF, Branch of Astrogeologic Studies
2255 N. Gemini Drive
Flagstaff, AZ 86001
(602) 779-3311 Ext. 1504

Washington University
RPIF, Box 1169
Dept. of Earth & Planetary Sciences
St. Louis, MO 63130
(314) 889-5679

RPIF – European Facilities:

Sourthern Europe RPIF
Inst. Astrofisica Spaziale, Reparto Planetol.
Viale Dell' Universita, 11
00185 Roma, ITALY

Reg. Planetary Image Facility, DFVLR
Oberpfaffenhofen
NE-OE-PE
8031 Wessling, WEST GERMANY
(089) 520-2417

Regional Planetary Image Facility
University of London Observatory
33/45 Daws Lane, Observatory Annexe
London, NW7 4SD UNITED KINGDOM

Regional Planetary Image Facility
Lab. de Geologie Dynam. Interne (bat. 509)
University Paris – Sud
F-91, 405 Orsay Cedex, FRANCE

NSSDC Address:

NSSDC
Goddard Space Flight Center – Code 633.4
Greenbelt, MD 20771
(301) 286-6695

*These are identifiable by the numbers in parentheses (e.g. AS15-..), and can be obtained from NSSDC.
**Crater locations were taken from *Annual Gazetteer of Planetary Nomenclature*, H. Masursky et al., IAU Working Group: Open-File Report 84-692, Department of Interior, U.S. Geological Survey (1986).
***A lunar far-side map, *The Earth's Moon*, has also been published as Nat. Geog. Mag. **135**, 245 (1969).

312

313

EARTHSIDE LUNAR MAP

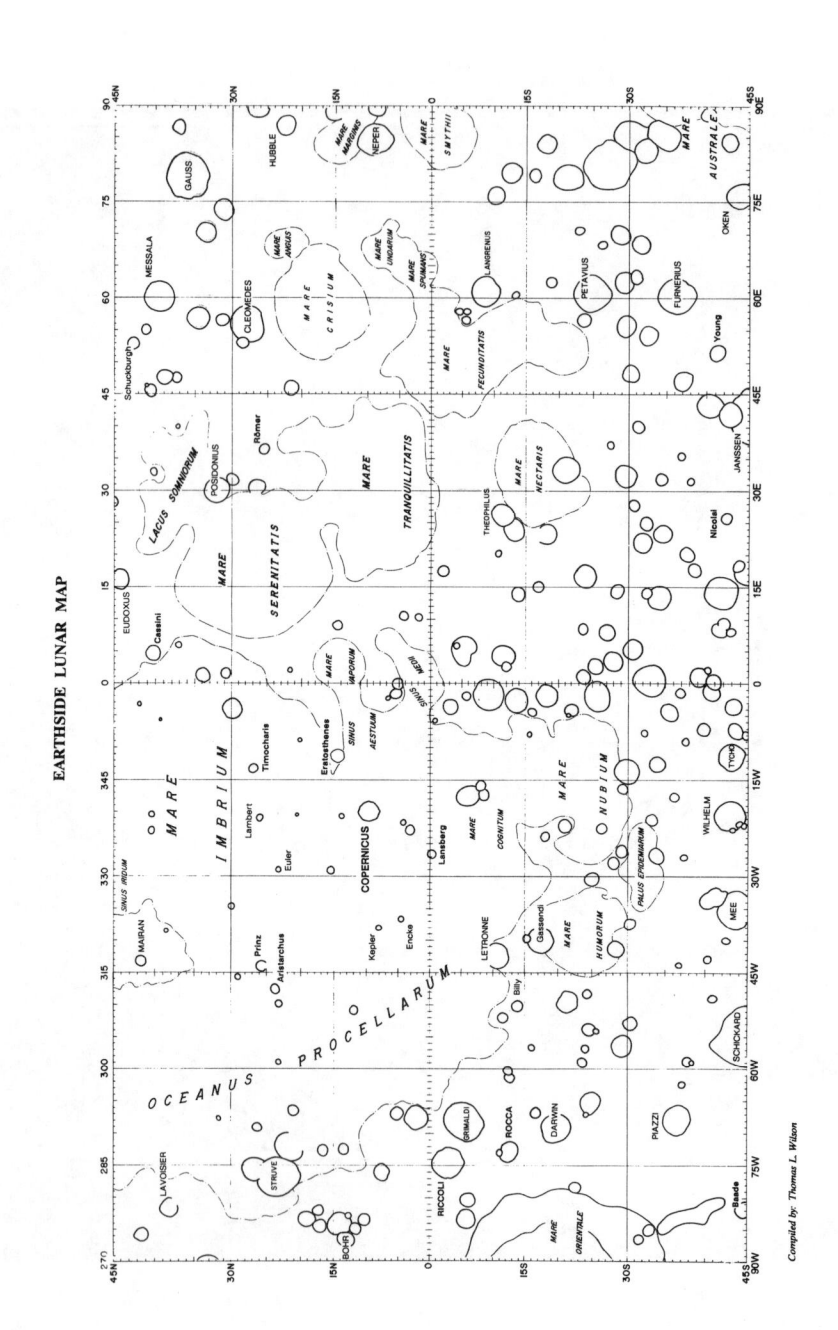

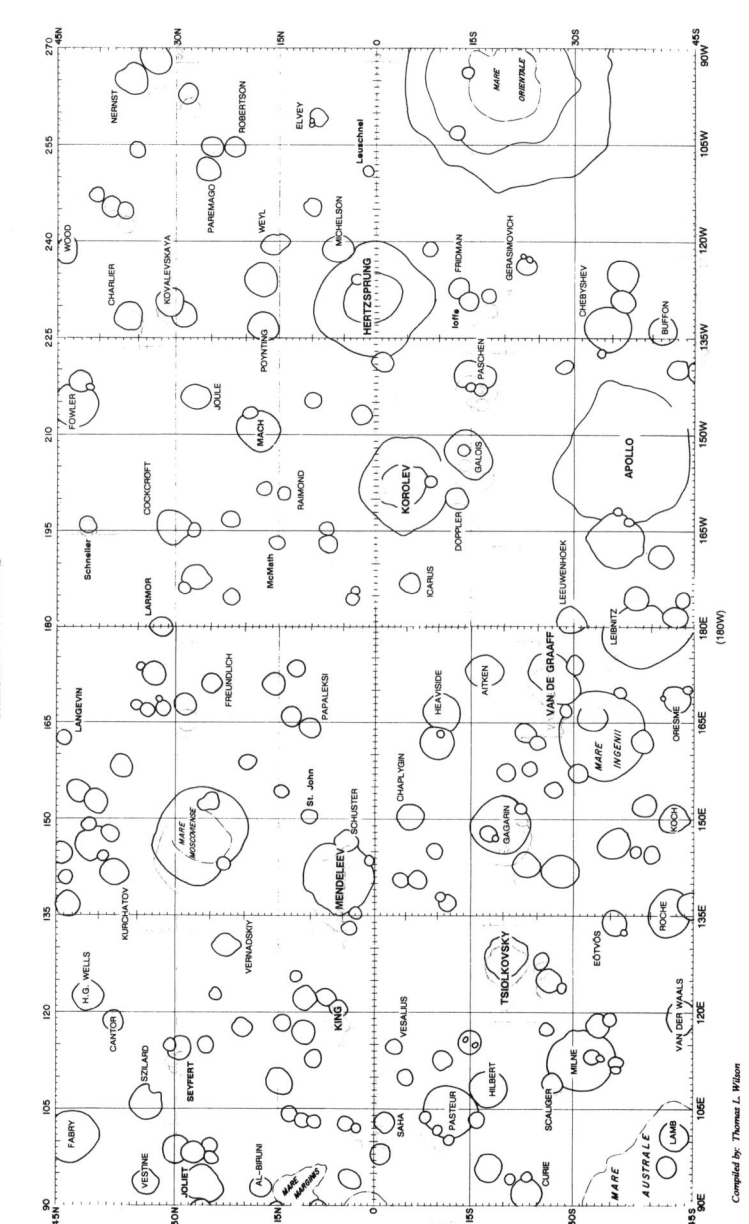

FARSIDE LUNAR MAP

Compiled by: Thomas L. Wilson

NORTH POLAR REGION

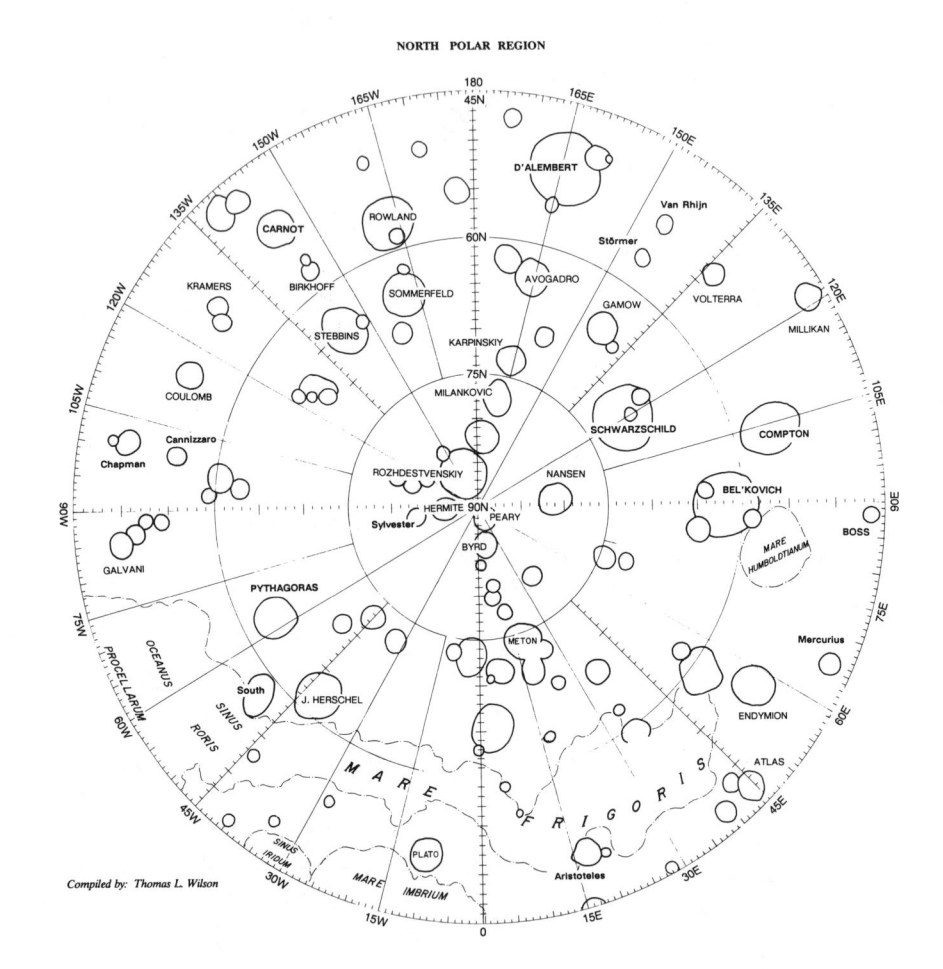

317

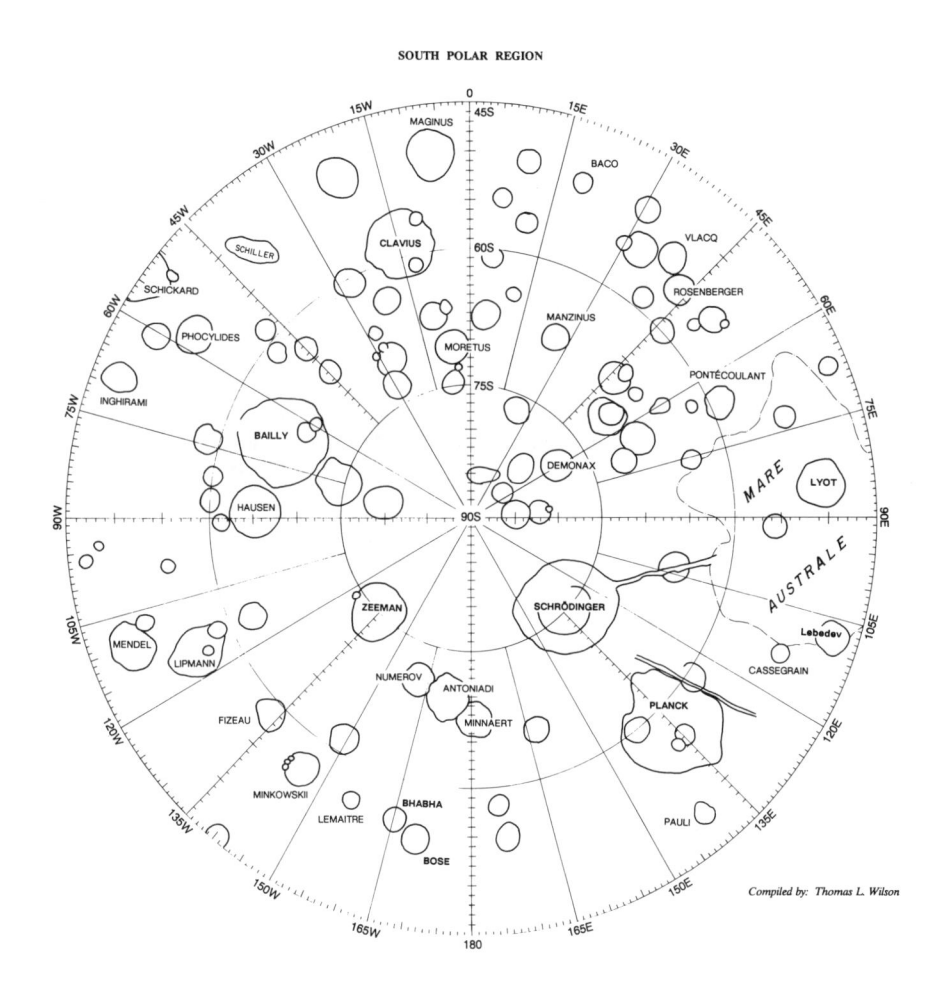

SOUTH POLAR REGION

Compiled by: Thomas L. Wilson

Lunar Physics Constants

Moon Mass
$\mathcal{M}_{Moon} = \mathcal{M}_{Earth}/81.30059$ 7.347×10^{25} g   [Earth: 5.973×10^{27} g]
= 0.7347×10^{20} metric tons
Mean density [1]   3.3437 g cm^{-3}

Moon Gravity
Surface gravity [2]   162.2 cm s^{-2} [Earth: 980.665 cm s^{-2}]
Escape velocity, surface   2.38 km s^{-1}

Moon Figure
Mean radius [1]   1737.53 km
     = 0.27252 Earth radii ($R_E = 6371$km)
Volume 2.199×10^{25} cm^3
Semi-diameter at mean distance (geocentric)   15' 32".6
(topocentric)   15' 48".3

Moon Orbit [2]
Mean Earth–Moon distance 3.84401×10^8 cm   ($\simeq 60\ R_E$)
Eccentricity of lunar orbit   0.0549
Inclination of orbit, with respect to ecliptic 5^0 8' 43"
of equator to ecliptic 1^0 32'.5
to orbit 6^0 41'
L2 distance behind Moon relative to Earth

Moon Rotation [3,2]
Sidereal period (fixed star to fixed star) $27.32166140 + 0.0^616T$ ephemeris days
Synodical month (new moon to new moon) $29.5305882 + 0.0^616T$ ephemeris days
     [T in centuries since 1900.0]

Moon Temperatures
Surface, equatorial (Apollo 17) [4] 102^0 K to 384^0 K during lunar day
Surface, polar [5] 84^0 K
Subsurface (at 30 cm) [4] 250^0 K (Steady, $\pm 2^0$K noon to dawn)

Moon Magnetic Field
Surface, equatorial (Apollo 16) 3×10^{-3} gauss (300 γ)   [Earth: $3 \times 10^4\ \gamma$]
External, in solar wind at 1 A.U. 5×10^{-5} gauss (5 γ)
External, in Earth's geomagnetic tail 1×10^{-4} gauss (10 γ)

Moon Regolith
Thermal conductivity [6] 10^{-4} W cm^{-1} ^{0}K^{-1}
Thermal diffusivity [6] 10^{-4} cm^2 s^{-1}
Regolith depth, most (range)   5–10 m (2–30 m)
Regolith, dust grain size [7]   40–268 μm

Moon's Atmosphere
Molecular concentration [8,9,10] 0.2–1×10^6 molecules cm^{-3}

[1] Ferrari, A.J. et al. JGR 85, 3939 (1980);[2] C.W. Allen, Astophysical Quantities, §68 (Athlone Press, 1973, 3rd ed.);[3]Landolt–Börnstein Tables, 3, 83 (1952);[4] Keihm, S.J., & M.G. Langseith, GCAS 4, 2503 (1973);[5] Mendell, W.W., & F.J. Low, JGR 75, 3319 (1970);[6]Langseth, M.G. et al., GCAS 7, 3143 (1976); [7] Heiken, G., RGSP 13, 567 (1975);[8] Johnson, F.S., RGSP 9, 813 (1971);[9]Hoffman, J.S. et al., GCAS 4, 2865 (1973); [10] Potter, A.E., & T.H. Morgan, Science 241, 675 (1988). JGR→J.Geophys.Res.; GCAS→Geochim.Cosmochim.Acta Suppl.(Lu.Sci.Conf.); RGSP→Rev.Geophys.Spa.Phys.

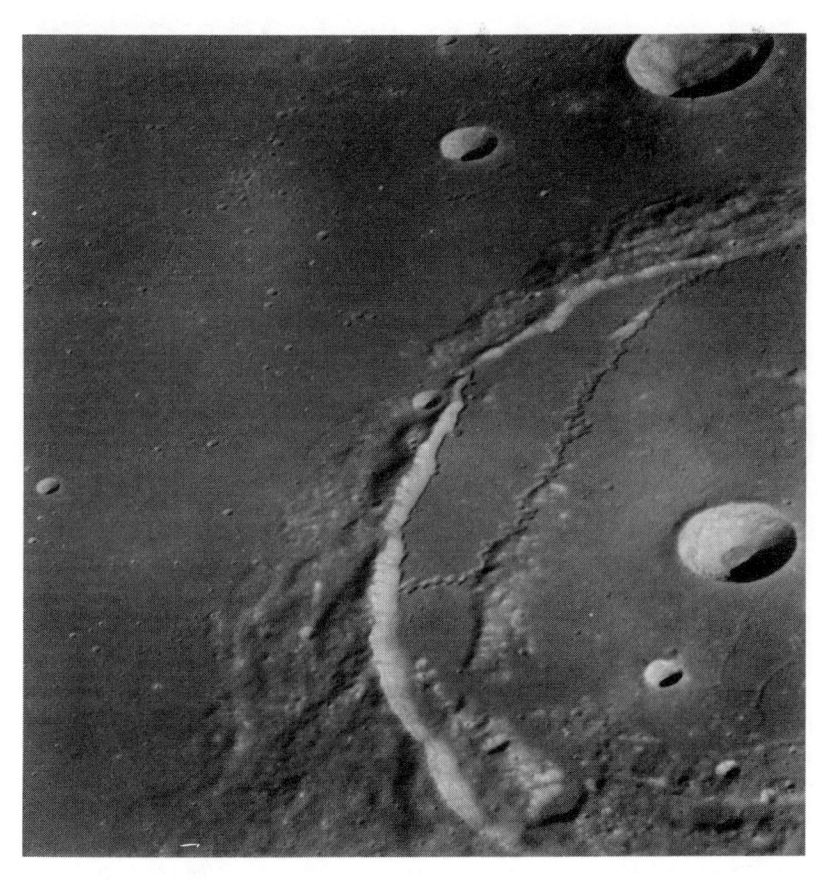

Western half of crater Posidonius [31.8° north, 29.9° east: 95 km diameter] on the eastern margin of Mare Serenitatis (Sea of Serenity). (AS15-91-12366).

High-oblique view during ascent from the lunar surface, looking to the southwest from southern Mare Imbrium. Eratosthenes [14.5^0 north, 11.3^0 west: 58 km diameter] is at left center, and Copernicus [9.7^0 north, 20.0^0 west: 93 km diameter] is on the right horizon. Montes Apenninus is to the left of Eratosthenes. (AS17-145-22285).

Crescent Earth, low over the lunar horizon, would become a common view for Lunar Orbital satellites should we establish a return to the Moon as our next major space initiative. Taken near latitude 24^0 S, longitude 99^0 E, the steep slopes of Humboldt loom ahead. (AS15-97-13268).

Earthrise, when viewed from just beyond the eastern limb of the Moon as seen from Earth, high above crater Pasteur [11.4^0 S latitude, 104.5^0 E longitude] in the foreground. Perhaps during the next great supernova, there will be witnesses of the event from a scene such as this. (AS14-66-9228).

This human footprint on the Moon represents an uncommon and indelible view of physics and astrophysics from the frontier of space. It symbolizes physics at its most fundamental level, without Earth's noise and interference, without weather and meteorology, without comparison in many Earth-based laboratories. It can remain untouched and unblemished for billions of years. In like fashion, long-duration, long-exposure, and large-array scientific experiments at a far-side lunar base stand in stark comparison to the limitations of some terrestrial physics. (AS11-40-5877).

AUTHOR INDEX

A

Atac, M., 243

B

Badhwar, Gautam D., 49
Bender, Peter L., 188
Bouquet, Alain, 227

C

Chaney, R. C., 243
Cline, David B., 88

F

Faller, James E., 153
Fenyves, Ervin J., 243
Freeman, John W., 9

H

Hoffman, J. H., 243
Hughes, E. Barrie, 243, 278

I

Ignatiev, Alex, 285

J

Johnson, Warren, 183

K

Koch, David G., 278

L

Learned, John G., 119

M

Mann, Alfred K., 128
Michel, F. Curtis, 3
Michelson, Peter F., 263

P

Park, J., 243
Pati, J. C., 297
Potter, Andrew, xi
Price, P. Buford, 42

R

Rosenblum, Arnold, 143
Rudaz, Serge, 217

S

Salam, Adbus, 297
Salamon, Michael H., 42
Salati, Pierre, 227
Silk, Joseph, 227
Smoot, George F., 205
Sreekantan, B. V., 297
Stanev, Todor, 29, 291
Stebbins, R. T., 188
Stenger, Victor J., 113
Swordy, Simon P., 211, 293

T

Tarle, G., 42
Tumer, O. T., 243

V

Vondrak, Richard, 17

W

Weber, Joseph, 159
White, S. R., 243
Wiedenbeck, Mark E., 33
Wilson, Thomas L., x, xi, 53
Wood, Kent S., 263

Z

Zhang, W., 128
Zych, A. D., 243

AIP Conference Proceedings

		L.C. Number	ISBN
No. 160	Advances in Laser Science–II (Seattle, WA, 1986)	87-71962	0-88318-360-9
No. 161	Electron Scattering in Nuclear and Particle Science: In Commemoration of the 35th Anniversary of the Lyman-Hanson-Scott Experiment (Urbana, IL, 1986)	87-72403	0-88318-361-7
No. 162	Few-Body Systems and Multiparticle Dynamics (Crystal City, VA, 1987)	87-72594	0-88318-362-5
No. 163	Pion–Nucleus Physics: Future Directions and New Facilities at LAMPF (Los Alamos, NM, 1987)	87-72961	0-88318-363-3
No. 164	Nuclei Far from Stability: Fifth International Conference (Rosseau Lake, ON, 1987)	87-73214	0-88318-364-1
No. 165	Thin Film Processing and Characterization of High-Temperature Superconductors	87-73420	0-88318-365-X
No. 166	Photovoltaic Safety (Denver, CO, 1988)	88-42854	0-88318-366-8
No. 167	Deposition and Growth: Limits for Microelectronics (Anaheim, CA, 1987)	88-71432	0-88318-367-6
No. 168	Atomic Processes in Plasmas (Santa Fe, NM, 1987)	88-71273	0-88318-368-4
No. 169	Modern Physics in America: A Michelson-Morley Centennial Symposium (Cleveland, OH, 1987)	88-71348	0-88318-369-2
No. 170	Nuclear Spectroscopy of Astrophysical Sources (Washington, D.C., 1987)	88-71625	0-88318-370-6
No. 171	Vacuum Design of Advanced and Compact Synchrotron Light Sources (Upton, NY, 1988)	88-71824	0-88318-371-4
No. 172	Advances in Laser Science–III: Proceedings of the International Laser Science Conference (Atlantic City, NJ, 1987)	88-71879	0-88318-372-2
No. 173	Cooperative Networks in Physics Education (Oaxtepec, Mexico 1987)	88-72091	0-88318-373-0
No. 174	Radio Wave Scattering in the Interstellar Medium (San Diego, CA 1988)	88-72092	0-88318-374-9
No. 175	Non-neutral Plasma Physics (Washington, DC 1988)	88-72275	0-88318-375-7

No. 176	Intersections Between Particle and Nuclear Physics (Third International Conference) (Rockport, ME 1988)	88-62535	0-88318-376-5
No. 177	Linear Accelerator and Beam Optics Codes (La Jolla, CA 1988)	88-46074	0-88318-377-3
No. 178	Nuclear Arms Technologies in the 1990s (Washington, DC 1988)	88-83262	0-88318-378-1
No. 179	The Michelson Era in American Science: 1870–1930 (Cleveland, OH 1987)	88-83369	0-88318-379-X
No. 180	Frontiers in Science: International Symposium (Urbana, IL 1987)	88-83526	0-88318-380-3
No. 181	Muon-Catalyzed Fusion (Sanibel Island, FL 1988)	88-83636	0-88318-381-1
No. 182	High T_c Superconducting Thin Films, Devices, and Application (Atlanta, GA 1988)	88-03947	0-88318-382-X
No. 183	Cosmic Abundances of Matter (Minneapolis, MN 1988)	89-80147	0-88318-383-8
No. 184	Physics of Particle Accelerators (Ithaca, NY 1988)	87-07208	0-88318-384-6
No. 185	Glueballs, Hybrids, and Exotic Hadrons (Upton, NY 1988)	89-83513	0-88318-385-4
No. 186	High-Energy Radiation Background in Space (Sanibel Island, FL 1987)	89-083833	0-88318-386-2
No. 187	High-Energy Spin Physics (Minneapolis, MN 1988)	89-083948	0-88318-387-0
No. 188	International Symposium on Electron Beam Ion Sources and their Applications (Upton, NY 1988)	89-084343	0-88318-388-9
No. 189	Relativistic, Quantum Electrodynamic, and Weak Interaction Effects in Atoms (Santa Barbara, CA 1988)	89-084431	0-88318-389-7
No. 190	Radio-frequency Power in Plasmas (Irvine, CA 1989)	89-045805	0-88318-397-8
No. 191	Advances in Laser Science–IV (Atlanta, GA 1988)	89-085595	0-88318-391-9
No. 192	Vacuum Mechatronics (First International Workshop) (Santa Barbara, CA 1989)	89-045905	0-88318-394-3
No. 193	Advanced Accelerator Concepts (Lake Arrowhead, CA 1989)	89-045914	0-88318-393-5

No. 194	Quantum Fluids and Solids—1989 (Gainesville, FL, 1989)	89-81079	0-88318-395-1
No. 195	Dense Z-Pinches (Laguna Beach, CA, 1989)	89-46212	0-88318-396-X
No. 196	Heavy Quark Physics (Ithaca, NY, 1989)	89-81583	0-88318-644-6
No. 197	Drops and Bubbles (Monterey, CA, 1988)	89-46360	0-88318-392-7
No. 198	Astrophysics in Antarctica (Newark, DE, 1989)	89-46421	0-88318-398-6
No. 199	Surface Conditioning of Vacuum Systems (Los Angeles, CA, 1989)	89-82542	0-88318-756-6
No. 200	High T_c Superconducting Thin Films: Processing, Characterization, and Applications (Boston, MA, 1989)	90-80006	0-88318-759-0
No. 201	QED Stucture Functions (Ann Arbor, MI, 1989)	90-80229	0-88318-671-3